Men's Ideas/Women's Realities

THE ATHENE SERIES
An International Collection of Feminist Books
General Editors: Gloria Bowles and Renate Duelli-Klein
Consulting Editor: Dale Spender

The ATHENE SERIES assumes that all those who are concerned with formulating explanations of the way the world works need to know and appreciate the significance of basic feminist principles.

The growth of feminist research has challenged almost all aspects of social organization in our culture. The ATHENE SERIES focuses on the construction of knowledge and the exclusion of women from the process—both as theorists and subjects of study—and offers innovative studies that challenge established theories and research.

ON ATHENE – When Metis, goddess of wisdom who presided over all knowledge was pregnant with ATHENE, she was swallowed up by Zeus who then gave birth to ATHENE from his head. The original ATHENE is thus the parthenogenetic daughter of a strong mother and as the feminist myth goes, at the "third birth" of ATHENE she stops being Zeus' obedient mouthpiece and returns to her real source: the science and wisdom of womankind.

Volumes in the Series
MEN'S STUDIES MODIFIED
The Impact of Feminism on the Academic Disciplines
edited by Dale Spender

MACHINA EX DEA
Feminist Perspectives on Technology
edited by Joan Rothschild

WOMAN'S NATURE
Rationalizations of Inequality
edited by Marian Lowe and Ruth Hubbard

SCIENCE AND GENDER
A Critique of Biology and Its Theories on Women
Ruth Bleier

WOMAN IN THE MUSLIM UNCONSCIOUS
Fatna A. Sabbah

MEN'S IDEAS/WOMEN'S REALITIES
Popular Science, 1870-1915
edited by Louise Michele Newman

NOTICE TO READERS

May we suggest that your library places a standing/continuation order to receive all future volumes in the Athene Series immediately on publication? Your order can be cancelled at any time.

Also of interest

WOMEN'S STUDIES INTERNATIONAL FORUM*
Editor: Dale Spender

*Free sample copy available on request

Men's Ideas/ Women's Realities

Popular Science, 1870-1915

Edited by
Louise Michele Newman

Pergamon Press

New York • Oxford • Toronto • Sydney • Paris • Frankfurt

Pergamon Press Offices:

U.S.A.	Pergamon Press Inc., Maxwell House, Fairview Park, Elmsford, New York 10523, U.S.A.
U.K.	Pergamon Press Ltd., Headington Hill Hall, Oxford OX3 0BW, England
CANADA	Pergamon Press Canada Ltd., Suite 104, 150 Consumers Road, Willowdale, Ontario M2J 1P9, Canada
AUSTRALIA	Pergamon Press (Aust.) Pty. Ltd., P.O. Box 544, Potts Point, NSW 2011, Australia
FRANCE	Pergamon Press SARL, 24 rue des Ecoles, 75240 Paris, Cedex 05, France
FEDERAL REPUBLIC OF GERMANY	Pergamon Press GmbH, Hammerweg 6, D-6242 Kronberg-Taunus, Federal Republic of Germany

For my parents, Marsha and Herbert Newman

Copyright © 1985 Pergamon Press Inc.

Library of Congress Cataloging in Publication Data

Main entry under title:

Men's ideas/women's realities.

   (The Athene series)
   Bibliography: p.
   Includes index.
   1. Feminism--United States--History--Addresses, essays, lectures. 2. Women's rights--United States--History--Addresses, essays, lectures. 3. Women--United States--Social conditions--Addresses, essays, Lectures. I. Newman, Louise Michele. II. Popular science (New York, N.Y.) III. Series.
HQ1426.M46 1984   305.4′2′0973   84-1072
ISBN 0-08-031930-0
ISBN 0-08-031929-7 (pbk.)

All Rights reserved. No part of this publication may be reproduced, stored in a retrieval system or transmitted in any form or by any means: electronic, electrostatic, magnetic tape. mechanical, photocopying, recording or otherwise, without permission in writing from the publishers.

Printed in the United States of America

CONTENTS

List of Tables and Documents	ix
Acknowledgments	xi
Foreword—Ruth Hubbard, Professor of Biology, Harvard University	xiii
A Note on Vocabulary	xix
Introduction: *The Popular Science Monthly* on the "Woman Question"	xxi

1: *The Problem of Biological Determinism (1870–1890)*   1
  Darwin's Theories of Natural and Sexual Selection   2
  Spencer's Contribution to the Theory of Sexual Differences   5
  Women's Responses to Darwin and Spencer   8
  Overcoming the Problem of Biological Determinism   10
  Selections: Herbert Spencer, "Psychology of the Sexes,"
        November 1873   17
        Frances Emily White, "Woman's Place in Nature,"
        January 1875   25
        Miss M. A. Hardaker, "Science and the Woman
        Question," March 1882   33
        Nina Morais, "A Reply to Miss Hardaker on the
        Woman Question," May 1882   39
        Mary T. Bissell, "Emotions Versus Health in Women,"
        February 1888   48

2: *The Evils of Education (1870–1900)*   54
  Overview of Women's Education in the United States   54
  The Debate Over Women's Colleges   56
  The Debate Over Coeducation   59
  Selections: Editor's Table, "The Higher Education of
        Woman," April 1874   69
        Editor's Table, "Normal Co-Education," January 1875   72
        Editor's Table, "Who Shall Study the Babies?"
        June 1876   74
        Henry Maudsley, "Sex in Mind and in Education,"
        June 1874   77

James Rowland Angell, "Some Reflections Upon the
    Reaction from Coeducation," November 1902   87
David Starr Jordan, "The Higher Education of
    Women," December 1902   96

3: *The Birth Rate Question (1890–1905)*   105
  Selections: Grant Allen, "Plain Words on the Woman Question,"
    December 1889   125
  Alice B. Tweedy, "Is Education Opposed to
    Motherhood?" April 1890   132
  An Alumna, "Alumna's Children," May 1904   137
  Another Alumna, "Alumna's Children Again,"
    July 1904   143
  A. Lapthorn Smith, "Higher Education of Women
    and Race Suicide," March 1905   147
  Olivia R. Fernow, "Does Higher Education Unfit
    Women for Motherhood?" April 1905   152

4: *The Professional Homemaker (1880–1910)*   156
  Selections: Editor's Table, "Improved Domestic Economy,"
    January 1879   168
  Mary Roberts Smith, "Education for Domestic Life,"
    August 1898   170
  Charlotte Smith Angstman, "College Women and the
    New Science," September 1898   174
  Laura Clarke Rockwood, "Food Preparation and its
    Relation to the Development of Efficient Personality
    in the Home," September 1911   181

5: *The Suffrage Debate (1890–1900)*   192
  Overview of the Suffrage Movement, 1848–1920   194
  The Arguments   197
  Selections: Edward D. Cope, "The Relations of the Sexes to
    Government," October 1888   210
  Frank Cramer, "The Extension of the Suffrage to
    Women," January 1889   216
  Therese A. Jenkins, "The Mental Force of Woman,"
    April 1889   217
  George F. Talbot, "The Political Rights and Duties of
    Women," May 1896   220
  Alice B. Tweedy, "Woman and the Ballot,"
    June 1896   228

  Grace A. Luce, "Occupations, Privileges, and Duties
   of Woman," September 1896   236
  Editor's Table, "Women and Politics," August 1896   239

6: *Working for Wages (1875–1915)*   243
 Selections: Ely Van de Warker, "The Relations of Women to
   the Professions and Skilled Labor," February 1875   259
  Emily Blackwell, "The Industrial Position of Women,"
   July 1883   271
  Editor's Table, "Progress and the Home," July 1883   281
  Elfrieda Hochbaum Pope, "Women Teachers and
   Equal Pay," July 1913   286
  D. R. Malcolm Keir, "Women in Industry,"
   October 1913   292

7: *The "New Woman" (1890–1915)*   298
 Selections: Editor's Table, "Individuality for Woman,"
   September 1891   305
  Mrs. Burton Smith, "The Mother as a Power for
   Woman's Advancement," March 1895   307
  Editor's Table, "'The New Woman' and the Problems
   of the Day," November 1896   312
  Ellen Coit Elliott, "Let Us Therewith Be Content,"
   July 1897   316
  Earl Barnes, "The Celibate Women of Today,"
   June 1915   323

Index   331
About the Author   337

LIST OF TABLES AND DOCUMENTS

2.1.	Secondary School Graduates From Public and Nonpublic Schools, by Sex and as a Proportion of the 17-Year-Old Population, 1890–1920	62
2.2.	Women Enrolled in Institutions of Higher Learning, 1870–1920	62
2.3.	Women Enrolled in Institutions of Higher Learning According to Type of Institution, 1870–1920	62
2.4.	Advanced Degrees Conferred, by Sex, 1870–1920	63
2.5.	Some Colleges Wholly or Partly Open to Women in the Nineteenth Century	64
3.1.	Birth Rates of American Women, 1800–1900	108
3.2.	Birth Rates of College Men, 1700–1900	112–113
3.3.	Birth Rates of College Women, 1870–1920	114–115
5.1.	Constitutional Amendments, XIV, XV, XIX	203–204
5.2.	Calendar of Suffrage Campaigns	204–206
6.1.	The Employed Population, 10 Years and Over, of the United States, 1880–1920	251
6.2.	Employed Women, 16 Years and Over, by Marital Status, Race, and Nativity, 1900	252–253
6.3.	Selected Occupations of Women, 1870–1920	254–255

ACKNOWLEDGMENTS

Some of the preliminary research for this book was begun in 1977 by Mary Sue Henifin, who compiled a bibliography of the scientific literature on women for a course that Ruth Hubbard, Professor of Biology at Harvard University, was preparing for the NSF-Chautauqua short course program. Their discovery of the debates in *The Popular Science Monthly* sparked their interest in editing a collection of articles. In 1980, Joan Cindy Amatniek read and arranged the more interesting material under thematic headings similar to those I have used in this book. I became involved in 1981 as a co-editor with Ms. Amatniek and gradually assumed sole responsibility for the writing and completion of this collection.

I would like to thank the staff of the New York Public Library, who patiently furnished me with most of the sources listed in my notes, particularly Mr. Walter Zervas who granted me permission to use the Allen Room. I would also like to thank a number of other libraries and historical societies that provided the biographical details appearing in the headnotes: Olin Library of Cornell University; Dartmouth College Library; California State Historical Society; State Historical Society of Wisconsin; Lawrence University Library; Mills College Library; Stanford University Library; Dewitt Historical Society, Ithaca; Maine Historical Society; Iowa State Historical Society; and the University of Iowa Library. I wish to express my deep gratitude to Radcliffe College and the Murray Research Center for providing financial support to Professor Hubbard during the early stages of the project.

Several friends and teachers read portions of the manuscript, and I would like to thank them for their helpful suggestions. Among them are Brian Meyer, Ann Douglas, Kathy Evans, William Heffernan, and Joan Cindy Amatniek. I am especially indebted to Brian Meyer for his astute criticisms. I would also like to thank Warren Fields, Jr., for helping me type the manuscript.

I was fortunate to have three exceptional editors, Ruth Hubbard of Harvard University; and Gloria Bowles and Renate Duelli Klein, editors of the Athene series. They are, to a large extent, responsible for whatever is of merit in this work. My thanks also go to Phyllis Hall of Pergamon Press, who gave much timely assistance and advice.

Finally, I am deeply grateful to Rogers Brubaker, who inspired me by his own commitment to scholarship and gave generously of his emotional and intellectual support.

FOREWORD

The idea that woman's mind was limited by her body was as old as antiquity. But at no time was that idea ever more fervently held or more highly elaborated than it was in America after the Civil War. In those years many, perhaps most, Americans feared that changes taking place in the traditional division of sex roles posed a serious threat to a social fabric already weakened by war. . . . Because the home no longer defined the limits of female activity and women were joining the men in the outside world, however marginally, many Americans believed that the need to draw a clear line between appropriately male and female activities had become acute.

Rosalind Rosenberg,
Beyond Separate Spheres: Intellectual Roots of Modern Feminism
(New Haven: Yale University Press, 1982), p. xv.

That women and men are profoundly different was not a subject of debate in the nineteenth and early twentieth century, but the social and political implications of that difference were. Science played an important role in the debate because all sides tried to base their arguments on "scientific facts," adduced from theories of biological and social evolution and of heredity, and from physiology and medicine. *The Popular Science Monthly* served as an important forum for the debate; it printed numerous editorials on the subject as well as articles by various "experts," such as physicians and scientists (including Herbert Spencer), and responses from readers. The present collection, culled from *The Popular Science Monthly* between 1870 and 1915, offers a rich menu for anyone interested in the history of the debate about woman's nature, about her capacity for education and professional employment, and her right to vote and participate in politics.

It is unfortunate that primary sources usually are accessible only to scholars, for until we read people's own writings, we remain at arm's length (or much further) from what they believed and said. I marvel at the positiveness with which evolutionary connections were (and still are) cited to prove that the sexual division of labor was rooted in the ways cave dwellers lived in the prehistoric past. There seemed little room for doubt that in those distant times women and men divided tasks exactly as they had among the well-to-do classes in America and Britain before social reformers began to stir things up. During the period covered by this collection, women and men argued vehemently. There is much illogic and self-serving rationalization in their arguments, but also much reason and passion.

What is so interesting about these writers is the eagerness with which they sought and embraced "scientific proof." This tendency seems to be an intrinsic part of the republican, optimistic, and individualistic traditions of the nineteenth century in Britain and the United States, and the vestiges are with us still. For educated people, most of whom had left the authority of the church behind them, science was the ultimate authority. Yet, of course, what they meant by "science" (and what is usually meant by it today) were the interpretations of the workings of nature and society made by people like themselves that did not contradict or challenge too seriously the coherence of their world view.

Miss Hardaker wrote confidently:

> If it were possible to collect all the results of the muscular activity of men, from the beginning of civilization until the present, and likewise all the results of the muscular activity of women for the same period, we should reason instantaneously, from these phenomena, to the superior quality of masculine muscle. (p. 34)

To which Nina Morais countered: "[T]rue science says that, if woman's power is to be judged by her work, she must be given a fair field for its display.... A just trial is the whole demand of the reform philosophy" (p. 46). Both asked science to be judge, but clearly they expected opposite verdicts.

Meanwhile the distinguished British physician, Henry Maudsley, offered his expert testimony:

> When we thus look the matter honestly in the face, it would seem plain that women are marked out by Nature for very different offices in life from those of men, and that the healthy performance of her [sic] special functions renders it improbable she will succeed, and unwise for her to persevere, in running over the same course at the same pace with him. For such a race she is certainly weighted unfairly.... Women cannot rebel successfully against the tyranny of their organization.... This is not the expression of prejudice nor of false sentiment; it is the plain statement of a physiological fact. (pp. 78–79)

But though he believed women's concern with domesticity to be firmly rooted in evolution and physiology, he, as well as other writers of the period, made plain the practical reasons why this had to be. Most of the authors assumed that if women were given a choice in the matter, they would prefer to devote themselves to other than domestic tasks, with unequivocally disastrous effects for the "race." "Whatever aspirations of an intellectual kind [women] may have," wrote Maudsley,

> they cannot be relieved from the performance of [their foreordained work as mothers and nurses of children] so long as it is thought necessary that mankind should continue on earth. . . . [I]t would be an ill thing, if . . . we got the advantages of a quantity of female intellectual work at the price of a puny, enfeebled, and sickly race. (p. 81)

Not surprisingly, he concluded that women and men must be educated differently.

Another physician, Van de Warker, claiming to speak "as a gynaecologist, leaving out of consideration the social aspects of the case" (p. 259), argued that since eminent women usually didn't marry and therefore didn't have daughters, the laws of inheritance explained the "average" woman's limited achievements as compared with men. Van de Warker believed that the menstrual cycle handicapped women and pinned the danger alternately on ovulation (which "is constantly liable to accidents" and lays women open to "hysteria" [p. 266], on menstruation, or on its absence ("aemenorrhoea, a disease to which every woman is liable who follows an intellectually rather than physically active life" [p. 266]).

The physician, Emily Blackwell, opposed these claims and pointed out that:

> Most women who have been engaged in any new departure would testify that the difficulties of the undertaking lay far more in [the] artificial hindrances and burdens than in their own health, or in the nature of the work itself. (p. 279)

Science and politics intermingled closely and inevitably, and it is fascinating to see how the different writers, sometimes consciously, sometimes inadvertently, went back and forth between them.

One thing that strikes me is how much woman-blaming existed in this debate. Both sides made women responsible for the future of the "race." The question was how best to prepare women to fulfill that responsibility. Louise Newman, in her commentary, points out that much of the discussion (for example, about women's education and its effect on the birth rate) essentially ignored that men also have a part in this. This is not very different from current discussions about reproductive hazards in the work place, which focus on the dangers to women's reproductive capabilities while ignoring similar dangers for men, or about the effect of a woman's age on her children's health. Much of the discussion still is framed as though making babies and rearing children involve only women. Another interesting point is that marriage and reproduction were generally presented as women's duty, with the clear implication that, given a choice (as might be offered by "manly" education or work), women wouldn't marry or have children. So, depending on the writers' temperaments, they threatened, persuaded or cajoled. "We ought frankly to recognize," wrote Grant Allen, "that most women must be wives and mothers; that most women should therefore be trained, physically, morally, socially, and mentally, in the way best fitting them to be wives" (p. 127). We need devoted mothers, claimed Lapthorn Smith, because "we want our children, especially our boys, to be good and happy" (pp. 149–150).

So, once the earlier "scientific" question, whether girls' physiology *could* stand up to their being educated more like boys, was answered in the affirmative, it gave way to the more frankly political question of whether they *should* be. Yet that, too, was discussed in quasi-scientific terms, focusing on evolution, birth rates, the betterment of the "race."

The rise of the domestic science movement was in part intended to make domesticity more attractive to educated women, but it also was meant to provide employment for qualified women who were not being hired to teach other subjects at the college level. The domestic science movement, too, squarely placed on women the responsibility for the health and happiness of future generations. In doing so, it glossed over a basic contradiction: the earlier argument that women were fitted by nature (evolution, physiology, heredity) to *be* wives and mothers was replaced by the opposite notion that women needed to *learn to be* proper wives and mothers. The high incidence of childhood sickness and death among the poor was blamed on the ignorance of immigrant and other poor women who gave their children unnutritious foods and didn't know how to keep house. "The appalling mortality of children that are born fairly normal and vital," argued Mary Roberts Smith,

> is chiefly to be accounted for by the ignorance of mothers . . . (p. 172). [D]omesticity—that is the care of household and children—is in itself a profession for which the best training and the fullest development attainable are not too much. (p. 170)

And Laura Clarke Rockwood, invoking scientific authority, wrote:

> "A good stomach kept in a healthy condition is the foundation of all true greatness," says Dr. Tyler, professor of biology at Amherst College.
>
> The housemother's first duty then after bearing her family is to provide them with food for their growing needs. . . . But . . .closely connected with the necessity for food are the other hygienic necessities for survival—air, water, sunshine, shelter, rest, all in due season; and depending upon these vital necessities are the opportunities for the development of the mental, moral and social personality or the completely social man. Because woman is specialized for home work, these wider responsibilities for individual growth have, in large measure, become hers also. (p. 183)

No small order! And not so different from the present tendency in some quarters to blame health problems on "life style."

So, while higher education prepares men to be doctors, lawyers, politicians and business men, women would use theirs to improve the "race" by becoming yet better wives and mothers.

Charlotte Smith Angstman rhapsodized that

[w]ith that largeness of vision and special understanding born of her special opportunities, yet true to her woman's instinct, which nothing can eradicate, [woman] has seen what might be bettered, and is bettering it in that place which is most potent for all that is good and evil in life—the home. (p. 181)

It is important to look at what people were saying in the context of the social limitations and opportunities with which they were living, and this collection offers a rich fare of original sources set in Louise Newman's vivid sketches of the times. It includes demographic data and offers a wide range of sources for further exploration.

Having whetted your appetite, I want to end on a personal note. As I sit on a New England beach in brilliant sunshine and watch the seagulls sail on the wind, I am keenly aware that the social history I have been describing is not my own. I was born in post-World War I Vienna, the capital of what, until recently, had been a continental empire. Although the Vienna in which I grew up was proudly forging new forms of social democracy, and although my parents were struggling, Jewish, first-generation professionals, my childhood milieu retained an aristocratic tradition in which college-educated women (as upper-class women before them) were not expected—or expecting—to devote their lives to being wives and mothers. Of course, most of them planned to marry and have children, but they would hire maids to do housework and nursemaids to care for the children, while they themselves would enter the professions. (This did not mean that these women were feminists—most of them weren't.)

My own mother got her medical degree around 1920 and all her women friends were professionals. I grew up expecting to work outside the home as a professional of some sort. When I came to the United States in early adolescence, I was too preoccupied with language and social problems to realize that home and childcare were considered my destiny. Later, as a wife and mother in the United States, I had to do the usual juggling of conflicting demands in order to reconcile work and family obligations, but I never assumed that I was the one best equipped "by nature" to care for my household or children. On the contrary, I was quite sure that I could find other women (and of course they were women!) who were better at doing both than I was. This did not mean to me that I was surrendering responsibility for my children or that they wouldn't know who was their mother, but only that they would find out that other people could care for them, too.

Not until the women's movement, when I began to talk about such things with other mothers, did I realize how much easier my lack of Anglo-American socialization had made my efforts to combine motherhood with scientific work. My problems and conflicts were practical—how to be in several places at the same time—not moral—how to blend my professional work with my responsibility "to make the household sweet, wholesome, dignified, a place of growth" (to quote Mary Roberts Smith, p. 173).

This collection will lead you into a debate that has significantly shaped the social forms in which we live. Many of the arguments are still with us, although they are usually worded differently. The introductions will pull you in, the resources will tell you where to find more. So, read on, gentle reader, read on!

May 1984

Ruth Hubbard
Woods Hole, Massachusetts

A NOTE ON VOCABULARY

Because the vocabulary used today to describe women's quest for equality can often be confusing, a brief discussion of the historical terms appearing in this book may prove helpful. During the nineteenth century in the United States, the singular terms, the "woman movement" or the "woman's movement," were used in a broad sense to refer to all of women's new public activities, not just those activities explicitly aimed at attaining suffrage or political and legal rights. The segment of the woman movement devoted to securing the franchise for women was called the "woman('s) suffrage movement." Sometimes the suffrage movement was also called the "woman's rights movement," but as I mention in Chapter 5, suffrage was just one of many issues that the woman's rights movement supported.

Today, Americans tend to use the plural variant, the "women's movement," or another term, borrowed from the French early in the twentieth century, the "feminist movement," to refer to women's struggle for equal rights. For a brief period in the late 1960s and early 1970s, the "women's liberation movement" was a popular designation, but it now rarely appears in the current works of feminist historians and writers. Thus, following recent usage, I speak of the women's movement. At times, however, I purposely return to the singular terms, woman question, woman movement, and woman suffrage, particularly when I want to emphasize the historical antecedents of the modern women's movement. But for the most part, and at the risk of gross generalization, I use "women" and the plural possessive variations: women's education, women's nature, women's profession, women's sphere, women's work, and so on. Whenever possible I try to clarify which women I (or others) mean in order to avoid giving the false impression that women were a homogenous class with identical interests.

At the outset, I should mention that this book is not about all women; it is a book about how some people, generally well-educated, white middle-class men and women, *conceived of white middle-class women* (most of whom were of Anglo-Saxon Protestant descent). Poor women, immigrant women, and black women received only the briefest mention in the pages of *The Popular Science Monthly*, the journal in which the selections reprinted in this collection appeared. Yet, many of the observations about women were made originally without any qualifications. It was assumed rather than stated explicitly that the observations pertained to white middle-class women, and few writers questioned whether these statements also pertained to other groups of women. Where information is readily available, I point out the differences for immigrant and black women. But I have not discussed them in detail, not because these women are undeserving of mention, but because they merit much closer analysis than I was able to provide here.

Introduction
THE POPULAR SCIENCE MONTHLY ON "THE WOMAN QUESTION"

> "O come and be my mate!" said the Eagle to the Hen;
> "I love to soar, but then
> I want my mate to rest
> Forever in the nest!"
> Said the Hen, "I cannot fly
> I have no wish to try,
> But I joy to see my mate, careering through the sky!"
> They wed, and cried, "Ah, this is Love, my own!"
> And the Hen sat, the Eagle soared, alone.
>
> Charlotte Perkins Gilman, "Wedded Bliss,"
> *In This Our World and Other Poems*
> (San Francisco: James H. Barry and John H. Marble, 1895), p. 87.

This book seeks to gain insights into the ideological background of the women's movement in the United States by presenting and discussing articles that were published originally in an influential journal of the late nineteenth century, *The Popular Science Monthly*. All the selections contained in this collection appeared during the years 1873–1915, when the United States was, as Richard Hofstadter called it, "the Darwinian Country."[1] *The Popular Science Monthly* was, in its turn, the Darwinian journal, and its founding editor, Edward Livingston Youmans (1821–1887), enamored of social-Darwinian theory, was Herbert Spencer's most devout American disciple. As the story goes, Youmans founded *The Popular Science Monthly* in 1872 because he was in desperate need of an American magazine to print a work he had commissioned from Spencer, but for several years before this Youmans had recognized the need for a journal that would familiarize Americans with current discoveries in evolutionary theory and that would spread the growing influence of science.[2] Science, and evolutionary theory in particular, had much to say regarding women's nature and their appropriate place in society. It was an article by Spencer, "Psychology of the Sexes" (November 1873)—the first essay in this collection—that sparked the debate on the woman question in *The Popular Science Monthly*.

The magazine covered the woman question from various perspectives, publishing the views of men and women, conservatives and liberals, scientists and housewives. Frequently, the contributors addressed each other directly in continuing dialogues that spanned several months and occasionally several years. Thus, many of the selections speak to one another and illustrate clearly the lines of argument and counter argument. For this reason, the debates that follow have an unusual cohesiveness. Lack of space forced me to omit a large number of excellent selections, but the bibliographies at the end of each chapter introduction list all the relevant articles published in *The Popular Science Monthly* during this period (1873–1915).

Although the earliest article in this anthology appeared in 1873, the debate on the woman question began long before then. Concern over women's position in America had its origin at least as far back as the eighteenth century, when Mary Wollestonecraft and Abigail Adams, among others, criticized society for its neglect of women's education and political rights.[3] Neither does the debate on the woman question end in 1915, the date of the last selection, although after women achieved suffrage in 1920, the women's movement dissipated and the debates in the pages of *The Popular Science Monthly* subsided. The year 1915 proved a convenient cutoff point both because World War I was to become a more pressing concern and because *The Popular Science Monthly* was reorganized under a new title, *The Scientific Monthly*.

The selections are arranged in thematic rather than in chronological order to illustrate as clearly as possible the arguments about women's nature, education, fertility, domesticity, suffrage, and employment. One difficulty in this arrangement is that it may suggest a separation among the debates that did not exist in the minds of the debators. Any particular conception of women's nature had specific implications for their education and their domestic, economic, and civic roles. What women were and what they might become were essential factors in determining what type of schooling they should receive, whether they were capable of voting, and what type of work they could and should do. Thus, the chapter introductions are meant to integrate the various issues as well as point out the incongruities and contradictions that emerged in the different phases of the debate on the woman question.

The chapters follow in approximate chronological order and reflect the changing emphases of the debate. The first two chapters, which treat the period from 1870 to 1890, focus on biological sexual differences and their significance for women's education. The next two chapters examine various reactions to the increasing availability of higher education for women during the late nineteenth century. Chapter 5 covers the suffrage movement from its inception in 1848 until 1920, but the selections are drawn from two discussions, one occurring in 1889, the other in 1896. The last two chapters deal with the later years, 1900–1915, a time when earlier attitudes and practices began to break down. Chapter 6 describes early twentieth century attitudes toward employed women; Chapter 7, which serves as the conclusion to this book, analyzes the ideal of the "new woman" that emerged in the 1890s.

In preparing the articles from *The Popular Science Monthly* for publication in this collection, I corrected typographical errors and omitted a few extraneous footnotes, changes that will not be apparent to the reader. I also deleted repetitions, digressions, and obscure allusions, indicating these omissions with ellipses. I renumbered remaining footnotes, and on the occasions when I added an editorial comment, I bracketed my remarks [if they appeared within the text] or followed them with parentheses (Ed.) if they were given as a footnote. Full citations for all the selections appear in the bibliographies at the end of each chapter introduction.

During the late nineteenth century, men and women participated in emotional debates in an effort to reassess women's nature and their proper place in society. This reassessment occurred in response to fundamental changes in the lives of Americans brought about by the industrial development of the mid-nineteenth century, a process that accentuated the division between men's and women's spheres of activities. As industrial production replaced household manufactures, men and women worked in separate locations: men, in the work place; women, at home. In the social rhetoric of the day and commonly in practice, women's tasks were confined to childrearing and housework.

The narrowness of women's sphere gave impetus to a small number of educated women, who agitated for increased educational and professional opportunities for women. It is this element of the women's movement that we frequently think of when we consider women's struggle for equality in the United States. However, the woman movement in the nineteenth century included many diverse groups, whose only common aim was to "advance" the position of women. These groups often had different, even conflicting, ideals. Proponents of women's education, for example, comprised both those who believed women needed special training for their profession as homemakers and those who wanted to give women an education identical to that available to men. Suffragists included both those who considered women the intellectual equals of men and those who, conceding women's intellectual inferiority, based their demand for the franchise on women's moral superiority. Hence, significant differences of opinion existed within the woman movement concerning the steps needed to "advance" women's position: Would an education designed for men benefit women or render them infertile? Would suffrage help women gain political and legal equality or distract them from familial responsibilities? These questions were the subjects of vehement public debate, both among the various groups of the woman movement and between supporters and opponents of women's rights.

The term *the woman question* appeared in American periodicals in the early 1860s and referred specifically to woman suffrage. The adoption of the Fourteenth (1868) and Fifteenth (1870) Amendments to the United States Constitution (see Document 5.1, p. 203) clarified the political status of freed male slaves, but not that of women, white or black. The debate over whether women were "equal" citizens, entitled to vote, continued relentlessly and was not resolved until the adoption of the Nineteenth Amendment in 1920. The suffrage question led Americans to reconsider other aspects of women's equality, and the woman question ultimately encompassed not only political rights, but also a range of social and legal issues, involving education, familial responsibilities, and employment.

Some of the questions debated in the nineteenth century are no longer issues today. Women vote and hold office. They go to college and enter professions that were once (and not that long ago) closed to them, although many obstacles still exist. But one critical, unresolved problem remains: How can women reconcile the inherent conflicts between their personal and professional lives, between having a family and a career? Writers in the nineteenth century rarely phrased this question in these terms, but as women became more active in public activities, they began to comprehend the difficulty of integrating motherhood with activities outside the home. This difficulty was not a central aspect of women's lives prior to the industrialization of the nineteenth century when most men and women jointly supported families by farming. In colonial, agricultural America, women could farm and remain physically close to children. Men worked nearby. Although men's work differed from women's work, both were engaged in agricultural pursuits, and in this respect men and women shared a common experience.

In the decades following the Civil War, for the first time in American history, men on a large scale did not work at home. With industrial development, there occurred a physical separation and an ideological division between the home and the work place. As the home and work place separated physically, they took on complementary functions. The work place gradually became the source of the family's economic support; the home became a so-called haven from the competition and cruelty of the work place. The participants in the work place, men, were viewed as "producers," the providers of the family. Women, who remained at home, were seen as "consumers," and their productive role in the home was frequently overlooked, although many women performed tasks that were essential for the family's survival. By the late nineteenth century, many people, men and women alike, spoke of women's work in the home as ancillary to men's work in the work place.

At the same time, however, control of the family's spiritual and moral development fell increasingly to women. Childrearing took on added significance. Motherhood was considered a holy office, the most noble of professions, a full-time occupation. Society idealized, romanticized, and sentimentalized mothers. Nothing could be as pure or as good as a mother's love for her child. No knowledge or wisdom deserved as much respect as women's "maternal instincts."

Domesticity—homemaking and childbearing—became the critical defining element of women's identities. While some women complained of the narrowness of the domestic sphere, the large majority accepted their position, even welcomed it. To begin to understand women's acceptance of their domestic role, we must remember that women's relegation to the home could be seen as much an indication of their privileged status as of their subordination. The special moral and social responsibilities of mothers, which as it happened made it impossible for them to leave the home, were a source of pride to women. The majority of women did not view these obligations as a means of enslaving them, despite the claims of feminists to the contrary. Many supporters of women's education and suffrage did not conceive nor represent their efforts as attempts to allow women to escape domestic duties, but as a means of making women better mothers and homemakers. Hence, the first "right" won by women in the beginning of the nineteenth century was the opportunity for young women to receive schooling to prepare them for their future role as wives and mothers. Later, in the 1840s and 1850s, married women gained control of their own property. It was not until late in the century that women began to demand the right to participate in public activities, the right to support themselves, and the right to an identity other than that of wife and mother. But even these aspects of women's demand for equality were not construed by most participants in the woman movement as a challenge to the basic social institutions, marriage and the family. Charlotte Perkins Gilman (1860–1935), with lucid insight, was among the few to realize that the separation of the home (in which women still spend much more time than men) from all public and paid activity was the major obstruction to women's freedom.[4] The failure of the woman movement in the nineteenth century and of parts of the women's movement today lies in the inability to acknowledge the problems of integrating women into an economic order that denies them the possibility of satisfactorily integrating their dual roles as mother and employee. The demands of the job market force women to choose between their own need for economic self-sufficiency and professional success and their children's need for close and loving supervision and care.

Science, common sense, and social custom interacted smoothly in the late nineteenth century to make the different parental roles of men and women unchallangable. The woman movement strove to enlarge women's sphere, but without resolving the conflict between women's familial roles and the industrial organization of their society. For most supporters of the woman movement, the problem was how to make the different spheres equal, without making them identical, and without challenging the parental division of labor. They failed to realize that given these constraints, the spheres could never be made equal because the female sphere would continue to contain intrinsic tensions that the male sphere did not have. Financial success and the commitment necessary to achieve that success was compatible with fatherhood, as the essence of a good father was that he be a good provider. Moreover, economic and social realities gave men more opportunities than women to assert their individuality and attain independence. The fact that men received money for their work, while housework was unpaid and women's paid work received low wages, not only reflected society's different evaluation of men's and women's work, but more importantly, gave men the freedom to marry or not, just as they pleased. Nor did men have to ask anyone's permission to spend the money they had earned, and they could use financial as well as other resources to support and strengthen an industrial order that favored them.

As we begin to explore nineteenth-century assumptions about women's nature, the first question we must ask is, which women? Do we mean poor, rich, white, black or immigrant women, educated or uneducated women, or poor white women, educated black women, rich immigrant women? The category "women" is impossibly broad, and yet, as we shall see, much of the discussion concerning women does not differentiate among the multitude of groups. In fact, this is one of the most remarkable and troubling assumptions about women: the persisting belief that there is some essence of womanhood that applies equally to all women.

Female biology gives women the potential to bear and nurse children. But assumptions about womanhood extend far beyond this. For example, women traditionally have been considered more "natural" and better caretakers of young children than men, and childrearing has remained a significant portion of women's work. From women's close relationships with children, we often draw other conclusions about women's nature. We say women are loving, gentle, nurturing, sympathetic, that they have an "instinct" for childrearing. But are these qualities the products of biology, environment, conditioning, social norms, or parental expectations? Here we are led to ask another, more fundamental question: Is it possible to differentiate among these determinants? Can we distinguish the effects of nature and nurture, when biological and social components are inextricably mixed up with one another? How and where women live, what they eat, and what they do affects their size, their physiological functions, and their intellectual capacities. We often refer to biology, physiology, and psychology as if they constituted stable and distinct determinants of constitution and behavior, when they are variable, interdependent factors that both influence and are influenced by social structures, natural environments, cultural beliefs, and personal expectations. The conclusions of nineteenth-century scientists regarding women's nature were founded on false premises, for they believed that some underlying unchanging biological reality governed the character and behavior of women.

This is a book about what people believed and said. The writers included in this collection expressed personal views about the position, rights, and duties of women in their society, but these views were also shared by many others. In the chapter introductions, I attempt to place their statements in their historical context. Although I recognize the impossibility of determining with any precision the relation of historical events to changing attitudes, I nonetheless felt it my obligation to examine the ideological, social, economic, and political settings of the debates presented here. I make no claim to resolve the problems of causation and interpretation that perpetually challenge historians. Instead I present the evidence upon which I base my conclusions in the hope that my readers will formulate their own.

NOTES

1. Richard Hofstadter, *Social Darwinism in American Thought* (Boston: Beacon Press, 1955), pp. 4–5.
2. See Frank Luther Mott, *A History of American Magazines 1865–1885* (Cambridge: Belknap Press, 1957), 3: 495–499. For Eliza Youmans' account of the origin of *The Popular Science Monthly* see *The Popular Science Monthly* 30 (March 1887): 688.
3. See Mary Wollestonecraft, *A Vindication of the Rights of Woman With Strictures on Political and Moral Subjects* (New York: W. W. Norton, 1967; 1st ed., 1792) and Abigail Adams' letter to John Adams, "Remember the Ladies," reprinted in Miriam Schneir, ed., *Feminism: The Essential Historical Writings* (New York: Vintage Books, 1972), pp. 3–4. Historian Dale Spender cautions against dating the beginning of women's protests against male power and argues that as far back as historical records take us, there is evidence of such protest. See Spender, *Women of Ideas and What Men Have Done to Them: From Aphra Behn to Adrienne Rich* (Boston: Ark Paperbacks, 1983), especially pp. 44–56.
4. See Charlotte Perkins Gilman, *Women and Economics* (New York: Harper Torchbooks, 1966; 1st ed., 1898) and Charlotte Perkins Gilman, *The Home: Its Work and Influence* (Urbana: University of Illinois Press, 1972; 1st ed., 1903).

SUGGESTIONS FOR FURTHER READING

Anthologies

Agonito, Rosemary. *History of Ideas on Woman: A Sourcebook*. New York: G. P. Putnam, 1977.
Cott, Nancy F., ed. *Root of Bitterness: Documents of the Social History of American Women*. New York: E. P. Dutton, 1972.
Cott, Nancy F., and Pleck, Elizabeth H., eds. *A Heritage for Her Own: Toward A New Social History of American Women*. New York: Simon and Schuster, 1979.
Epstein, Cynthia Fuchs, and Goode, William J. *The Other Half: Roads to Women's Equality*. Englewood Cliffs, NJ: Prentice-Hall, 1971.
Hartman, Mary S., and Banner, Lois, eds. *Clio's Consciousness Raised: New Perspectives on the History of Women*. New York: Harper & Row, 1974.
Kraditor, Aileen S., ed. *Up From the Pedestal: Selected Writings in the History of American Feminism*. Chicago: Quadrangle Books, 1968.
Lerner, Gerda, ed. *The Female Experience: An American Documentary*. Indianapolis: Bobbs-Merrill, 1977.

Rossi, Alice S., ed. *The Feminist Papers: From Adams to de Beauvoir.* New York: Bantam Books, 1976.
Schneir, Miriam, ed. *Feminism: The Essential Historical Writings.* New York: Vintage Books, 1972.
Welter, Barbara, ed. *The Woman Question in American History.* Hinsdale, IL: Dreyden Press, 1973.

Social Histories of Women in the United States

Banner, Lois W. *Women in Modern America: A Brief History.* New York: Harcourt Brace Jovanovich, 1974.
Beard, Mary Ritter. *Women as a Force In History.* New York: Macmillan, 1946.
Degler, Carl N. *At Odds: Women and the Family in America from the Revolution to the Present.* New York: Oxford University Press, 1980.
Hymowitz, Carol, and Weissman, Michaele. *A History of Women in America.* New York: Bantam Books, 1978.
Lerner, Gerda. *The Woman in American History.* Menlo Park, CA: Addison-Wesley, 1971.
Rothman, Sheila M. *Woman's Proper Place: A History of Changing Ideals and Practices, 1870 to the Present.* New York: Basic Books, 1978.
Wilson, Margaret Gibbons. *The American Woman in Transition: The Urban Influence, 1870–1920.* Westport, CT: Greenwood Press, 1979.

History of Feminism and the Women's Movement

Banks, Olive. *Faces of Feminism: A Study of Feminism as a Social Movement.* New York: St. Martin's Press, 1981.
Degler, Carl. "Revolution Without Ideology: The Changing Place of Women in America," *Daedalus* 93 (September 1964): 653–670.
Flexner, Eleanor. *Century of Struggle. The Woman's Rights Movement in the United States.* Cambridge, MA: Belknap Press, 1959.
Hersh, Blanche Glassman. *The Slavery of Sex: Feminist-Abolitionists in America.* Urbana: University of Illinois Press, 1978.
O'Neill, William L. *Everyone Was Brave: The Rise and Fall of Feminism in America.* Chicago: Quadrangle Books, 1969.
Rosenberg, Rosalind. *Beyond Separate Spheres: Intellectual Roots of Modern Feminism.* New Haven: Yale University Press, 1982.
Sinclair, Andrew. *The Better Half: The Emancipation of the American Woman.* New York: Harper & Row, 1965.
Spender, Dale. *Women of Ideas and What Men Have Done to Them.* Boston: Ark Paperbacks, 1983.

Men's Ideas/Women's Realities

Chapter 1
THE PROBLEM OF BIOLOGICAL DETERMINISM (1870–1890)

The older physiologists not only studied nature from the male standpoint—as, indeed, they must chiefly, being generally men—but they interpreted fact by the accepted theory that the male is the representative type of the species—the female a modification preordained in the interest of reproduction, and in that interest only or chiefly.

Antoinette Brown Blackwell, "Sex and Education,"
The Sexes Throughout Nature
(New York: G. P. Putnam's Sons, 1875), p. 17.

In nineteenth-century America, most people believed that salient physical and psychological traits distinguished men from women: The average man was taller, heavier, stronger, more aggressive, intelligent and creative than the average woman, while women were thought to be delicate, emotional creatures, with more acute powers of intuition and perception as well as an instinct for childrearing.[1]

Emerging scientific theories, particularly those of Charles Darwin (1809–1889) and Herbert Spencer (1820–1903) reinforced these common sense notions and lent the weight of scientific evidence to the belief in female distinctiveness.[2] These theories—the foundation of social-Darwinism—were called upon both by conservatives, who opposed any change in social conventions, and by supporters of the women's rights and reform movements, who advocated an extension of the female sphere. Conservatives appealed to social-Darwinism to defend the status quo, claiming that social institutions and practices had evolved in accordance with the laws of nature and were leading inevitably toward progress. The "Woman-Question agitators," as Grant Allen called them,[3] also embraced evolution's explanation of sexual differences because it affirmed their belief in the uniqueness of women and offered reassurance that proposed changes in women's activities would benefit society.

According to the theories of Darwin and Spencer, sexual differences or "secondary sexual characters," which included mental as well as physical traits, had evolved in response to the activities and functions of each sex, ultimately becoming rooted in biological structures. Thus, men and women had undergone divergent paths of evolution; men and women had fundamentally different natures—some would say diametrically opposed natures—because they had done different work. In the 1870s and 1880s, most men and women agreed on what these essential sexual differences were. Hence the point of contention was not so much what women's nature *was*, but what that nature *could* and *might* become. The controversy concerned the potential for modifying the feminine character and the probable effect such modification might have on the future development of society.

DARWIN'S THEORIES OF NATURAL AND SEXUAL SELECTION

Darwin's theories of natural and sexual selection, published in *On the Origin of Species* (1859), represented an attempt to explain the noticeable variations among individuals of a species and among different species. Natural selection, the broader of the two theories, derived from three observations, of which two had been treated by Malthus.[4] The first was the potential of all organisms to produce numerous offspring. Yet, despite this potential (and herein lay the second observation), the numbers of any given species remained fairly constant. From these two observations, Darwin concluded that organisms compete with each other for survival in a "struggle for existence." This deduction, along with a third observation that all organisms vary, led to the theory of natural selection or "the survival of the fittest"[5]: Some of the variations among individuals are advantageous in the struggle for survival, others are disadvantageous. A higher proportion of the individuals with favorable variations survive and reproduce, transmitting these traits to offspring, while a higher proportion of those with unfavorable variations die before they can reproduce. In this fashion natural selection was said to eliminate those characteristics that were unfavorable or harmful to individuals and to preserve those traits that assisted individuals in adapting to their surroundings.[6]

But as Darwin observed, not all existing variations seemed essential for the survival of the individual. The theory of sexual selection attempted to explain some of these nonessential variations. Operating in the same manner as natural selection to preserve favorable variations and eliminate unfavorable ones, sexual selection depended "not on a struggle for existence, but on a struggle between males for possession of the females." The result of sexual selection was not "death to the unsuccessful competitor, but few or no offspring."[7] Under natural selection, there was a limit to the amount of advantageous modification that could occur so long as the conditions of life did not change. Sexual selection, however, could produce many more modifications, thus accounting in part for the "frequent and extraordinary amount of variability presented by secondary sexual characters."[8]

Natural and sexual selection encompassed two distinct processes: The first was the initial appearance of variations (acquired traits); the second was the appearance, or lack of appearance, of variations in offspring (inherited traits). Although Darwin acknowledged that heredity was not well understood, he attempted to set forth his understanding of it.[9] First, Darwin believed that the majority of acquired traits were inheritable; second, he distinguished between the "transmission of characters" and the "development of characters." This distinction was particularly important for understanding the development of sexual differences. Sexual characteristics could be *transmitted* through both sexes but *developed* in one sex alone. In this fashion, women could pass along their fathers' muscular builds to sons, but not develop this build themselves. Third, the law of equal transmission (the transmission of characters to both sexes) did not always hold, although it was the most common form of inheritance. Consequently, Darwin postulated that inheritance could be limited by sex. Some characteristics would be transmitted only to offspring of the same sex. Darwin conceded that he did not know why the law of equal transmission governed the inheritance of some characters while the law of partial inheritance governed the transmission of other traits. But his observations led him to believe that characters acquired early in the life of an individual (in particular, physical attributes such as size and muscular structure) were acquired through natural selection and tended to be transmitted to both sexes, though not necessarily developed in both sexes. Characters that appeared later in life (in particular, mental attributes such as intelligence and reason) were acquired through sexual selection and seemed to be transmitted only to the offspring of the same sex.[10]

Darwin formulated these laws governing variation and inheritance with respect to all species and genera. In *Descent of Man* (1872), he reviewed these principles with specific reference to their operation in human beings,* emphasizing how natural and sexual selection, but particularly the latter with its associated law of partial inheritance, accounted for men's physical and mental superiority over women.

*I use gender-neutral words such as "human beings" and "people" where Darwin used "man" when the context suggests that the claim applies to both men and women. When a specific statement refers to one sex alone, I use "men" or "women" as appropriate.

Most of the significant "male" physical traits, such as greater size and strength, men had inherited from their "half-human male ancestors," these characteristics having been:

> preserved or even augmented during the long ages of man's savagery, by the success of the strongest and boldest men, both in the general struggle for life and in their contests for wives. . . . With civilised people the arbitrament of battle for the possession of the women has long ceased; on the other hand, the men, as a general rule, have to work harder than the women for their joint subsistence, and thus their greater strength will have been kept up.[11]

Moreover, because male savages had to fight for their mates, sexual selection had resulted in men acquiring greater courage, perseverance, intelligence, and imagination than women:

> But merely bodily strength and size would do little for victory, unless associated with courage, perseverance, and determined energy. With social animals, the young males have to pass through many a contest before they win a female, and the older males have to retain their females by renewed battles. They have, also, in the case of mankind, to defend their females, as well as their young, from enemies of all kinds, and to hunt for their joint subsistence. But to avoid enemies or to attack them with success, to capture wild animals, and to fashion weapons, requires the aid of the higher mental faculties, namely observation, reason, invention, or imagination. These various faculties will thus have been continually put to the test and selected during manhood; they will, moreover, have been strengthened by use during this same period of life. Consequently in accordance with the principle often alluded to [sexual selection and its corresponding law of partial inheritance], we might expect that they would at least tend to be *transmitted chiefly to the male offspring* at the corresponding period of manhood.[12]

Modern women trailed far behind modern men in mental ability because their female ancestors, having been protected and supported by men, had not needed and thus had not acquired higher mental faculties. They had not inherited these "sexual" characters from male ancestors, because, as it happened, these traits were "transmitted chiefly to male offspring." Darwin believed that in order for a woman to "reach the same standard as man, she ought, when nearly adult, to be trained to energy and perseverance, and to have her reason and imagination exercised to the highest point; and then she would probably transmit these qualities chiefly to her adult daughters."[13] But even then many women, over numerous generations, would have to be educated in this fashion, and then would have to reproduce more than other women before women, as a sex, could augment their mental ability. And because men still underwent a more severe struggle in order to maintain themselves, their wives and their children, they would be able "to keep up or even increase their mental powers, and, as a consequence, the present inequality between the sexes [would persist]."[14]

For Darwin then, it was clear that the realities of sexual and natural selection had resulted in men becoming "superior" to women, primarily because primitive men had acquired the physical and mental power needed to conquer other males in order to possess the opposite sex, transmitting their skill "more fully" to male than to female offspring. Women, never having done physical battle with each other, had not acquired the advanced secondary sexual characters that distinguished men. The only mitigating factor in all of this was that the "law of the equal transmission of characters to both sexes prevails with mammals." This was fortunate indeed, otherwise it would have been "probable that man would have become as superior in mental endowment to woman, as the peacock is in ornamental plumage to the peahen."[15]

SPENCER'S CONTRIBUTION TO THE THEORY OF SEXUAL DIFFERENCES

Along with Darwin, Spencer believed that physical organs and mental faculties changed in response to activities performed by the individual and that these modifications would be passed on to offspring. Hence, some of the so-called "feminine" traits such as maternal instinct, tenderness, and a love of the helpless were more pronounced in women than in men because women, as primary childrearers, had acquired these traits as a result of their daily interaction with children. Over generations, these characteristics had become fixed in biological structures. In Spencer's terminology, there appeared a "constitutional modification *produced* by excess of function."[16] Other traits, Spencer believed, derived from the history of the relation of women to men. In order to survive in primitive societies, women had learned to live with brutal men. Consequently, they had developed the art of pleasing, had learned to disguise their feelings, and had become skilled at quickly intuiting the feelings of those around them.[17]

For Spencer, as well as Darwin, social conditioning and sexual division of labor resulted in the evolution of different adaptive features between the sexes. Men, who had hunted and fought, had grown strong, intelligent, and courageous, transmitting these traits to their sons. Women, who had cared for children and adapted themselves to the "stronger sex," had become tender and perceptive, passing these traits on to their daughters. Both Darwin and Spencer considered sexual differences "natural." Natural in its most narrow sense meant simply that the traits had developed according to the laws of natural and sexual selection and were thus inevitable. As a consequence of Darwin's assumption that preserved characteristics had endured because they were advantageous to the organism and of Spencer's equation of evolution with social progress, natural also meant beneficial. To say, then, that the maternal instinct was natural in women meant that it was a beneficial and inevitable feature of womanhood. A woman without maternal instinct (a contradiction in terms) was an aberration, but also a threat to the future existence of society.

Spencer, adapting Hermann Helmholtz's theory of the conservation of energy to human development, argued that individuals had access to a limited fund of energy ("vital force"), which could be applied either to growth and development ("individual evolution") or to reproduction. Since the female reproductive system required more vital force than the male—something that apparently was sufficiently obvious to require no proof or explanation—women had less energy for physical and mental development than men. In short, the exacting demands of their reproductive system precluded women's full individual evolution. This view of the female reproductive system was reinforced by nineteenth-century physicians, who believed that the uterus was connected to the central nervous system and warned that shocks to the nervous system could impair the reproductive organs and vice versa: Changes in the reproductive cycle could adversely affect the emotional and mental states of women. Doctors attributed many mental and emotional disorders, from headaches to hysteria, to malfunctioning of the female reproductive system. Women's highly sensitive reproductive system remained precariously in balance only so long as no other strain was placed upon them. Thus, too much "brain work," especially during adolescence, was believed to lead to reproductive disorders and to endanger women's capacity to have children.[18]

Spencer suspected that mere physical ("bodily") labor rendered women less prolific, but admitted that the evidence for this assertion was inconclusive. But there was better evidence, Spencer argued, to support the claim that "absolute or relative infertility is generally produced in women by mental labour carried to excess." Thus, the "deficiency of reproductive power" among upper-class girls could be

> reasonably attributed to the overtaxing of their brains—an overtaxing which produces a serious reaction on the physique. This diminution of reproductive power is not shown only by the greater frequency of absolute sterility; nor is it shown only in the earlier cessation of child-bearing; but it is also shown in the very frequent inability of such women to suckle their infants. In its full sense, the reproductive power means the power to bear a well-developed infant, and to supply that infant with the natural food for the natural period. Most of the flat-chested girls who survive their high-pressure education, are incompetent to do this. Were their fertility measured by the number of children they could rear without artificial aid, they would prove relatively very infertile.[19]

This relationship between mental labor and fecundity did not hold true for men, Spencer believed, because "the cost of reproduction to males . . . [is] so much less than it is to females." To back up this pronouncement, Spencer resorted to the following argument:

> Special proofs that in men, great cerebral expenditure diminishes or destroys generative power, are difficult to obtain. It is, indeed, asserted that intense application to mathematics, requiring as it does extreme concentration of thought, is apt to have this result . . . [I]t is a matter of common remark how frequently men of unusual mental activity leave no offspring. But facts of this kind admit of another interpretation. The reaction of the brain on the body is so violent . . . that the incapacities caused . . . are probably often due more to *constitutitional* disturbance than to the direct deduction which excessive action entails.[20]

Thus, while the end result might have been the same for both sexes, with educated men and women both having lower fertility rates than uneducated men and women respectively, Spencer argued that the problem was a "constitutional" disturbance in men, while in women, it was a "reproductive" one.

Furthermore, because the onset of puberty occurred earlier in women and because women finished developing sooner than men, Spencer supposed that female brains and mental faculties had not had enough time to develop fully, these being "the latest products of human evolution."[21] Other scientists' observations of the differences between the brains of men and women were taken as confirmation of Spencer's hypotheses. Male brains weighed more and contained more convolutions; hence the greater size and complexity of men's brains seemed to account for their superior mental abilities. Darwin, who postulated a hierarchy of mental functions, beginning with instinct, intuition, and emotion and rising to imitation, imagination, and reason, also subscribed to the view that the less highly evolved female brain was better adapted to the lower mental functions: "It is generally admitted that with woman the powers of intuition, of rapid perception, and perhaps of imitation, are more strongly marked than in man; but some, at least, of these faculties are characteristic of the lower races, and therefore of a past and lower state of civilisation."[22] Until the twentieth century, many scientists accepted, but never proved, the link between brain size and intelligence.[23]

Within the lifetime of a single woman, Spencer believed, the potential for socially induced variability was small. Modifications in the size of a woman's brain or in the functioning of her reproductive system in response to changes in her activities occurred too slowly to make any significant difference in the mental characteristics of an individual. Darwin, although he did not put much faith in the conjecture, could conceive of the possibility for long-term evolutionary change, hypothesizing that if greater numbers of better educated women reproduced, eventually women would inherit more advanced mental faculties. Spencer's theory precluded the possibility of long-term evolutionary change because the female reproductive system required so much energy to function properly. There simply was not enough vital force left over for women to develop their intellects. Women who ignored the laws of nature by carrying mental labor to excess were subject to a greater risk of becoming infertile and would, Spencer believed, give birth to proportionately fewer children (and fewer daughters) than uneducated women. This "fact" of evolution implied that women, as a sex, could not advance in intellectual ability, since the greater fertility of uneducated women would inevitably lead to larger numbers of intellectually limited women.

Proponents of Darwin and Spencer acknowledged that these theories applied to the average woman, not to any particular individual who might be the exception that proved the rule. But social-Darwinists noticed very little variation among women, much less than among men. Men were taken to be more advanced than women in the evolutionary hierarchy because adult males seemed to differ significantly from children, whereas women seemed to resemble children in their emotional reactions and "illogical" thought. The greater degree of uniformity among women, as well as their similarity to children, were considered additional evidence of women's inferiority.[24]

WOMEN'S RESPONSES TO DARWIN AND SPENCER

Women who commented on the writings of Darwin and Spencer largely accepted the evolutionary account for the genesis of sexual traits. They agreed with social-Darwinists that women were unlike men. Evolution, which explained women's uniqueness in terms of their adaptation to the domestic sphere, seemed to them a valid and logical theory. The few women who ventured criticisms accepted most of Darwin's and Spencer's assumptions regarding women's biological and psychological characteristics, but they pointed out that these two scientists quickly dismissed the unique evolutionary development of women, overlooking or giving cursory treatment to the beneficial feminine traits that had appeared in women as a result of sexual selection. Dr. Frances White, in "Women's Place in Nature" (see pp. 25-32) observed that Darwin was primarily interested in how males competed for females and neglected to describe how sexual selection worked in the case of women. White pointed out that some of women's secondary sexual characters, notably their beauty, sweet low voice, mild character, and "devotional sentiment," were advantageous to mankind and deserved recognition.

Another woman to reply to Darwin and Spencer was Antoinette Brown Blackwell (1825-1921), the first female minister ordained in the United States. In an article "Sex and Evolution," published in 1875, Blackwell posited that an "equilibrium" or "equivalence" existed between the sexes of all species. She also criticized Darwin for not giving sufficient value to feminine traits, claiming that "woman's intuition" was not inferior to "man's reason," nor less developed than it, but an equally important mental faculty that was necessary for the advancement of society.[25]

Women also attempted to sever the connection between absolute brain size and intelligence. Nina Morais, in a response to Miss Hardaker's contention that men's larger brain accounted for their greater intellectual ability, argued that the association of brain size with mental capability was arbitrary and unfounded: The data (brain measurements) did not explain how the brain functioned and thus could not be taken as a determinant of intelligence. She countered with another observation: The *relative* brain size, the proportion of brain weight to total body weight, was greater in women than in men. (See "Science and the Woman Question" and "A Reply to Miss Hardaker," (pp. 33–47.)

Some female physicians, Dr. Mary Putnam Jacobi in particular, challenged the medical community's conception of the female reproductive system in an effort to change the widespread belief that menstruation was an inherently pathological and debilitating function. Male physicians commonly believed that the feminine reproductive cycle, from puberty to menopause, sapped women's strength and energy. In 1900, Dr. George J. Engelmann, president of the American Gynecology Society, dramatized the crippling nature of feminine sexuality:

> Many a young life is battered and forever crippled in the breakers of puberty; if it cross [sic] these unharmed and is not dashed to pieces on the rock of childbirth, it may still ground on the ever-recurring shallows of menstruation, and lastly, upon the final bar of the menopause ere protection is found in the unruffled waters of the harbor beyond reach of sexual storms.[26]

Doctors assumed that women would need to withdraw for a week every month in order to rest and advised them not to engage in physical activity—not to walk, dance, shop, ride or attend parties—during their menstrual periods.

In *The Question of Rest for Women During Menstruation* (1877), Jacobi argued that menstrual disorders were not "inherent" in the process and that physical and mental exertion did not by themselves lead to a greater incidence of pain or greater probability of malfunctioning. In fact, the amount of exercise and level of education correlated with less susceptibility to pain, findings that conflicted with the general beliefs of the medical community. (See Chapter 6.)

Another woman physician, Dr. Mary Taylor Bissell, disputed the claim that emotional fragility was a "natural"—meaning an inevitable and ineradicable—feature of femininity. Although Bissell believed that women were more susceptible to nervous disorders than men, she pointed out that social factors contributed to the emotional "weakness" of women. To strengthen women, both physically and emotionally, Bissell recommended that young girls be allowed to play outdoors, unrestrained by uncomfortable clothing. She also argued that nervous disorders were caused in part by boredom and advised women to make greater use of their intellectual faculties. (See "Emotions Versus Health in Women," pp. 48–53.)

OVERCOMING THE PROBLEM OF BIOLOGICAL DETERMINISM

The central features of Darwin's and Spencer's theories on evolution were the roles they assigned to the environment and to function in shaping the physical, biological constitution of the individual. Blackwell, White, Jacobi, and Bissell welcomed this aspect of social-Darwinian theory because it implied that changes in women's activities could improve women's intellectual and physical natures. Darwin, Spencer, and their proponents were more apt to view women's nature as a fixed entity, having become unmodifiable after thousands of years of evolutionary development. Evolution, intended as a theory of change and interpreted as such by women's rights advocates, was used by conservatives to oppose changes in social institutions and practices. Conservatives inverted the original proposition, women's nature has evolved in response to social functions, into: Social functions have developed naturally in accordance with women's nature. It was considered dangerous to the continuation of social progress to alter the female sphere because women's activities and women's nature were now perfectly suited to each other.

The question of whether women's nature was biologically fixed or socially modifiable underlay all other debates on the woman question. For, as opponents of women's education argued, what sense would it make to provide women with an education if they were physiologically unable to learn? Antisuffragists warned against the danger of enfranchising a sex that was incapable of rational thought. Interwoven with these biological questions were moral and political questions: Opponents of women's education greatly feared that extensive schooling would interfere with women's foremost social responsibility: bearing and raising children. And because many women shared society's conviction that women had a moral obligation to become mothers, motherhood continued to be considered women's ideal occupation. In the 1870s, the educational reformers who insisted women *could* be educated came under attack for overlooking the biological truths of womanhood. But those educational reformers who took up the motherhood crusade, arguing that women *should* be educated so that they would be better able to care for their children, were applauded and supported.[27] The woman movement partially succeeded in overcoming women's supposed mental and physical inferiority by appealing to what was thought to be distinctly feminine in women's nature and by convincing others that the proposed reforms were compatible with conventional views of motherhood and would even strengthen and promote these views. In this manner, the woman movement perpetuated and strengthened the ideal of conventional motherhood. Despite the social changes that ensued —the increase in availability of higher education for women and the appearance of a few women in the "male" professions—women's identity continued to be based in their biology, and having children and caring for families remained their foremost social and moral duties.

NOTES

1. These notions concerning women's nature were prevalent thoughout the scientific, medical, and popular literature in the late nineteenth and early twentieth centuries. See Charles Darwin, *Descent of Man and Selection in Relation to Sex* (New York: D. Appleton, 1906; 1st ed., 1871), pp. 568–569, 576 and G. Stanley Hall, *Adolescence: Its Psychology and Its Relations to Physiology, Anthropology, Sociology, Sex, Crime, Religion, and Education* (New York: D. Appleton, 1905), 2: 561–569. For a critique, see Helen Bradford Thompson, *The Mental Traits of Sex* (Chicago: Chicago University Press, 1903). Several excellent articles appeared in *The Popular Science Monthly* that could not be included in this collection for lack of space. For example, see Van de Warker, "The Genesis of Woman," 5: 269–277; Patrick, "The Psychology of Woman," 47: 209–225; and Brooks, "The Condition of Women," 15: 145–155; 347–356. An excellent review of the medical literature is contained in Carroll Smith-Rosenberg and Charles E. Rosenberg, "The Female Animal: Medical and Biological Views of Woman and Her Role in Nineteenth-Century America," *Journal of American History* 60 (September 1973): 332–356.
2. See Rosalind Rosenberg "In Search of Woman's Nature, 1850–1920," *Feminist Studies* 3 (Fall 1975): 142–143.
3. Grant Allen, "Plain Words on the Woman Question," *The Popular Science Monthly* 36 (December 1889): 170–181. This article appears in Chapter 4, pp. 125–131.
4. See Thomas Malthus, *An Essay on the Principles of Population as it Affects the Future Improvement of Society with Remarks on the Speculations of Mr. Godwin, M. Condorcet and other Writers* (London: J. Johnson, 1798).
5. "Survival of the fittest" was coined by Herbert Spencer in *The Principles of Biology* (New York: D. Appleton, 1864), p. 444. In a later edition of *The Principles of Biology* (New York: D. Appleton, 1898), p. 530, Spencer commented on the history of the term: "It will be seen that the argument naturally leads up to this expression—Survival of the Fittest—which was here used for the first time [in 1864]. Two years later (July, 1866) Mr. A. R. Wallace wrote to Mr. Darwin contending that it should be substituted for the expression 'Natural Selection.' Mr. Darwin demurred to this proposal. Among reasons for retaining his own expression he said that I had myself, in many cases, preferred it—'continually using the words Natural Selection.' (*Life and Letters*, Vol. III, pp. 45–6.) Mr. Darwin was quite right in his statement, but not right in the motive he ascribed to me. My reason for frequently using the phrase 'Natural Selection,' after the date at which the phrase 'Survival of the Fittest' was first used above, was that disuse of Mr. Darwin's phrase would have seemed like an endeavour to keep out of sight my own indebtedness to him, and the indebtedness of the world at large. The implied feeling has led me ever since to use the expressions Natural Selection and Survival of the Fittest with something like equal frequency."
6. Charles Darwin, *On The Origin of Species by Means of Natural Selection or the Preservation of the Favoured Races in the Struggle for Life* (London: John Murray, 5th ed., 1869; 1st ed., 1859) p. 96. Often it is not realized that Spencer developed his own evolutionary theory concurrently with Darwin, publishing "A Theory of Population, Deduced From the General

Law of Animal Fertility," *Westminster Review* 57 (April 1852): 468–501 seven years prior to the publication of Darwin's *On the Origin of Species* (1859). In "A Theory of Population," Spencer argued that the pressure of subsistence upon human population (caused by an excess of fertility) had a beneficial effect, resulting in the selection of the most skilled and intelligent individuals for survival and ultimately leading to the highest state of civilization. Thus, evolutionary development, in Spencer's mind, was equated with social progress.

7. Darwin, *Origin of Species*, p. 100.
8. Darwin, *Descent of Man*, p. 230.
9. See Darwin, *Origin of Species*, p. 14 Mendel's work on genetics, although originally published in 1865, was not well known until the twentieth century. Darwin, following Jean Baptiste Lamarck, believed in pangenesis: Characteristics acquired through interaction with the environment were transmitted through the blood stream in "gemmules" to the reproductive organs and thus passed on to offspring. This formulation of inheritance, however, seemed to Darwin somewhat inadequate. In the text, I discuss some of Darwin's speculations on inheritance.
10. Darwin, *Descent of Man*, pp. 231–238.
11. Ibid., p. 575.
12. Ibid., p. 576 (emphasis added).
13. Ibid., pp. 577–578.
14. Ibid., p. 578.
15. Ibid., p. 577.
16. Herbert Spencer, *The Principles of Biology* (New York: D. Appleton, 1864), p. 252; emphasis in original. This work also contains Spencer's views on variation (pp. 257–272) and heredity (pp. 239–256).
17. Spencer's most detailed exposition on biological sexual differences is contained in "Psychology of the Sexes," which first appeared in *The Popular Science Monthly* 4 (November 1873): 30–38, and was later published as the conclusion to chapter fifteen in *The Study of Sociology* (International Science Series, 1873). Unless otherwise noted, my discussion draws from "Psychology of the Sexes," which appears on pp. 17–24. However, Spencer's views on women are spread throughout his major works. For example, see *Education: Intellectual, Moral and Physical* (New York: D. Appleton, 1860), pp. 266–273 for his views on the education of girls and pp. 295–299 for the ill effects of brain-forcing on young women. In *Social Statics* (New York: D. Appleton, 1865; 1st ed., 1855), pp. 173–191 Spencer argued that women should have the same political rights as men, but in later editions of this work, he argued much less forcefully for this. See *Social Statics, Abridged and Revised* (New York: D. Appleton, 1892), pp. 73–80. In *The Principles of Ethics* (New York: D. Appleton, 1893), 2: 157–166, Spencer reversed himself on woman suffrage, arguing against the enfranchisement of women on the grounds that their impulsiveness would have an injurious effect upon legislation. *The Principles of Sociology* (New York: D. Appleton, 1896), 1: 725–744 contains an evolutionary and anthropological discussion of woman's position in society and pp. 767–770 give reasons why the political status of women must gradually improve. Lastly, *The Principles of Biology* (New York: D. Appleton, 1867), 2: 484–486 presents Spencer's views on women's reproductive system, which I discuss on pp. 6–7.
18. Excessive mental stimulation or "brain-forcing" was thought to endanger young women's health in other ways as well, causing anaemia (severe paleness), stunted growth or undue thinness, diathesis (nervousness), and hysteria. See for example, S. Weir Mitchell, *Wear and Tear, or Hints for the Overworked* (Philadelphia: J. B. Lippincott, 1874; 1st ed., 1871), pp. 33–41; and T. S. Clouston, "Female Education from a Medical Point of View," *The Popular Science Monthly* 24 (December 1883): 223–224.
19. Herbert Spencer, *The Principles of Biology* (New York: D. Appleton, 1867), 2: 485–486.

20. Ibid., pp. 486–487; emphasis added.
21. Quote is from "Psychology of the Sexes," see p. 19. Although Spencer believed that women were mentally inferior to men, in the 1850s and 1860s, he did not think this was a sufficient reason to deny women equal political rights. By the 1890s, however, Spencer's views on the rights of women had altered. See *Social Statics, Abridged and Revised* (New York: D. Appleton, 1892), pp. 73–79; and *The Principles of Ethics* (New York: D. Appleton, 1893), 2: 157–166.
22. Darwin, *Descent of Man*, p. 576. Darwin's ranking of mental faculties can be found in *Descent of Man*, pp. 65–75. See also Rosalind Rosenberg, *The Dissent From Darwin, 1890–1930: The New View of Women Among American Social Scientists* (Ph.D. dissertation, Stanford University, 1975), p. 13.
23. For a discussion of how scientists' and physicians' conceptions of women's nature were challenged successfully at the beginning of the twentieth century, see Rosalind Rosenberg, "In Search of Woman's Nature, 1850–1920," *Feminist Studies* 3 (Fall 1975): 141–153 and Rosalind Rosenberg, *Beyond Separate Spheres: Intellectual Roots of Modern Feminism* (New Haven: Yale University Press, 1982), especially pp. 84–113.
24. A leading proponent of the variation hypothesis was Havelock Ellis. See Ellis, "Variation in Man and Woman," *The Popular Science Monthly* 62 (January 1903): 237–253. For an excellent analysis of this theory, see Stephanie A. Shields, "The Variability Hypothesis: The History of a Biological Model of Sex Differences in Intelligence," *Signs: Journal of Women in Culture and Society* 7 (Summer 1982): 769–797.
25. Antoinette Brown Blackwell, "Sex and Evolution," *The Sexes Throughout Nature* (New York: G. P. Putnam's Sons, 1875), pp. 11–137. In response to Spencer's "Psychology of the Sexes," Blackwell wrote an article, "The Alleged Antagonism between Growth and Reproduction," which was published in *The Popular Science Monthly* 5 (September 1874): 606–610. Lack of space resulted in the omittance of this article from this collection. It can also be found in *The Sexes Throughout Nature*, pp. 138–148.
26. George J. Engelmann, "The American Girl of Today," President's Address, *American Gynecology Society*, cited in G. Stanley Hall, *Adolescence* 2: 588.
27. Carl N. Degler makes a similar argument regarding the tactics employed by suffragists. See Degler, *At Odds: Women and the Family in America from the Revolution to the Present* (New York: Oxford University Press, 1980), pp. 357–358.

ARTICLES ON BIOLOGICAL DETERMINISM APPEARING IN *THE POPULAR SCIENCE MONTHLY*

(Arranged by Date of Publication)

Herbert Spencer. "Psychology of the Sexes," (November 1873) 4:30–38.
Sara S. Hennell. "Mr. Spencer and the Women," (April 1874) 4: 746–747.
Henry Maudsley. "Sex in Mind and in Education," (June 1874) 5: 198–215.
Ely Van de Warker. "The Genesis of Woman," (July 1874) 5: 269–277.
Antoinette Brown Blackwell. "The Alleged Antagonism Between Growth and Reproduction," (September 1874) 5: 606–610.
Frances Emily White. "Woman's Place in Nature," (January 1875) 6: 292–301.
Ely Van de Warker. "Sexual Cerebration," (July 1875) 7: 287–301.

Ely Van de Warker. "The Relations of Women to Crime," (November 1875 and January 1876) 8: 1–16; 334–344.
Ely Van de Warker. "The Relations of Sex to Crime," (April 1876) 8: 727–736.
"The Question of Rest for Women During Menstruation," (December 1877) 12: 241–244. [Book review]
Mary Putnam Jacobi. "The Question of Rest for Woman," (February 1878) 12: 492–494.
T. Clifford Allbutt. "On Brain-Forcing," (June 1878) 13:217–230.
"Biology and 'Woman's Rights,'" (December 1878) 14: 201–213.
W. K. Brooks. "The Condition of Women from a Zoological Point of View," (June 1879 and July 1879) 15: 145–155; 347–356.
Benjamin Ward Richardson. "Dress in Relation to Health," (June 1880) 17: 182–199.
William Henry Flower. "Fashion in Deformity," (October 1880) 17: 721–742.
G. Delauney. "Equality and Inequality in Sex," (December 1881) 20: 184–192.
Miss M. A. Hardaker. "Science and the Woman Question," (March 1882) 20: 577–584.
Nina Morais. "A Reply to Miss Hardaker on the Woman Question," (May 1882) 21: 70–78.
Mrs. Z. D. Underhill. " A Premature Discussion," (July 1882) 21: 376–377 [response to Hardaker].
Charles Roberts. "Bodily Deformities in Girlhood," (January 1883) 22: 322–328.
William A. Hammond. "The Relations Between the Mind and the Nervous System," (November 1884) 26: 1–20.
W. K. Brooks. "Influences Determining Sex," (January 1885) 26: 323–330.
Lucy M. Hall. "Physical Training of Girls," (February 1885) 26: 495–498.
Eugene L. Richards. "The Influence of Exercise Upon Health," (July 1886) 29: 322–335.
Ambrose L. Ranney. "The Care of the Brain," (July 1886) 29: 386–393.
A. B. Rosenberry. "The Physiology of Exercise," (October 1886) 29: 841.
William A. Hammond. "Brain-Forcing in Childhood," (April 1887) 30: 721–732.
Frances Emily White. "Hygiene as a Basis of Morals," (May 1887) 31: 67–78.
Helen H. Gardener. "Sex and Brain-Weight," (June 1887) 31: 266–268.
Joseph Simms. "Human Brain-Weights," (July 1887) 31: 355–359.
George J. Romanes. "Mental Differences of Men and Women," (July 1887) 31: 383–401.
William A. Hammond. "Men's and Women's Brains," (August 1887) 31: 554–558.
Helen H. Gardener. "More About Men's and Women's Brains," (September 1887) 31: 698–700.
William A. Hammond. "An Explanation," (October 1887) 31: 846.
Mary T. Bissell. "Emotions Versus Health in Women," (February 1888) 32: 504–510.
Mary T. Bissell. "A Correction," (May 1888) 33: 125–126.
Herr Dr. Bilsinger. "Modern Nervousness and Its Cure," (November 1892) 42: 90–93.
C. E. Brewster. "The Symmetrical Development of Our Young Women," (December 1892) 42: 217–223.
Mary T. Bissell. "Athletics for City Girls," (December 1894) 46: 145–153.
G. T. W. Patrick. "The Psychology of Woman," (June 1895) 47: 209–225.
William Hirsch. "Epidemics of Hysteria," (August 1896) 49: 544–549.
Sophia Foster Richardson. "Tendencies in Athletics for Women in Colleges and Universities," (February 1897) 50: 517–526.
Rebecca Sharpe. "Random Remarks of a Lady Scientist," (March 1901) 58: 548–550.
Otis T. Mason. "Environment in Relation to Sex in Human Culture," (February 1902) 60: 336–345.
Havelock Ellis. "Variation in Man and Woman," (January 1903) 62: 237–253.
J. Madison Taylor. "Conservation of Human Energy, Preservation of Beauty," (September 1904) 65: 397–413.
Nellie Comins Whitaker, "The Health of American Girls," (September 1907) 71: 234–245.
Pearce Bailey. "Hysteria as an Asset," (June 1909) 74: 568–574.
Otto Charles Glaser. "The Constitutional Conservatism of Women," (September 1911) 79: 299–302.

SUGGESTIONS FOR FURTHER READING

Primary Sources

Blackwell, Antoinette Brown. *The Sexes Throughout Nature.* New York: G. P. Putnam's Sons, 1875.
Darwin, Charles. *The Descent of Man and Selection in Relation to Sex.* New York: D. Appleton, 1906; 1st ed., 1871.
Darwin, Charles. *On the Origin of Species by Means of Natural Selection, Or the Preservation of the Favoured Races in the Struggle for Life.* London: John Murray, 1869; 1st ed., 1859.
Duffey, Eliza. *The Relations of the Sexes.* New York: Wood and Holbrook, 1876.
Ellis, Havelock. *Man & Woman: A Study of Human Secondary Sexual Characters.* London: A. and C. Black, 1930; 1st ed., 1894.
Geddes, P. and Thomson, J. A. *The Evolution of Sex.* New York: Humboldt, 1890.
Jacobi, Mary Putnam. *The Question of Rest for Women During Menstruation.* New York: G. P. Putnam's Sons, 1877.
Lamarck, Jean Baptiste. *Zoological Philosophy.* London: Macmillan, 1913; 1st ed., 1809.
Malthus, Thomas. *An Essay on the Principles of Population as it Affects the Future Improvement of Society with Remarks on the Speculations of Mr. Godwin, M. Condorcet and Other Writers.* London: J. Johnson, 1798.
Spencer, Herbert. *The Principles of Biology.* 2 vols. New York: D. Appleton, 1867; Vol. 1 first published in 1864.
Spencer, Herbert. *The Study of Sociology.* New York: D. Appleton, 1896.
Thompson, Helen Bradford. *The Mental Traits of Sex.* Chicago: Chicago University Press, 1903.

Secondary Sources

Darwin, Spencer, and Evolutionary Theory

Bannister, Robert C. *Social Darwinism, Science and Myth in Anglo-American Social Thought.* Philadelphia: Temple University Press, 1979.
Bleir, Ruth. *Science and Gender: A Critique of Biology and Its Theories on Women.* New York: Pergamon Press, 1984.
Eisely, Loren. *Darwin's Century.* New York: Doubleday, 1958.
Hofstadter, Richard. *Social Darwinism in American Thought.* Boston: Beacon Press, 1955.
Hubbard, Ruth, "Have Only Men Evolved?" *Women Look at Biology Looking at Women.* Edited by Ruth Hubbard, Mary Sue Henifin, and Barbara Fried. Cambridge: Schenkman, 1981, pp. 7–36. Also in *Biological Woman—The Convenient Myth.* Edited by Ruth Hubbard, Mary Sue Henifin, and Barbara Fried. Cambridge: Schenkman, 1982, pp. 17–46.
Lowe, Marian, and Hubbard, Ruth, eds. *Woman's Nature: Rationalizations of Inequality.* New York: Pergamon Press, 1983.
Mosedale, Susan, "Science Corrupted: Victorian Biologists Consider The Woman Question," *Journal of the History of Biology* 11 (Spring 1978): 1–55.
Sayers, Janet. *Biological Politics: Feminist and Anti-Feminist Perspectives.* London and New York: Tavistock Publications, 1982.
Shields, Stephanie, A. "The Variability Hypothesis: The History of a Biological Model of Sex Differences in Intelligence," *Signs: Journal of Women in Culture and Society* 7 (Summer 1982): 769–797.

Women's Nature, Health and Sexuality

Cott, Nancy F. "Passionlessness: An Interpretation of Victorian Sexual Ideology, 1790–1850," *Signs: A Journal of Women in Culture and Society* 4 (Winter 1978): 219–236. Reprinted in Cott, Nancy F., and Pleck, Elizabeth H., eds. *A Heritage of Her Own: Toward a New Social*

History of American Women. New York: Simon and Schuster, 1979, pp. 162–181.

Degler, Carl N. "What Ought to Be and What Was: Women's Sexuality in the Nineteenth Century," *American Historical Review* 79 (December 1974): 1467–1490.

DuBois, Ellen Carol, and Gordon, Linda. "Seeking Ecstasy on the Battlefield: Danger and Pleasure in Nineteenth-Century Feminist Sexual Thought," *Feminist Studies* 9 (Spring 1983): 7–25.

Ehrenreich, Barbara, and English, Deirdre. *For Her Own Good: 150 Years of the Experts' Advice to Women*. New York: Anchor Press, 1979.

Gordon, Michael. "From an Unfortunate Necessity to a Cult of Mutual Orgasm: Sex in American Marital Education Literature 1830–1940," *Studies in the Sociology of Sex*. Edited by James M. Henslin. New York: Meredith, 1971, pp. 53–77.

Haller, John S., and Haller, Robin M. *The Physician and Sexuality in Victorian America*. Urbana: University of Illinois Press, 1974.

Klein, Viola. *The Feminine Character: History of an Ideology*. New York: International Universities Press, 1949.

McGovern, James R. "The American Woman's Pre-World War I Freedom in Manners and Morals," *Journal of American History* 55 (September 1968): 315–333.

Rosenberg, Charles E. "Sexuality, Class and Role in 19th-Century America," *American Quarterly* 25 (May 1973): 131–153.

Rosenberg, Rosalind. *Beyond Separate Spheres: Intellectual Roots of Modern Feminism*. New Haven: Yale University Press, 1982.

Rosenberg, Rosalind. "In Search of Woman's Nature, 1850–1920," *Feminist Studies* 3 (Fall 1975): 141–153.

Smith-Rosenberg, Carroll. "The Female World of Love and Ritual: Relations Between Women in Nineteenth-Century America," *Signs: A Journal of Women in Culture and Society* 1 (Autumn 1975): 1–29. Reprinted in Cott, Nancy F., and Pleck, Elizabeth H., eds. *A Heritage of Her Own: Toward a New Social History of American Women*. New York: Simon and Schuster, 1979, pp. 311–342.

Smith-Rosenberg, Carroll. "The Hysterical Woman: Sex Roles and Role Conflict in 19th-Century America" *Social Research* 39 (Winter 1972): 652–678.

Smith-Rosenberg, Carroll. "Puberty to Menopause: The Cycle of Femininity in Nineteenth-Century America," *Feminist Studies* 1 (Fall 1973): 58–72. Reprinted in Hartman, Mary S., and Banner, Lois, eds. *Clio's Consciousness Raised: New Perspectives on the History of Women*. New York: Harper & Row, 1974, pp. 23–37.

Smith-Rosenberg, Carroll, and Rosenberg, Charles E. "The Female Animal: Medical and Biological Views of Woman and her Role in Nineteenth-Century America," *Journal of American History* 60 (September 1973): 332–356.

Wood, Ann Douglas. " 'The Fashionable Diseases': Women's Complaints and Their Treatment in Nineteenth-Century America," *Journal of Interdisciplinary History* 4 (Summer 1973): 25–52. Reprinted in Hartman, Mary S., and Banner, Lois, eds. *Clio's Consciousness Raised: New Perspectives on the History of Women*. New York: Harper & Row, 1974, pp. 1–37.

Herbert Spencer (1820-1903), an English philosopher, did more than any other person of his age to integrate his own and others' scientific theories into popular cultural thought. From reading Lyell, Spencer became an advocate of Lamarckian principles, believing that function exists prior to structure, that the results of the use and disuse of organs eventually become incarnate in structural adaptions, and thus organisms may acquire traits through the exercise of function, which may then be passed on through hereditary transmission. Coining the phrase, "Survival of the Fittest," Spencer argued that the pressure of subsistence upon a population would result in the survival of the most skilled individuals and ultimately lead to beneficial changes in the society. In the following article, Spencer investigated the origin of sexual differences and presented an account that had much in common with Darwin's theory of sexual selection.

PSYCHOLOGY OF THE SEXES

HERBERT SPENCER

November 1873

ONE further instance of the need for psychological inquiries as guides to sociological conclusions may be named — an instance of quite a different kind, but one no less relevant to questions of the time. I refer to the comparative psychology of the sexes. Women, as well as men, are units in a society, and tend by their natures to give that society certain traits of structure and action. Hence the question, Are the mental natures of men and women the same? is an important one to the sociologist. If they are, an increase of feminine influence is not likely to affect the social type in a marked manner. If they are not, the social type will inevitably be changed by increase of feminine influence.

That men and women are mentally alike, is as untrue as that they are alike bodily. Just as certainly as they have physical differences which are related to the respective parts they play in the maintenance of the race, so certainly have they psychical differences, similarly related to their respective shares in the rearing and protection of offspring. To suppose that along with the unlikenesses between their parental activities there do not go unlikenesses of mental faculties, is to suppose that here alone in all Nature there is no adjustment to special powers to special functions.[1]

---

[1] The comparisons ordinarily made between the minds of men and women are faulty in many ways, of which these are the chief:
Instead of comparing either the average of women with the average of men, or the *élite* of women with the *élite* of men, the common course is to compare the *élite* of women with the average of men. Much the same erroneous impression results as

Two classes of differences exist between the psychical, as between the physical, structures of men and women, which are both determined by this same fundamental need—adaptation to the paternal and maternal duties. The first set of differences is that which results from a somewhat earlier arrest of individual evolution in women than in men, necessitated by the reservation of vital power to meet the cost of reproduction. Whereas, in man, individual evolution continues until the physiological cost of self-maintenance very nearly balances what nutrition supplies, in woman, an arrest of individual development takes place while there is yet a considerable margin of nutrition: otherwise there could be no offspring. Hence the fact that girls come earlier to maturity than boys. Hence, too, the chief contrasts in bodily form: the masuline figure being distinguished from the feminine by the greater relative sizes of the parts which carry on external actions and entail physiological costs—the

would result if the relative statures of men and women were judged by putting very tall women side by side with ordinary men.

Sundry manifestations of nature in men and women are greatly perverted by existing social conventions upheld by both. There are feelings which, under our predatory *régime*, with its adapted standard of propriety, it is not considered manly to show; but which, contrariwise, are considered admirable in women. Hence, repressed manifestations in the one case, and exaggerated manifestations in the other; leading to mistaken estimates.

The sexual sentiment comes into play to modify the behavior of men and women to one another. Respecting certain parts of their general characters, the only evidence which can be trusted is that furnished by the conduct of men to men, and of women to women, when placed in relations which exclude the personal affections.

In comparing the intellectual powers of men and women, no proper distinction is made between receptive faculty and originative faculty. The two are scarcely commensurable; and the receptivity may, and frequently does, exist in high degree where there is but a low degree of originality, or entire absence of it.

Perhaps, however, the most serious error usually made in drawing these comparisons is, that of overlooking the limit of normal mental power. Either sex under special stimulations is capable of manifesting powers ordinarily shown only by the other; but we are not to consider the deviations so caused as affording proper measures. Thus, to take an extreme case, the mammae of men will, under special excitation, yield milk: there are various cases of gynaecomasty on record, and in families, infants whose mothers have died have been thus saved. But this ability to yield milk, which, when exercised, must be at the cost of masculine strength, we do not count among masculine attributes. Similarly, under special discipline, the feminine intellect will yield products higher than the intellects of most men can yield. But we are not to count this as truly feminine if it entails decreased fulfillment of the maternal functions. Only that mental energy is normally feminine which can coexist with the production and nursing of the due number of healthy children. Obviously a power of mind which, if general among the women of a society, would entail disappearance of the society, is a power not to be included in an estimate of the feminine nature as a social factor.

limbs, and those thoracic viscera which their activity immediately taxes. And hence, too, the physiological truth that, throughout their lives, but especially during the child-bearing age, women exhale smaller quantities of carbonic acid, relatively to their weights, than men do; showing that the evolution of energy is relatively less as well as absolutely less. This rather earlier cessation of individual evolution thus necessitated, showing itself in a rather smaller growth of the nervo-muscular system, so that both the limbs which act and the brain which makes them act are somewhat less has two results on the mind. The mental manifestations have somewhat less of general power or massiveness; and beyond this there is a perceptible falling short in those two faculties, intellectual and emotional, which are the latest products of human evolution — the power of abstract reasoning and that most abstract of the emotions, the sentiment of justice — the sentiment which regulates conduct irrespective of personal attachments and the likes or dislikes felt for individuals.[2]

After this quantitative mental distinction, which becomes incidentally qualitative by telling most upon the most recent and most complex faculties, there come the qualitative mental distinctions consequent on the relations of men and women to their children and to one another. Though the parental instinct, which, considered in its essential nature, is a love of the helpless, is common to the two; yet it is obviously not identical in the two. That the particular form of it which responds to infantine helplessness is more dominant in women than in men, cannot be questioned. In man the instinct is not so habitually excited by the very helpless, but has a more generalized relation to all the relatively weak who are dependent upon him. Doubtless, along with this more specialized instinct in women, there go special aptitudes for dealing with infantine life — an adapted power of intuition and a fit adjustment of behavior. That there is here a mental specialization, joined with the bodily specialization, is undeniable; and this mental specialization, though primarily related to the rearing of offspring, affects in some degree the conduct at large.

The remaining qualitative distinctions between the minds of men and women are those which have grown out of their mutual relation as stronger and weaker. If we trace the genesis of human character, by considering the conditions of existence through which the human race passed in early barbaric times and during civilization, we shall see that the weaker sex has naturally acquired certain mental traits by its dealings with the stronger. In the course of the struggles for existence among wild tribes, those tribes survived in which the men were not only powerful and courageous, but aggressive, unscrupulous, intensely egoistic. Necessarily, then, the men of the conquering races

---

[2]Of course it is to be understood that in this, and in the succeeding statements, reference is made to men and women of the same society, in the same age. If women of a more-evolved race are compared with men of a less-evolved race, the statement will not be true.

which gave origin to the civilized races, were men in whom the brutal characteristics were dominant; and necessarily the women of such races, having to deal with brutal men, prospered in proportion as they possessed, or acquired, fit adjustments of nature. How were women, unable by strength to hold their own, otherwise enabled to hold their own? Several mental traits helped them to do this.

We may set down, first, the ability to please, and the concomitant love of approbation. Clearly, other things equal, among women living at the mercy of men, those who succeeded most in pleasing would be the most likely to survive and leave posterity. And (recognizing the predominant descent of qualities on the same side) this, acting on successive generations, tended to establish, as a feminine trait, a special solicitude to be approved, and an aptitude of manner to this end.

Similarly, the wives of merciless savages must, other things equal, have prospered in proportion to their powers of disguising their feelings. Women who betrayed the state of antagonism produced in them by ill-treatment would be less likely to survive and leave offspring than those who concealed their antagonism; and hence, by inheritance and selection, a growth of this trait proportionate to the requirement. In some cases, again, the arts of persuasion enabled women to protect themselves, and by implication their offspring, where, in the absence of such arts, they would have disappeared early, or would have reared fewer children. One further ability may be named as likely to be cultivated and established—the ability to distinguish quickly the passing feelings of those around. In barbarous times, a woman who could, from a movement, tone of voice, or expression of face, instantly detect in her savage husband the passion that was rising, would be likely to escape dangers run into by a woman less skilled in interpreting the natural language of feeling. Hence, from the perpetual exercise of this power, and the survival of those having most of it, we may infer its establishment as a feminine faculty. Ordinarily, this feminine faculty, showing itself in an aptitude for guessing the state of mind through the external signs, ends simply in intuitions formed without assignable reasons; but when, as happens in rare cases, there is joined with it skill in psychological analysis, there results an extremely remarkable ability to interpret the mental states of others. Of this ability we have a living example never hitherto paralleled among women, and in but few, if any, cases exceeded among men.

Of course, it is not asserted that the specialties of mind here described as having been developed in women, by the necessities of defense in their dealings with men, are peculiar to them: in men also they have been developed as aids to defense in their dealings with one another. But the difference is, that, whereas, in their dealings with one another, men depended on these aids only in some measure, women in their dealings with men depended upon them almost wholly—within the domestic circle as well as without it. Hence, in virtue of that partial limitation of heredity by sex, which many facts throughout Nature show us, they have come to be more marked in women than in men.[3]

One further distinctive mental trait in women springs out of the relation of the sexes as adjusted to the welfare of the race. I refer to the effect which the manifestation of power of every kind in men has in determining the attachments of women. That this is a trait inevitably produced will be manifest, on asking what would have happened if women had by preference attached themselves to the weaker men. If the weaker men had habitually left posterity when the stronger did not, a progressive deterioration of the race would have resulted. Clearly, therefore, it has happened (at least since the cessation of marriage by capture or by purchase has allowed feminine choice to play an important part) that, among women unlike in their tastes, those who were fascinated by power, bodily or mental, and who married men able to protect them and their children, were more likely to survive in posterity than women to whom weaker men were pleasing, and whose children were both less efficiently guarded and less capable of self-preservation if they reached maturity. To this admiration for power, caused thus inevitably, is ascribable the fact that sometimes commented upon as strange, that women will continue attached to men who use them ill, but whose brutality goes along with power, more than they will continue attached to weaker men who use them well. With this admiration of power, primarily having this function, there goes the admiration of power in general, which is more marked in women

---

[3] As the validity of this group of inferences depends on the occurrence of that partial limitation of heredity of sex here assumed, it may be said that I should furnish proof of its occurrence. Were the place fit, this might be done. I might detail evidence that has been collected showing the much greater liability there is for a parent to bequeath malformations and diseases to children of the same sex, than to those of the opposite sex. I might cite the multitudinous instances of sexual distinctions, as of plumage in birds and coloring in insects, and especially those marvelous ones of dimorphism and polymorphism among females of certain species of Lepidoptera, as necessarily implying (to those who accept the Hypothesis of Evolution) the predominant transmission of traits to descendants of the same sex. It will suffice, however, to instance, as more especially relevant, the cases of sexual distinctions within the human race itself, which have arisen in some varieties and not in others. That in some varieties the men are bearded, and in others not, may be taken as strong evidence of this partial limitation of heredity; and, perhaps, still stronger evidence is yielded by that peculiarity of feminine form found in some of the negro races, and especially the Hottentots, which does not distinguish to any such extent the women of other races from the men. There is also the fact, to which Agassiz draws attention, that, among the South American Indians, males and females differ less than they do among the negroes and the higher races; and this reminds us that among European and Eastern nations the men and women differ, both bodily and mentally, not quite in the same ways and to the same degrees, but in somewhat different ways and degrees—a fact which would be inexplicable were there no partial limitation of heredity by sex.

than in men, and shows itself both theologically and politically. That the emotion of awe aroused by contemplating whatever suggests transcendent force or capacity, which constitutes religious feeling, is strongest in women, is proved in many ways. We read that among the Greeks the women were more religiously excitable than the men. Sir Rutherford Alcock tells us of the Japanese that "in the temples it is very rare to see any congregation except women and children; the men, at any time, are very few, and those generally of the lower classes." Of the pilgrims to the temple of Juggernaut, it is stated that "at least five-sixths, and often nine-tenths, of them are females." And we are also told of the Sikhs, that the women believe in more gods than the men do. Which facts, coming from different races and times, sufficiently show us that the like fact, familiar to us in Roman Catholic countries, and to some extent at home, is not, as many think, due to the education of women, but has a deeper cause in natural character. And to this same cause is in like manner to be ascribed the greater respect felt by women for all embodiments and symbols of authority, governmental and social.

Thus the *a priori* inference, that fitness for their respective parental functions implies mental differences between the sexes, as it implies bodily differences, is justified; as is also the kindred inference that secondary differences are necessitated by their relations to one another. Those unlikenesses of mind between men and women, which, under the conditions, were to be expected, are the unlikenesses we actually find. That they are fixed in degree, by no means follows: indeed, the contrary follows. Determined as we see they some of them are by adaptation of primitive women's natures to the natures of primitive men, it is inferable that as civilization readjusts men's natures to higher social requirements, there goes on a corresponding readjustment between the natures of men and women, tending in sundry respects to diminish their differences. Especially may we anticipate that those mental peculiarities developed in women, as aids to defense against men in barbarous times, will diminish. It is probable, too, that, though all kinds of power will continue to be attractive to them, the attractiveness of physical strength and the mental attributes that commonly go along with it will decline, while the attributes which conduce to social influence will become more attractive. Further, it is to be anticipated that the higher culture of women, carried on within such limits as shall not unduly tax the *physique* (and here, by higher culture, I do not mean mere language-learning and an extension of the detestable cramming-system at present in use), will in other ways reduce the contrast. Slowly leading to the result everywhere seen throughout the organic world, of a self-preserving power inversely proportionate to the race-preserving power, it will entail a less early arrest of individual evolution, and a diminution of those mental differences between men and women which the early arrest produces.

Admitting such to be changes which the future will probably see wrought out, we have meanwhile to bear in mind these traits of intellect and feeling which distinguish women, and to take note of them as factors in social phenomena—much more important factors than we commonly suppose.Considering them in the above order, we may note, first, that the love of the helpless, which in her maternal capacity woman displays in a more special form than man, inevitably affects all her thoughts and sentiments; and, this being joined in her with a less developed sentiment of abstract justice, she responds more readily when appeals to pity are made than when appeals are made to equity. In foregoing chapters we have seen how much our social policy disregards the claims of individuals to whatever their efforts purchase, so long as no obvious misery is brought on them by the disregard; but, when individuals suffer in ways conspicuous enough to excite commiseration, they get aid, and often as much aid if their sufferings are caused by themselves as if they are caused by others—often greater aid, indeed. This social policy, to which men tend in an injurious degree, women tend to still more. The maternal instinct delights in yielding benefits apart from deserts; and, being partially excited by whatever shows a feebleness that appeals for help (supposing antagonism has not been aroused), carries into social action this preference of generosity to justice, even more than men do. A further tendency, having the same general direction, results from the aptitude which the feminine intellect has to dwell on the concrete and proximate rather than on the abstract and remote. The representative faculty in women deals quickly and clearly with the personal, the special, and the immediate; but less readily grasps the general and the impersonal. A vivid imagination of simple direct consequences mostly shuts out from her mind the imagination of consequences that are complex and indirect. The respective behaviors of mothers and fathers to children sufficiently exemplify this difference: mothers thinking chiefly of present effects on the conduct of children, and regarding less the distant effects on their characters; while fathers often repress the promptings of their sympathies with a view to ultimate benefits. And this difference between their ways of estimating consequences, affecting their judgments on social affairs as on domestic affairs, makes women err still more than men do in seeking what seems an immediate public good without thought of distant public evils. Once more, we have in women the predominant awe of power and authority, swaying their ideas and sentiments about all institutions. This tends toward the strengthening of governments, political and ecclesiastical. Faith in whatever presents itself with imposing accompaniments is, for the reason above assigned, especially strong in women. Doubt, or criticism, or calling in question of things that are established, is rare among them. Hence in public affairs their influence goes toward the maintenance of controlling agencies, and does not resist the extension of such agencies; rather, in pursuit of immediate promised benefits, it urges on that extension; since the concrete good in view excludes from their thoughts the remote evils of multiplied restraints. Reverencing power more than men do, women, by implication, respect freedom less—freedom, that is, not of the nominal kind, but of that real kind which consists in the ability of each to carry on his own life without hindrance from others, so long as he does not hinder them.

As factors in social phenomena, these distinctive mental traits of women have ever to be remembered. Women have in all times played a part, and, in modern days, a very notable part, in determining social arrangements. They act both directly and indirectly. Directly, they take a large, if not the larger, share in that ceremonial government which supplements the political and ecclesiastical governments; and as supporters of these other governments, especially the ecclesiastical, their direct aid is by no means unimportant. Indirectly, they act by modifying the opinions and sentiments of men—first, in education, when the expression of maternal thoughts and feelings affects the thoughts and feelings of boys, and afterward in domestic and social intercourse, during which the feminine sentiments sway men's public acts, both consciously and unconsciously. Whether it is desirable that the share already taken by women in determining social arrangements and actions should be increased, is a question we will leave undiscussed. Here I am concerned merely to point out that, in the course of a psychological preparation for the study of Sociology, we must include the comparative psychology of the sexes; so that, if any change is made, we may make it knowing what we are doing. . . .

Frances Emily White (1832–1903) was well known during her lifetime as one of a few women physicians in the United States. She graduated from the Woman's Medical College in Philadelphia in 1872 and served as professor of physiology and hygiene there from 1876 to 1903. White was a frequent contributor to The Popular Science Monthly, *writing on a variety of topics. In the following selection, she reviewed and criticized Darwin's and Spencer's account of women's evolutionary development.*

WOMAN'S PLACE IN NATURE

FRANCES EMILY WHITE, M.D.

January 1875

. . . .

In Darwin's "Descent of Man" we have an accumulation of statements of facts gathered from vast fields of observation by many of the foremost naturalists of the age; and his deductive interpretations of these facts seem to have been accepted by a majority of the leading naturalists and physicists of the day. Such being the case, we are warranted in making this work the basis of our inquiry, thus looking at the subject from the side of natural history. Should some additional deductions and interpretations be brought out, it is hoped that they will not be found either forced or imaginary.

In order to a clear understanding of the line of reasoning employed, we must distinguish between the terms "natural selection" and "sexual selection," as used by Darwin. The traits resulting from these two processes are under a different law of heredity—those arising through natural selection being transmitted alike to the young of both sexes, while the results of sexual selection are inherited mainly by the adults of the corresponding sex. It will be seen that these are important laws, and that they furnish a key to our inquiry into the conditions and influences which have resulted in the woman of to-day. Under the operation of this second law (quoting from the "Descent of Man"), "it is the male which, with rare exceptions, has been chiefly modified—the female remaining more like the young of her own species, and more like the other members of the same group. The cause of this seems to lie in the males of almost all animals having stronger passions than the females. Hence it is that the males fight together, and sedulously display their charms before the females; and those which are victorious transmit their superiority to their male offspring." The question naturally arises, How have the males of the lower animals acquired this greater strength of passion? Says Darwin: "It would be no advantage, and some loss of power, if both sexes were mutually to search for each other; but why should the male almost always be the seeker?" Reasoning from the lower forms of life, he points out the fact that the ovules, developed in the female organs of plants, must be nourished for a time after fertilization; hence the pollen is necessarily brought to them—being conveyed to the stigma by insects, by winds, or by the spontaneous movements of the stamens themselves, upon which the pollen grows. "With lowly-organized animals permanently affixed to the same spot, and having their sexes separate, the male element is invariably brought to the female; and we can see the reason; for the ova, even if detached before being fertilized, and not requiring subsequent nourishment or protection, would be, from their larger relative size, less easily transported than the male element. . . . In case of animals having little power of locomotion, the fertilizing element must be trusted to the risk of at least a short transit through the waters of the sea. It, would, therefore, be a great advantage to such animals, as their organization became perfected, if the males, when ready to emit the fertilizing element, were to acquire the habit of approaching the female. The males of various lowly-organized animals having thus aboriginally acquired the habit of seeking the females, the same habit would naturally be transmitted to their more highly-developed male descendants; and, in order that they should become efficient seekers, they would have to be endowed with strong passions. The acquirement of such passions would naturally follow from the more eager males leaving a larger number of offspring than the less eager."

I have quoted thus at length upon this point, in accordance with the principle already laid down, that the lower is a type of the higher.

Following Darwin's argument—"the greater eagerness of the male has thus indirectly led to the more frequent development of secondary sexual characters in the male than in the female"—secondary sexual characters being those not directly concerned in reproduction. Among these are the greater size, strength, courage, and pugnacity of the male, which most naturalists admit to have been acquired or modified by sexual selection—not depending on any superiority in the general struggle for life, but on certain individuals of one sex, generally the male, having been successful in conquering other males, and thus having left a larger number of offspring to inherit their superiority.

In the human species, the differences between the sexes are marked. The greater size and strength of man are apparent. His broader shoulders, more powerful muscles, greater physical courage and pugnacity, may be plainly claimed, by Darwin and his adherents, as man's inheritance from a long line of ancestry, of which the vanishing-point is in the remote past, among the lowest forms of life.

Whether or not this relationship be accepted, the same principles which have prevailed among lower animals must have been operative in the progress and development of the human race.

During the long ages when man was in a condition of barbarism, it must have been the strongest and boldest hunters and warriors who would succeed best in the struggle for existence, thus improving the race through the operation of natural selection, and the survival of the fittest; while the stronger passions accompanying these traits would lead to their success in securing the wives of their choice.

They would necessarily, by means of the same advantages, leave a more numerous progeny than their less successful rivals. It is here that the laws of sexual selection and heredity come in to maintain and increase the differences between the sexes. Who can doubt that a difference in mental characteristics would result from such causes? The greater necessity for exertion on the part of men would inevitably result in the development of more robust intellects. "Mere bodily size and strength would do little for victory unless associated with courage, perseverance, and determined energy.

"To avoid enemies or to attack them successfully, to capture wild animals, and to invent and fashion weapons, require the aid of the higher mental faculties, namely: observation, reason, invention, or imagination. These various faculties will thus have been continually put to the test and selected during manhood; they will, moreover, have been strengthened by use during this same period of life.

"Consequently, in accordance with the principle often alluded to, we might expect that they would at least tend to be transmitted chiefly to the male offspring at the corresponding period of manhood. . . . These faculties will have been developed in man partly through sexual selection, that is, through the contests of rival males, and partly through natural selection, that is, from success in the general struggle for life. . . .

"Thus," continues Darwin, "man has ultimately become superior to woman." We will say, rather, thus have men and women come to differ mentally as well as physically. We will take further testimony, and inquire what sexual selection has been accomplishing for women during these long periods of man's physical and mental development, before accepting the unmodified dictum of superiority.

The authority so frequently quoted tells us that "the equal transmission of characters to both sexes is the commonest form of inheritance," and that "this form has commonly prevailed throughout the whole class of mammals." Hence the advantages primarily gained by man have been bestowed upon his descendants of both sexes, though, as has been shown, in a somewhat less degree upon the female. Let us now glance at the converse of these vivid pictures of the advantages accruing to man through habits and conditions arising from primary sexual characters, and endeavor to learn whether the habits and conditions necessarily attaching to the female have been the source of any gain either to herself or to the race as a whole.

The less degree of hardship and exposure to which she has been subjected have doubtless tended to develop in her the physical beauty in which she is generally acknowledged to be man's superior; while the fact that women have long been selected and prized for their beauty will have tended, on the principle of sexual selection, to increase the differences originally acquired through natural selection.

The "sweet low voice" which has so long been accounted "an excellent thing in woman," has undoubtedly been gained in a similar manner. In the pursuit of her more quiet avocations there would be less likelihood of the development of large and powerful vocal organs, as it is during the excitements of battle and the chase that the fiercest yells and wildest shouts are produced. The perception of musical cadences, and a sensitiveness to the influence of rhythm, manifested even by many of the lower animals, naturally associating themselves with the rhythm of motion, would tend to early development, on the part of the female, in the care and nursing of her young; while sexual selection has probably played a still more important part in the origin of music.

"Although," says Darwin, "the sounds emitted by animals of all kinds serve many purposes, a strong case can be made out that the vocal organs were primarily used and perfected in relation to the propagation of the species."

Many of the lower animals are mute except during the breeding-season, and the calls, melodious or frightful, of most animals have either a social, an amatory, or a maternal meaning. Thus, through the principle of inherited associations, music asserts its sway over the deepest emotions of the nature—spoken of by Herbert Spencer as arousing "dormant sentiments of which we had not conceived the possibility, and do not know the meaning," and apostrophized as follows by the more impassioned Richter: "Away! thou tellest me of that which I have not and never can have; which I forever seek, and never find!" Its mysterious influence is explained by Darwin as consisting in its power of exciting sensations and ideas which "appear from their vagueness, yet depth, like mental reversions to the emotions and thoughts of a long-past age."

Woman, unable to obtain an influence by those means so readily at the command of man, will have naturally resorted to milder measures, both for securing any desired object, and in self-defense; and music, appealing as it does to the gentler and more tender emotions, will have been often employed in arousing the better nature of him at whose mercy her inferior strength has placed her. Thus she will have held the ruder passions of man in check, and, in taming his wilder nature, will have developed an increasing gentleness both of feelings and of manners in the entire race.

During the battles of rival males, the female will have occupied the less active but more dignified position of arbiter and judge. Not being in the heat of the conflict, she will have had opportunity to observe the strategy of each, and to weigh their comparative merits. By this exercise of the faculties of observation, comparison, judgment, and reason, her intellectual powers will have been "continually put to the test and selected during" womanhood. Unfairness in the conduct of the battle will doubtless have roused her indignation, and compelled her better feelings in favor of the more honorable combatant. Sympathy for the vanquished will sometimes have taken the place of exultation in the superior prowess of the victor, and admiration for mere muscular power will have had to contend with these finer emotions.

While man has been engaged in contests with the common enemy, during which his fiercest passions will have been aroused, woman has been subjected to the discipline of family life. To meet emergencies successfully, to provide for the sick, to maintain order and discipline in the household, which, at an early period in human history, included slaves as well as children, will have required mental powers of a high order. At the same time she will have developed a milder character through the exercise of the beneficent traits of maternal love, and solicitude for the absent husband and father. These feelings of tenderness and love will have gradually prepared the way for the development of the devotional sentiment, and will have thus furnished a basis for the deeper religious nature which has become a part of woman's birthright.

Darwin says that the foundation of the moral qualities lies in the social instincts, including in this term the family ties — the more important elements being love and sympathy.

Thus it appears that while sexual selection and intellectual development have gone hand-in-hand, it is no less true that the moral and emotional sides of human nature have been developed by the operation of the same laws mainly through the female portion of the race. Though Darwin scarcely does more than touch upon this phase of the subject, he says: "Woman seems to differ from man in mental disposition, chiefly in her greater tenderness and less selfishness;" and again: "It is indeed fortunate that the law of equal transmission of characters to both sexes has commonly prevailed throughout the whole class of mammals; otherwise it is probable that man would have become as superior in mental endowments to woman as the peacock is in ornamental plumage to the peahen."

I shall refrain from indulging in any "would-have-beens" upon the moral aspects of this picture, in the contingency to which Darwin alludes, since we are concerned only with what *is*.

Our authority continues; "That there is a tendency to the equalizing of the sexes is undoubted in many of the secondary sexual characteristics; woman bestows these superior qualities on her offspring of both sexes."

Applying the principles, to the operation of which he imputes man's mental superiority, we will add — though in a greater degree upon her adult female offspring, since it is during her maturity that these qualities of greater tenderness and less selfishness are most called into exercise.

Although Darwin states that man has been more modified than woman by the law of heredity in connection with sexual selection, he admits its force in the development of both sexes by many statements which might be quoted, were it necessary. The principal argument against its equal force in the two cases is found in the fact that the young of both sexes in many animals, including the human, most resemble the mother. While this is true in a limited sense, the points of greater resemblance being mainly of a physical character, as superior softness and smoothness of skin, greater delicacy of muscles, muscular tissue, etc., it is not applicable to the qualities of tenderness and unselfishness, the cruelty and selfishness of children, especially boys, being proverbial.

Still quoting from the same work: "Although men do not now fight for the sake of obtaining wives, and this form of selection has passed away, yet they generally have to undergo, during manhood, a severe struggle in order to maintain themselves and their families, and this will tend to keep up, or even increase, their mental powers, and, as a consequence, the present inequality between the sexes. . . . In order that woman should reach the same standard as man, she ought, when nearly adult, to be trained to energy and perseverance, and to have her reason and imagination exercised to the highest point; and then she would probably transmit these qualities chiefly to her adult daughters. The whole body of women, however, could not be thus raised, unless, during many generations, the women who excelled in the above robust virtues were married, and produced offspring in larger numbers than other women."

Though the writer appears to see no incompatibility in these two conditions of the intellectual elevation of women, doubtless many of my readers will, particularly such of them as have borne and raised large families.

Herbert Spencer says: "Taking degree of nervous organization as the chief correlative of mental capacity, and remembering the physiological cost of that discipline whereby high mental capacity is reached, we may suspect that nervous organization is very expensive; the inference being that bringing it up to the level it reaches in man, whose digestive system, by no means large, has at the same time to supply materials for general growth and daily waste, involves a great retardation of maturity and sexual genesis." This is a general statement, applicable to the race as a whole, but it follows that, in so far as reproduction is a greater physical tax upon woman than upon man, so far she labors under a natural disability to equal man intellectually, there being a necessary antagonism between self-evolution and race-evolution, since energy expanded in one direction is not available in another.

Darwin, in suggesting a method — evidently impracticable, however — by which women may become the intellectual peers of men, fails to provide for the elevation of man to a moral equality with woman, although he admits that "the moral faculties are generally esteemed, and with justice, as of higher value than the intellectual powers." He says also that "the moral nature of man has reached the highest standard as yet attained, partly through the advancement of the reasoning powers and consequently of a just public opinion, but especially through the sympathies being rendered more tender and widely diffused."

In regard to the future progress of the race, Herbert Spencer asks "in what particular ways this further evolution, this higher life, this greater coördination of actions may be expected to show itself;" and concludes[1] that it will not be in the direction of increased muscular strength, but somewhat in an increase of mechanical skill, largely in intelligence, but most largely in morality.

Thus these high authorities assign to woman a place in the production of those influences which have developed and must continue to develop mankind, coextensive in importance with the moral interests of the race.

But, if I have read their teachings aright, neither man nor woman can justly take any individual pride, the one in his intellectual, the other in her moral superiority; rather they must see themselves as "Parts and proportions of a wondrous whole;" as the accompanying movements which make up the harmony of the grand diapason of the human race.

And there is that just adaptation of the different parts which is essential to and constitutes harmony. Bacon says that the causes of harmony are equality and correspondence; and Pope completes our argument with the line— "All discord, harmony not understood." There can be, then, no real conflict of interests between man and woman, since there is a mutual dependence of each upon the other, bringing mutual good. Neither can it be a misfortune to be a woman, as so many at the present day would have us believe, although her position may be in some respects subordinate to that of man.

In fact, the subordination of man to woman, different in kind from its converse, is equally apparent; both seem to be matters of common consciousness. It may be readily seen how, in early times, when muscular strength and general physical power were held in the highest esteem, that the position of woman should have been a subordinate one. Animal courage, endurance of physical hardships, the strength, cunning, and agility, which enabled men to cope with wild beasts and with each other, were the traits of character most prized, because most conservative of life in those barbarous times; hence the idea that, woman's position is naturally a subordinate one, has acquired the force of a primal intuition, and might almost be claimed as a "datum of consciousness." But, as the necessities of existence have been gradually modified by civilization, both the character and degree of her subordination have notably changed.

Those qualities, regarded as preëminently feminine, have risen in common estimation, and mere muscular superiority, and even intellectual power, are now put to the test of comparison with the higher moral qualities.

It is true that the laws of most countries still discriminate in a manner unfavorable to women. Legislation has been largely upon the ideal basis of every woman being under the protection of some man, and of all men being the

[1] For argument, *see* "Principles of Biology," vol. ii, p. 495.

true defenders of all women, and this is evidently traceable to the conviction, already alluded to, that a subordinate position belongs naturally to woman. Lecky says that "the change from the ideal of paganism to the ideal of Christianity was a change from a type which was essentially male to a type which was essentially feminine." As the race shall continue to approach the level of its lofty ideal, the subordination of woman, as well as that of man, will continue to lessen, since both have their chief foundation in the lower traits of character, the force in the one case being superior strength combined with power of will, and, in the other, superior beauty with the desire to fascinate. As these influences are gradually losing their power of despotic sway, woman, in place of acting as the slave, the toy, or the tyrant of man, is becoming not only his companion, but the custodian of the moral and religious interests of society, man looking at her as the natural critic and judge of the moral aspects of his conduct.

While the varying characteristics of the two sexes are thus seen to be inherent and inevitable (the secondary sexual characters having largely grown out of those which are primary and essential), it does not follow that they are necessarily indicative of the "sphere" of each for all time. While it is doubtless true, in a certain sense, that "that which has been is that which shall be," nevertheless, change (in accordance with law) underlies the very idea of evolution, and as it has been and is now, so it ever shall be, that the sphere of woman will be determined by the kind and degree of development to which she shall attain. Like man, she need know no other limitation; but when we look around upon the great industries of life, mining, engineering, manufacturing, commerce, and the rest, and consider how little direct agency woman has had in bringing them to their present stage of progress, we are compelled to believe that she must not look toward direct competition with man for the best unfolding of her powers, but rather, while continuing to supplement him, as he does her, in the varied interests of their common life, that her future progress, as in the past, will consist mainly in the development of a higher character of womanhood through the selection and consequent intension of those traits peculiar to her own sex.

The next two selections, Hardaker's "Science and the Woman Question" and Morais' reply, are representative of the debate over sexual differences that occurred in the 1880s. These articles addressed what was considered to be a critically important sexual difference between men and women: the larger size of man's brain. Many people believed this difference accounted for man's greater intelligence. Hardaker and Morais were frequent contributors to American magazines during the early 1880s, and they had clashed once before on the woman question. See Hardaker, "The Ethics of Sex," North American Review 131 (July 1880): 62–74 and Morais, "The Limitations of Sex," North American Review 132 (January 1881): 79–95.

SCIENCE AND THE WOMAN QUESTION

MISS M. A. HARDAKER

March 1882

THE reform philosophy which asks for the elevation of women admits an inferiority of position and power on their part, but at the same time claims that this inferiority is due to temporary causes. It bridges the broad gulf between masculine and feminine achievement by the excuse of a different environment. If the difference is, indeed, due to temporary conditions, it is, of course, removable by the removal of such conditions. It is the purpose of this paper to consider whether the difference (especially in intellectual power) of the two sexes is attributable to permanent or to temporary conditions. The nearest way of getting at this question is to attack it upon its physiological side.

Students of physiology see that a final and conclusive law can not yet be drawn from differences in brain-weights and measurements, because of the present imperfection of such data. But there is an even broader and better foundation from which to build up a conclusion, and I propose to stand on this more general ground. In order, however, that such physiological details may have due influence upon the general argument, I give a few of the best-established facts. Professor Bastian's work on the brain, published in 1880, sums up his studies of this organ as affected by sex. I condense or quote from him the following statements: "Difference of sex, in its influence over capacity of skull, is often greater than difference of race. . . . Difference of cranial capacity between the sexes increases with the development of the race, so that the male European excels much more the female than the negro the negress. The difference in the average capacity of the skulls of male and female among modern Parisians is almost double that between the skulls of the male and female inhabitants of ancient Egypt. . . . The general superiority, in absolute weight, of the male over the female brain exists at every period of development. In new-born infants, the brain was found by Tiedemann to weigh 14½ ounces to 15¾ ounces in the male, and from 10 ounces to 13½ ounces in the female. The maximum weight of the adult male brain, in a series of 278 cases, was 65 ounces; the minimum weight, 34 ounces. The maximum weight of the adult female brain, in a series of 191 cases, was 56 ounces; the minimum, 31 ounces. In a large proportion the male brain ranges between 46 and 53 ounces, and the female between 41 and 47 ounces. A mean average weight of 49½ ounces may be deduced for the male, and of 44 ounces for the female brain." It is further given, on the authority of Gratiolet and others, that the male brain can not fall below 37 ounces without involving idiocy; while the female may fall to 32 ounces without such a result. All accepted authorities agree that the average male brain exceeds the average female brain in weight by about ten per cent. Professor Thurnam also adds, "The brain-weight of the male negro is the same of that of the female European."

Of *qualitative* differences of brain we know next to nothing; we can not study quality from the physiological side, but are driven to an appeal to the concrete products of brain activity. Yet it is most probable that we may at some time establish an exact correspondence between brain-substance and intelligence, as the size and condition of the lungs yield an exact measure of the breathing power, and as the contractile muscle of the heart measures the quantity of blood ejected at each pulsation. In the case of every other organ of the body we know that there is an ascertainable correspondence between size and condition, and the amount of work which the organ can do. Is there any good reason for making an exception of the brain? . . . Now, if we are not yet certain of the kind and degree of structural differences in the brains of men and women, we still have overwhelming external evidence of the existence of such differences. We have as much external evidence of the superior quality of the masculine brain as of the superior breathing power of the masculine lungs, or of the superior absorbing power of the masculine stomach. We do not examine a muscle to asertain its internal structure. The weight which it will lift, or the distance which it will travel in a given time, is an unfailing index of its quality. If it were possible to collect all the results of the muscular activity of men, from the beginning of civilization until the present, and likewise all the results of the muscular activity of women for the same period, we should reason instantaneously, from these phenomena, to the superior quality of masculine muscle. We should need no resurrection of dead men and women to demonstrate the difference.

Let us now consider with more attention the general physiological law that quantity of power is in proportion to the size of the body. This brings us to the still more fundamental principle of the inseparability of matter and force. A large amount of matter represents more force than a small amount; and this law includes vital organisms as well as inorganic masses. Under proper conditions for test, the amount of power evolved by any vital organism is in direct ratio to the size and weight of that organism. This settles the question of *quantity* of power permanently in favor of man. The weight of all the men of civilized countries would exceed that of all the women by perhaps fifteen or twenty per cent.

Again, it is an accepted truth of modern science that all human energy is derived from the food, and is an exact equivalent of the amount of food consumed and assimilated. The amount of food assimilated by men exceeds the amount assimilated by women by about twenty per cent. This fact has popular recognition in the higher rate of board demanded for men. It inevitably follows that man, as a sex, representing more food-assimilation than woman, must represent more energy *of some kind*. If the collective weights and food-assimilating capacities of men should ever fall below those of women, there would follow a reversal of the present relations; but, while these two facts remain, it follows, with mathematical certainty, that the amount of power evolved by men must exceed that evolved by women. While man eats more, he will stand for more. It may be simply more muscle, but it must be more of something. Scientific students are rapidly coming to the conclusion that the human body is subject to the same laws of the conservation and transformation of energy which pertain to the whole material universe. Power is in direct proportion to size. The kind of power will depend upon the organization. Food converted into muscle will reappear as work; food converted into brain will reappear as thought and speech. We have a right to insist on the legitimacy of judging brain-power by brain-products. We value brains for thought as we value looms for manufacture. A barrel of brains is of no account, unless we can evolve from them a steam-engine or a poem. We have seen it hinted, in a recent essay by Dr. Bedell, of Chicago, that women probably possess a larger amount of actual nervous matter, in proportion to the size of the body, than men. Many external facts strengthen this theory. But nervous matter distributed over the body, in the form of threads and sensory ganglia, although it would certainly increase the amount of sensitiveness, by giving more nervous surface in proportion to the entire mass, would tend to create a preponderance of emotion over thought. Very strong probability is lent to this explanation by the liability of women to nervous excitement and disease.

Before dropping the argument from physiology, I wish to refer once more to the necessary relation between size and power. Let us suppose two cases for experiment. Let them be respectively that of a man weighing one hundred and fifty pounds, and that of a woman weighing one hundred and thirty pounds. These figures are not far from the average weight of man and woman. In the woman the digestive organs will be smaller; the absorbing surface will be less; the lung-surface will be less; the quantity of blood will be less (and here we may throw in the interesting physiological fact that the solid constituents of the blood of woman are rated about eight per cent less than in man—i.e., the watery ingredient is larger in woman). Now, if we suppose these two persons to be under similar and normal conditions, with each organ bearing a harmonious relation in size to every other—i.e., if the brain of the man have the same proportion to the weight of his body as the brain of the woman to the weight of her body—it inevitably follows (if both assimilate proportionate amounts of food in a given time) that more blood will be sent to the brain of the man within a given time than to the brain of the woman in the same time. Consequently, the man will do more thinking in an hour than the woman. Nor will the woman ever be able to overtake the man in the race of thought, any more than she can overtake him in the amount of circulation; for his brain is of exactly the same *quality* as hers, but it is larger.

In order to see more clearly the importance of the factor of *time* in this calculation, as well as to get a clearer understanding of the law of the transformation of energy, let us make the further supposition that a pound of bread is gradually converted into vital energy in both these organisms. In the first place, the capacity for mastication and deglutition will be smaller in the woman. She will not be able to swallow her pound of bread in the same time as will the man. But supposing that he gave her this advantage (since it would depend on his will), yet, when the pound of bread had been transferred to the respective stomachs, the smaller lining membrane of the woman's could not secrete as much gastric juice in a given time as the man's; and as she could make no gain in time from his generosity (for the bread would have passed beyond the control of his will), she could not possibly get her bread digested and absorbed in so short a time as he. His blood could furnish the intestinal juices, reabsorb them with the food in solution, and go about some other work, while hers, with just as much work to do and fewer hands to do it with, can not possible keep pace with his.

If, now, we may further suppose this pound of bread converted into its equivalent of thought, it is evident that a pound of bread will represent as much thought in a woman's brain as in a man's; but, as her smaller organs refuse to assimilate as fast as his, the larger organism will have a permanent advantage over the smaller one in the element of time. Any other conclusion than the one just stated implies a contradiction of the established relations of matter and force, and there is a general historic corroboration of this idea in the actual record of sex activity. Women have done something of nearly everything that men have done, but they have come later and with smaller offerings. . . . Unless woman can devise some means for reducing the size of man, she must be content to revolve about him in the future as in the past. She may resist her fate, and create some aberrations in her course, but she will be held to her orbit nevertheless.

The argument from physiology has still another element of strength. The perpetuation of the human species is dependent on the function of maternity, and probably twenty per cent of the energy of women between twenty and forty years of age is diverted for the maintenance of maternity and its attendant exactions. Upon the supposition that woman's mental endowment were exactly equal to man's, the amount diverted to maternity must be continually subtracted from it, so that any original equality of intellect would certainly be lost through maternity. This diversion of power would also occur in the years of highest physical vigor. This period in man is that of most active intellectual development, because the physical basis of intellectual energy is most abundant in these years. Consequently, his period of greatest intellectual gain corresponds to her period of greatest loss. . . .

By reflecting that no such original equality ever existed, but that, on the contrary, a considerable physical superiority has been the possession of man from the beginning; by remembering that, in the perpetual struggle for existence, man's physical and intellectual faculties have been stimulated to the utmost in gaining the means of life for himself and for his weaker mate and offspring; and by considering how large an amount of woman's energy must be diverted from intellectual pursuits to the function of maternity — we see that the conditions of intellectual development are vastly in man's favor. We also see that the main features of these original conditions are permanent conditions of human life on the earth. Woman's inferior size and power will forbid her becoming the successful rival of man in the struggle for existence. Consequently, she will miss the powerful intellectual stimulus which this competition creates among men. Lastly, while society continues to exist, she will always be obliged to expend a large proportion of her energy in the function of maternity. All these enumerated and inevitable facts bear upon her chances of intellectual growth, and have a tendency to widen the intellectual gulf between herself and man. Mr. Darwin, in his "Descent of Man" (vol. ii, p. 313), after enumerating the causes which strengthened the differences in mental power of the sexes, adds, "It is, indeed, fortunate that the law of the equal transmission of characters to both sexes has commonly prevailed throughout the whole class of mammals; otherwise, it is probable that man would have become as superior in mental endowment to woman as the peacock is in ornamental plumage to the peahen."

It is not unlikely that the still imperfectly known laws of heredity have increased the intellectual endowment of the male, for Mr. Darwin finds a general law of transmission in the line of sex. If, too, we accept Mr. Spencer's proof of the inverse ratio between individuation and multiplication, we see that intellectual mothers will have fewer daughters than unintellectual ones; so that the chances of transmission of intellectual qualities in the female line will be lessened as culture increases among mothers. Accordingly, the intellectual tendencies which might have been acquired by the short and easy method of heredity will have to be acquired by the slower processes of application. This, again, will require the expenditure of a proportionately larger amount of energy in women than in men, supposing that men have a higher average of intellect through heredity. Moreover, the probabilities of marriage, or, at least, of early marriage, are lessened in cases of intellectual women; so that the chances are, not only that intellectual women will have few daughters, and so be unable to add to the general average of female intellect by sex-transmission, but also that they will be unable to add anything whatever to the sum of hereditary intellect in either sex.

Man has two powerful advantages over woman: the admitted superiority in the size and weight of body and brain, and the certainty of the continuance of conditions which insure that superiority, for the conditions of masculine superiority are the very ones upon which the preservation of the species depends. The necessary outcome of an absolute intellectual equality of the sexes would be the extinction of the human race. For, if all food were converted into thought in both men and women, no food whatever could be appropriated to the reproduction of the species. But, as an actual fact, women do not consume so much food as men; nor can they do so while their average size remains so much smaller. Moreover, of this smaller amount of food consumed by women some must always be spared for the continuance of the race; so that the sum total of food converted into thought by women can never equal the sum total of food converted into thought by men. It follows, therefore, that *men will always think more than women.*

Nevertheless, if it could be shown that the energy derived from food in men were an energy of inferior *quality*, women might gain a compensating factor in quality of thought. By the consent of competent judges, the reasoning power and the creative imagination are the highest and most complex forms of brain-energy. We have the most abundant evidence that, while man possesses both these powers in large amount and of superior quality, woman possesses them in much smaller amount and of inferior quality; so that the distinction of exceptional women, of whom a list could be made, would add little to the general low average of feminine power. We hear the power of *intuition* quoted as a higher one than reason. Women possess this power in a higher degree than men, and are sometimes rated above them in consequence of it. Very little study has been bestowed on this faculty, which has been the occasion of so much self-congratulation to women. But there is considerable evidence that it is acquired by heredity, that it is closely akin to instinct, and that some modification of it is the common possession of women, children, and the lower animals. It is possessed in a lower degree by men, but has escaped registration as a masculine trait. In his classification of the mental powers of the two sexes, Mr. Darwin puts observation, reason, imagination, and invention as those especially selected in man, and likeliest to be transmitted to male offspring. He also says: "It is generally admitted that with woman the powers of intuition, of rapid perception, and perhaps of imitation, are more strongly marked than in man; but some, at least, of these faculties are characteristic of the lower races, and therefore of a past and lower state of civilization."

A REPLY TO MISS HARDAKER ON THE WOMAN QUESTION

NINA MORAIS

May 1882

....

IN the March number of "The Popular Science Monthly" Miss Hardaker invokes Science to testify to the natural and irrevocable mental inferiority of the female to the male. A statement of this kind, coming as it does when woman is struggling for every step in her intellectual advance, is peculiarly baneful to her. To cover ancient prejudice with the palladium of scientific argument is to unite the strength of conservatism and of progress in one attack. An examination of the accuracy of the paper, "Science and the Woman Question," may not, therefore, be ill-timed.

Two propositions underlie Miss Hardaker's argument. They are as follows:

1. A large amount of matter represents more force than a small amount. Hence man is superior to woman in body and brain (pp. 34-35).

2. "All human energy is an exact equivalent of the amount of food consumed and assimilated." Man, by reason of his larger organs, eats and assimilates more food than woman does. Each of his organs, including the brain, is therefore capable of acting with proportionally greater energy. Hence, "men will always think more than women" (p. 38).

Collaterally our author finds that the demands of maternity must cause a large subtraction from the smaller amount of mental energy which women would otherwise exert, and, as the result of her fundamental propositions, she draws the startling conclusion that "unless woman can devise some means for reducing the size of man, she must be content to revolve about him in the future as in the past," (p. 36).

Before entering upon the question by means of her own original and scientific method, Miss Hardaker makes the following statements: "Students of physiology see that a final and conclusive law can not be drawn from differences in brain-weights and measurements, because of the present imperfection of data." But the superior power of the male brain, like the superior power of the male muscle, is shown conclusively by its product (p. 33).

The figures which begin Miss Hardaker's argument are those which all speculations regarding the brain take into consideration. These figures are quite complete enough to indicate distinctly that the average male brain is always larger than the female. Miss Hardaker herself states that "all accepted authorities agree that the average male brain exceeds the average female brain in weight by about ten per cent" (p. 34). Now, if the principle that bulk is power were admitted, the measurements obtained would be nearly, if not quite, conclusive of the natural superiority of the male: It would not have been reserved for Miss Hardaker to make the discovery. Miss Hardaker can not afford to dismiss brain-measurements as incomplete evidence, for these statistics become the key-stone of her own logic when she endeavors to prove man's mental superiority because of his excess of brain.

The student, however, does not reason as Miss Hardaker reasons. He, as well as she, possesses the historic fact that the product of the masculine mind has always been greater than that of the feminine. He might, therefore, find that, as the male brain has been more productive, it is the better organ. Upon this point Miss Hardaker contends that not only can we reason to the general quality of organs from their respective products, but we can actually arrive at a knowledge of their structure by such processes of logic. "We do not examine a muscle," she says, "to ascertain its internal structure" (p. 34). If this were true, the occupation of the anatomist would be gone: the valvular arrangement of the heart, the cellular formation of the lungs, would have been disclosed by an observation of the externally perceptible operations of these organs. The truth is, that we can never reason from product to structure until after we have internal evidence of the functional relations between the structure and product of the class of organs to which those under test belong; nor can we without such knowledge even reason to the generaly quality of two organs by their different product, unless our comparisons are made under the same environment. For instance, take two pairs of lungs: let one respire at sea-level, the other at the top of Mont Blanc. Their absolute product would be no estimate of their relative capacity. Still, the physiologist would have little difficulty in eliminating the effect of difference of circumstances in his calculation, because his complete knowledge of the lungs and of the influence of atmospheric pressure enables him to allow for differences of environment. But no such allowance can be made in estimating the normal power of the male and female brain which have always acted in different mental atmospheres; for the relation of structure to function as regards brain has not been accurately determined.

It is because of this lack of knowledge regarding the precise connection between brain-structure and thought, and not because of imperfection in the data of measurements, that students refuse to draw therefrom the law of brain capacity; and thinkers will not infer the capacity of male and female brains from their products, until the different influences acting upon men and women can be eliminated. While anatomy is unable to solve for us the enigma of sexual brain-power, we may have recourse to comparison under similar environment as the key to our problem. This method of discovery Miss Hardaker, with a perversity remarkable in a disciple of modern science, is laboring zealously to prevent.

"We need not," she says, "ascertain the meaning of brain-size by experiment; we can arrive at it by analogy. All other organs (under the same conditions) work in proportion to their size. Is there any good reason for making an exception of the brain?" (p. 34). Now, even if all other organs work in proportion to their size, the fact that the brain is exceptional, in the nature and in the variety and complexity of its functions, would render the argument from biceps to brain as questionable as that from marble to zinc. There may be properties in common, but in the production of forces the similar effects of these common properties may be wholly vitiated by others peculiar to only one of the objects compared. Besides, size is not always a gauge of organic capacity. Does the large eye see better, the large ear hear more, the large nerve feel more keenly? And, if, all other conditions being equal, they might do so, the incalculable variation of condition renders the size test of no practical value at all. This, however, is a phase of the subject to be discussed later, when we shall endeavor to show that, although we agree with Miss Hardaker that a larger brain means something, it does not necessarily mean a "greater amount of thinking in a given time." And, here we throw in, as interesting facts, that woman's smaller heart beats faster than man's larger one; that her circulation is to his in swiftness as ten to nine; and that, according to Miss Hardaker's figures, and to some celebrated authorities, the proportion of brain to body is larger in woman than in man.

But to meet Miss Hardaker upon her own ground in the discussion of her fundamental propositions, we shall waive, as she has done, all sexual differences, of physical or local environment, and all analogical inferences, and proceed to compare the male and female brains upon a supposititious level of like conditions. She proposes to prove, on quite new and highly scientific grounds, that absolute weights and measurements are, after all, the ultimate tests of capacity. It may be deemed singular that the profound students who have preceded Miss Hardaker—some of whom were undoubtedly scientists—should have entirely overlooked the beautifully simple conclusion she formulates, thus: "If mass represents force, the larger the brain, the larger the power." The reason why students have been so blind to Miss Hardaker's discoveries is quite as simple as the discovery itself. It is because her premises are false.

A large amount of matter does not represent more force than a small amount, nor does it represent any force at all. There is an elementary law of physics which declares that the momentum of a body equals its size multiplied by its velocity, and this may lead to the supposition that matter itself is force. But matter in a state of inertia is not power; it becomes powerful only when acted upon. The same force acting upon different bodies imparts velocity in the inverse ratio of their masses; and, since velocity as well as size is a factor of power, it follows that a force which imparts a greater velocity to a smaller body gives it as great a momentum as a larger body obtains when acted upon by the same force; for the velocity in the latter case is feebler. Even admitting (what Miss Hardaker does not appear to claim) that *potential* energy may be proportionate to size of mass, we see that potential energy can only be evolved by an appropriate force acting through or upon the mass, and, to make the potential energy of a large mass do more work than that of a smaller one, the force applied must always be greater. Hence, not the size of the body, but the strength of the impelling force, is the ultimate test of its power. A glance at obvious facts will show that size is not the gauge, that weight may indeed be a direct impediment to the evolution of force. The avoirdupois of the fat boy is a clog to his energy; the fast runner wins by his light weight; the champion oarsman reduces his flesh.

In applying her theory to the brain, one fact which Miss Hardaker herself states is sufficient to tell very disastrously against her conclusion that larger brain-weight means larger thinking power. "According to Gratiolet, the male brain can not fall below thirty-seven ounces without involving idiocy, while the female may fall to thirty-two ounces without such result" (p. 34). Here are two brains precisely of the same quality, one thirty-seven ounces the other thirty-one, an absolute difference of six ounces. Yet these six ounces represent just nothing. Indeed, give the woman thirty-four ounces and leave the man thirty-seven, his three ounces more are simply a minus: thirty-four is rational thought, thirty-seven irrational. In this instance a *small amount of matter represents more power than a large amount*. It would seem that the true law must be sought elsewhere than in the grocer's scale.

But the impelling force which Miss Hardaker omits in her former statement is supplied in her next assumption: "All human energy is derived from food. Man eats more than woman because his larger size requires him to do so, a larger proportion of nourishment is sent to his brain; hence men think more than women." A look backward at our elementary law of physics will show that Miss Hardaker's second conclusion is as weak as her first. To repeat that portion of our law which bears upon this argument, we find that the same force acting on different bodies imparts velocity in the inverse ratio of their masses, and it is therefore clear that in order to make the large machine run *as fast* as the small one, fuel must be supplied to the former more freely. The explosive force that sends the tiny rifle-ball at the rate of twenty miles a minute could not overcome the inertia of the missile discharged from the Krupp gun; a proportional force to each would send each just the same distance. Now, granting all the premises of Miss Hardaker's second proposition, that male and female eat in certain fixed proportions, that a certain fixed amount of that proportion goes to nourish relatively proportioned brains, the only logical conclusion is that the *larger brain*, supplied with *more blood*, would in a given time do heavier work, but not more work, than the smaller one supplied with less blood. Under these circumstances the momentum of the larger brain would be greater than that of the small brain; their velocities equal. Without his extra supply of blood, man's brain *could never overtake woman's in velocity*; indeed, without this additional stimulus it might not be able to move at all. The theory that the smaller brain is propelled more easily, might explain the quickness of perception and of fancy which, according to Miss Hardaker, are womanly traits.

Such reasoning, however, is at best mere theorizing, for it applies the simple laws of mechanics to the intricate and so far inexplicable structure of the brain, making no allowance for complications which would divert the action of the law. It may be true that blood is the primary motor of the brain; but there are many other elements besides the size of brain and body, or even the amount of food assimilated, which measure the quantity of blood sent to the brain. The problem is by no means, as Miss Hardaker has tried to make it, an easy sum in simple proportion which the school-boy may solve standing on one foot. Omitting altogether a consideration of the superior blood-circulation of women as a class, overlooking entirely the probability (indicated by the data of the idiot question heretofore discussed) that proportion of brain to body is an element in the capacity of the former, the individual rapidity of circulation, the richness of food in brain-making material become important terms of our problem. The opium-eater, the wine-drinker, the consumer of brain-stimulants, certainly drive more than a proportional share of blood to the brain. At the same time there is always a personal equation to vary the proportional action of food-supply. The brains of Moses and Mohammed were stimulated by prolonged fasts. The circumstances of travel, temperament, companionship, wealth, the passions, music, art, dancing, machine-stitching, and a thousand others, which can never be averaged, often exert an adventitious influence on the appropriation of fuel for thought. These influences are entirely independent of food-consumption and brain-size; they defy the application of any law of mechanics.

But Miss Hardaker's scientific argument, if true, proves too much; for if men, the greater consumers, think more or even better because of the large size of their bodies and the larger power of their digestive organs than women do, then it must follow that the larger and healthier men as a class must think, if not more, at least more profoundly, than smaller and less robust men. Yet the bulk of the world's thought has not been done by men of superior physique or even of superior health. Aristotle, Napoleon, Jeffrey, Thiers, were short in person; Shakespeare, Buckle, Comte, were delicate in frame; Descartes and Bacon were always sickly; Heine wrote his best while in physical agony; Newton and Spinoza were slight in form and of medium height; Herbert

Spencer's health has always been precarious; Mrs. Browning was a life-long invalid; while, unfortunately for a theory based upon superior digestion, Goethe and Carlyle were confirmed dyspeptics.

The instances here cited are by no means exceptional. Indeed, the seeker for data under this head will find that, instead of larger and more healthy physiques evolving a larger average amount of mental power than smaller and less robust ones, the contrary result is emphatically true. As a matter of fact, the circumstance of superior muscular development seems unfavorable to great exertion of the mind. The demands of the body itself are in large men imperative. The waste of the system must be repaired, and the first draughts of energy must go to this purpose. Afterward, though the potential energy represented by the food consumed may still be stored up, there is little power or little inclination to apply that energy to thought. The college student who is most active in the field, who has the greatest height in his stockings, and the biggest biceps, is rarely at the head of his class. Not only does the larger body require more in proportion for its nourishment, but the forces which effect this nourishment are not easily turned in other directions, and it is, therefore, a natural sequence that the body must dwindle as the power of the mind increases. The savage Teutons, whose great bodies affrighted the Romans of Caesar, have become the civilized possessors of less bulk and more knowledge. Human energy appears not to be harmonious, but to run in grooves. Thought produces thought, and the energy once sent to the brain is the direct cause of a new demand for supplies. In like manner, the arm that is developed by work needs a larger amount of food for its maintenance. This is the explanation of the historic fact that physical and mental powers have never been proportionally cultivated, but always at the expense of each other. The profound thinker and the superior pugilist are rarely united.

But, even if it is true that the larger and healthier *physique* affords more blood for brain-use, it does not follow that the larger the supply the greater the amount of brain-work possible. The argument assumes that the brain has no limit to its activity except in the quantity of blood that can be prepared for it. But it needs no scientific education to know that there are other influences which limit the thinker's activity, and that these limitations are somewhere in the mysterious recesses of the brain, or in the forces of which the brain is the organ. The physical health of the brain-worker may be perfect, his digestion unimpaired, his power to assimilate food the same, and yet he may not be able to concentrate his thoughts or carry on a complicated train of reasoning. The defect is not in his body—that is as healthy as ever; nor is it in any of the processes of blood-making—these go on as before. The trouble lies in the brain itself, whose capacity for work is measured by some hidden standard of its own, and which gives warning when a cessation of brain-work is imperative. The body is a furnace whose power of consuming fuel is greater than the capability of its boiler—the brain—to generate power. To keep the latter in good working condition, something more is necessary than building and feeding the fires. A supplementary but important consideration is, whether the steam beyond a certain point will not be productive of unpleasant consequences in the form of an explosion.

In the discussion of the collateral question, that of the effect of maternity on brain-power, Miss Hardaker's scientific logic takes its most amusing form. "The necessary outcome of absolute intellectual equality of the sexes," she says, "would be the extinction of the human race. *For, if all food were converted into thought in both men and women, no food whatever could be appropriated for the reproduction of species*" (p. 38). What Miss Hardaker really means by this last highly scientific axiom it is impossible to guess. She can not mean that, as all food *is* converted into thought in men, women must cease to be mothers in order to imitate his food-conversion. Whatever Miss Hardaker may intend by her impossible supposition, the fact that maternity does make large draughts upon the energy of woman is not to be overlooked. But, unless it can be shown that the mental activity of man is ceaseless, that his manual labor diverts no blood from the brain, that his imaginative and reasoning powers keep steadily at work year in and year out, limited only by supply of food, it does not necessarily follow that women must fall behind men in the brain-work of a life-time. Both men and women need mental rest — no brain-worker can keep at the top of his speed for ever; and women whose duties as mothers divert their energy from the brain may overtake men in their voluntary holidays. This fact will have more concrete significance when we reflect that the professional brain-workers in both sexes are in the minority, and that women who are such are usually unmarried, or mothers of small families. At the same time, the labors of men who form the great masses of population are not more stimulating to brain-culture than the vocations of their wives. But, granting what is probably true, that woman as a whole can never show as much mental product as man, because some of her time and energy must be devoted to motherhood, still she may be quite as capable of production. Therefore, any reasoning which excludes women as a class from the advantages of equal mental training with men, on the ground that they must be the mothers of the race, is forcing the activity of women into one channel, and rendering all other efforts (such as the writing of a scientific article, perhaps) unnatural and unwomanly.

But suppose the whole of Miss Hardaker's argument to be founded on true premises, and all her conclusions to be just and accurate, it may yet be pertinently asked, *Cui bono*? Miss Hardaker would slam the educational doors in women's faces because, being smaller, they are unfit to enter the select retreats of Brobdingnag. But, if justice is to prevail in the rules of admission, the woman who possesses a brain of fifty-six ounces is entitled to precedence over the great majority of males whose brains weigh only forty-nine and a half. Should the environment be more favorable to the woman whose brain-weight is forty-four ounces, she can claim the advantage over the larger male brain whose environment is less favorable. Then, too, the applicants for entrance must be subjected to the test of an eating-match, and the dyspeptic must consent to suicide or rejection. All this must be done, for, although Justice carries her scales, she is blindfolded. She can only weigh brains, food, environment, but can not see the sex of suitors for admission into the new academy. Miss Hardaker must be aware that, were every element in her assumptions true, some women must be greatly superior to the average men, although the highest point reached by the male could not be obtained by the female. Miss Hardaker would, perhaps, object to having the doors of journalism closed against her, because she can never think as profoundly as Lord Bacon, or because in general woman's literary production has not made so fair a showing as man's. It is not long ago since this sort of reasoning militated strongly against the publication of any article that might be signed with a woman's name. But science—not the false science which answered Miss Hardaker's invocation, not the science which would confine the negro to slavery because of his small brain and small mental achievement—true science says that, if woman's power is to be judged by her work, she must be given a fair field for its display. To clear the race-course for the man, and to block woman's road at a certain point, because we feel intuitively that she can go no further, is by no means consistent with modern scientific methods. If the line of woman's power is marked, let her discover the fact, as Bacon thought— all scientific truth should be discovered by experiment. The discovery will not long be delayed; the law of the survival of the fittest will not be abrogated. But, if it should be found that the mental steamship of the female can, after all, store enough fuel to cross the ocean of reasoning, it would give woman the inestimable benefit of correcting the possible errors into which a professed enemy of her sex has fallen. It would demonstrate that, like Mr. Darwin's pea-hen, women have remained inferior to their mates, not because of natural defect, but by reason of external circumstances. A just trial is the whole demand of the reform philosophy.

In the Royal Society, many years ago, it is said Charles II asked an explanation of the fact that a fish in water had no weight; that water plus a fish was no heavier than water without a fish. The wise gentlemen of the Royal Society (presumably males of large bulk) were much agitated over the problem, and gave many scientific reasons for the remarkable phenomenon. It was a wiser man (though not of so scientific a turn of mind) who, instead of giving his reasons why the fish had no weight in its own element, tried the experiment and found, to the surprise of the scientific gentlemen, that a practical test was of more value than any quantity of learned but ill-founded speculation. Perhaps it will not be out of place, by way of parallel to Miss Hardaker's triumphant demonstration of "the reason why," to cite the testimony of a prominent instructor, whose evidence tends to show that her scientific impossibility may be affected by some elements which she has not considered. "So far as my observation and experience go," says President Magill, of Swarthmore College (a gentleman who for ten years has been the instructor of about three hundred students of both sexes), "there is absolutely no difference in the average intellectual capacity of the sexes, under the same training and external influences. The valedictorians of our classes have been almost equally divided between the sexes, with a slight and accidental preponderance in favor of the young women."

Mary Taylor Bissell (1854–1936) was born in Brooklyn and educated at Vassar College (A. B., 1875) and the Woman's Medical College of the New York Infirmary (M.D., 1881), run by the Blackwell sisters. (See "The Industrial Position of Women," pp. 271–280.) Bissell served as the medical director of physical training for the Berkeley Ladies Athletic Club (New York) and as professor of hygiene at the Woman's Medical College. She was also the executive secretary of the New York State Consumers League (1908–1911), where she investigated the health of women working in factories and stores. She wrote several books on women's health including, Physical Development and Exercise for Women (1891) and A Manual of Hygiene (1894).

In the following selection, Bissell argued that sex-based differences in temperament did not fully account for women's greater susceptibility to nervous disorders, and she challenged a widespread belief that fragile health was the "natural" condition of women. Many male physicians (and most people) assumed that women could do little to strengthen a weak body, lessen susceptibility to hysteria, or minimize indisposition caused by menstrual pain. Physical strength and robust health conflicted with middle- and upper-class ideals of feminine beauty emphasizing weakness and delicacy. Delicacy referred not only to physical characteristics (slender, frail figures, tiny hands and feet, finely sculptured facial features) but also to mental capacities (sensitive, but not rigorous intellects, a refined aesthetic sense, highly moral and emotional sensibilities). In Bissell's conception, physical and emotional weakness were dependent variables, influenced by social as well as biological factors. At the center of Bissell's thought lay the conviction that women were not born weak, but became weak because they did not exercise; that women were not born susceptible to hysteria, but became hysterical when subject to stressful and irresolvable conflicts; and that women were not born flat-chested or thin, but did not develop fully because they ate poorly and wore constricting apparel. In this remarkable piece, Bissell examines the effects of dress, exercise, upbringing, class, and occupation in an attempt to understand how women's health might be improved.

EMOTIONS VERSUS HEALTH IN WOMEN

MARY T. BISSELL, M.D.
February 1888

WOMAN is believed to have been endowed by Nature with a strongly emotional temperament. She is accepted as the fairest exponent of sentiments, which in turn lend her her chiefest charm. Tears and smiles, emotion and sensibility, are expected of her. It is permitted to her to be a bluestocking if she will, but sympathetic and tender she must be. If Hypatia has her admirers, all the world loves Juliet! It is precisely in that natural aptitude for emotion, in that type of mind which is exquisitely sensitive to impressions and generously swayed by sympathetic feeling, that one of the great dangers to the perfection of womanhood, physical and mental, may be said to reside.

Many and varied influences tend to increase this emotional excitability until it often becomes a fixed habit of mind; an undue sensibility of the supreme centers to emotional ideas is created, which can only be maintained at the expense of sound health of body and of mind. First among these are certain home influences that are brought to bear upon a little girl from her earliest childhood, which foster in her self-consciousness and introspection.

She is generally permitted narrower limits, within which she can play, can dress, can succeed, than are allowed to her brother, even when her physique is equally able. She is housed more closely, her out-of-door sports are fewer and less interesting, and her dress is too often a limitation to her freedom. Such restrictions of her liberty, and constant reference to the fact that her sex denies her this or that employment or pleasure, tend to make a child self-conscious and emotionally overactive. Methods of family discipline which depend upon appeals to the emotional natures of children have like unhealthy results, for they promote a condition of mental commotion and unrest harmful to children, who require an even atmosphere for the mind as well as for the body. There are often undue claims made upon little children for the demonstration of their affections, and this is especially true of girls. . . .

We need only briefly refer to the unhealthy influence exerted upon the minds of little girls by foolish indulgence in showy dress or in social dissipation. Dissipation, indeed, is a serious term to apply to the social pleasures of little children, but, when we hear of children's parties beginning at nine o'clock, in which toilets and manners only suitable for their mammas are encouraged, we easily conclude that, in the lack of simplicity in social customs, we may find an abnormal stimulus to the emotional natures of American girls.

Certain school influences have a large responsibility in this direction. What is called the hot-house pressure of public schools, and the elaborate system of examinations in our higher institutions of learning, have their evil not in the exercise of the calmer faculties of the mind, such as judgment, reason, memory, etc., but in their tendency to arouse that complex emotion called *worry*. These influence are exerted, it is true, upon girls and boys alike, but, as the facility of the girls for emotional disturbance is greater, they suffer more largely per consequence. The repeated stimulation of such complex emotions can not fail to agitate the mind of young girls, and insidiously disturb its calm.

As the girl grows to womanhood, the impression made by these influences upon her plastic child nature can not be entirely thrown off. If she be of a strong and womanly type, she will meet the physical and social trials of life with such character and self-possession as she may, but they will have for such a one a double force. Life offers only too many facilities for overtaxing the sympathies of the unduly sensitized individual. The appeals of misery, poverty, and sorrow sound in every ear. The woman who would maintain a just equilibrium between sentimental mourning and efficient sympathy for these facts of existence needs to be re-enforced, not weakened, by the education of her childhood. And if to the friction of any life we add the strain of an elaborate social system, if our young woman be a society girl, with all the demands of a high-bred life of fashion upon her time, temper, versatility, and self-control, we have one more influence which maintains her at constantly high emotional pressure.

It is evident that the sum of these and similar forces constantly exerted upon the mind of women must have their due effect. The normal result of the stimulation of any organ of the body is well known to be a final loss of health in that organ. When the faculties of the mind, called out in the display of the emotions, are overtaxed, we generally find either a lack of will-power or a deficiency in reason and judgment, and our common expression for that condition is that such an individual is not well balanced.

It is possible that some of us have heard it suggested that woman is a less reasonable being than man. It has, indeed, been whispered that she—regarding her as a type, not as an individual—is less logical, less temperate in her judgments, more easily controlled by appeals to the feelings. In the recent article by Ouida in the "North American Review," speaking of the character of a woman's mind, she says: "The female mind has a radical weakness, which is often also its peculiar charm; it is intensely subjective; it is only reluctantly forced to be impersonal." Such opinions are not entirely unfamiliar to any of us.

We are in no wise concerned for the final judgment of mankind upon the mind of woman, nor do we imagine that it requires championship. But it is easily apparent that this very grace of her nature may be turned to bad account through undue stimulation, and that, through inheritance and the influences we have briefly suggested, she may acquire a tendency toward an unduly subjective type of mind—a tendency which threatens the loss of a just intellectual sense of proportion, and which, therefore, can not conduce to sound mentality.

The old meaning of the word emotion—*commotion*—is opposed to the best mental growth and health. In repose, in the quiet harmonious performance of its functions, the mind grows into vigorous maturity, and the constant unrest and commotion of nerve-elements, which accompany violent emotional disturbances, and repeated strain upon other than its reasoning faculties, can not fail to disturb the quiet, natural evolution of its powers. Can this tendency in woman's training be shown to affect her bodily health? Physicians and metaphysicians answer, Yes!

. . . Says Maudsley: "It may be questioned whether there is a single act of nutrition which emotion may not affect, infecting it with feebleness, inspiring it with energy, and so aiding or hindering recovery from disease. It is certain that joy or hope exerts an animating effect upon the bodily life, quiet and equable when moderate, but when stronger, evinced in the brilliancy of the eye, in the quickened pulse, in an inclination to laugh and sing; grief, or other depressing passion has an opposite effect, relaxing the arteries, enfeebling the heart, making the eye dull, impeding digestion, and producing an inclination to sigh or weep." This exaggeration of the emotions, seen in many cases among women, may be considered a serious factor in inducing some of the most common diseases of the nervous system from which Americans, in particular, are suffering. . . .

As regards the occurrence of hysteria, while it is frequently found among those belonging to what we call the lower classes of society, it is more frequently manifested among the more highly cultivated. A French author who, indeed, speaks for his own country only, states that one out of four of all females are decidedly affected with hysteria, and that one-half present an undue impressionability which differs very little from it. Although these statistics are too high for America, they are significant as being possible anywhere, and not the less so as coming from a land where, if a woman is anything, she is emotional.

Among the frequent causes of hysteria, all writers mention the depressing passions, such as fear, anxiety, jealousy, and remorse. One says: "The chief mental characteristic of hysterical patients is an excessive emotional excitability unchecked by voluntary exertion." And again: "This excessive emotional activity necessarily induces exhaustion." The treatment of this affection recognizes, first and last, the influence of mind over body. We find that moral suasion, the employment of the individual in directions without herself, the cultivation of an intellectual purpose, of an objective quality of mind, are remedies that rank with the nervines and antispasmodics in the treatment of this disorder.

As regards nervous exhaustion, we find that affection is almost entirely confined to the more highly civilized portions of the community—indeed, is a disease of civilization. Among the causes of nervous exhaustion this same truth is manifest—that excessive demands upon the individual exciting the complex emotions of anxiety, *worry*, are largely responsible for inducing this affection. We believe, indeed, that hard work, unaccompanied by emotional excitement, seldom injures either man or woman. It is the man who, in addition to close application to work, is harrassed by fears of poverty, of loss of position, of anxieties for himself or his family, and the woman who bears the burden of domestic cares, of private griefs, or sustains the strain of a complex social system, who suffers from nervous exhaustion, not the hard-working mechanic, or the unemotional washer-woman. The experience of every school-girl testifies that mental anxiety produces a degree of physical exhaustion out of all proportion to the muscular work effected. The agitations of school politics, the over-emotional character often infused into school-girl friendships, the fears of failure and kindred commotions result in more physical weariness than hours of calm, steady work in the laboratory or in the class-room.

A college graduate confesses that one of the most exhausting experiences of her college life was a morning spent in absolute physical inactivity in a student's meeting, but in a state of mental commotion impossible to describe, over an absorbing issue in college politics. "After four hours of that experience," she said, "I was fit for bed, and for nothing else." It requires no great ingenuity to suggest that this tendency in the training of woman which affects her mental and physical health, may be met by remedies addressed to body and mind alike. The education which shall discipline, not eradicate, the emotional susceptibility of women must begin where the gentility of Dr. Holmes's ideal gentleman began, with our great-grandmothers.

Heredity may not be able to shoulder all of the sins of mankind, but, at least, it must bear its share. The coming woman must not only be well-born, she must be bred in more hygienic methods. She must not only possess inherited vigor, she must also be educated nearer to Nature. The genuine child of Nature is not a morbidly emotional child. The girl who lives in the open air, who knows every bird and flower and brook in the neighborhood, has neither time nor inclination to spend in reading the sentimental histories of departed child-saints, and takes small delight in morbid conversation.

Out-of-door life has never been made popular or interesting for little girls as it always has been for boys. Girls will voluntarily seek fresh air and sunshine if they appreciate the delightful occupations as well as the *fun* to be found in it. They are quite right in "hating to go out because there is nothing to do." Open wide to them the fascinating book of Nature; let them read the story of bird-life and animal-life, and find their first hints of the wonders of plants and rocks by sunlight, and at first hand, not from a printed page in unventilated libraries. Then, when out-of-door life and out-of-door sports have been made as attractive and popular for girls as for boys, and when they have accepted the creed that a nobly-developed and active body is as much their birthright and glory as it has always been the glory of their brothers, we shall find we have gone a long way toward reducing exaggerated emotions in women. And if our first antidote for this condition lies in physical activity and in the cultivation of a sound body, the second antidote will be found in the provision of constant, congenial employment for the mind.

When a young woman went to Henry Ward Beecher to ask him to prescribe for her disappointed affections, he promptly advised her to begin the study of higher mathematics! There is no doubt but that among the less apparent, but no less real causes of undue emotional development among women we may count the lack of congenial and effective work. There is nothing sanitary in intellectual idleness. Physiology forbids that the inactive brain should be a healthy one. The overworked individual may suffer from undue strain, but the mind which is denied congenial employment suffers even a worse penalty in the disability of its best powers, and the waste of purposeless energy.

Women who are receiving the so-called higher education, find in its discipline and opportunity the best remedy for any tendency to excessive emotional disturbance. "The worst enemy of the emotions is the intellect." There is no stronger argument for opening to women new avenues for the acquisition of knowledge than these facts of her constitution offer, justified as the experiment has been by those who have found life a better and a broader thing to them because of these opportunities.

Undoubtedly the actual erudition that is gained in a collegiate training for women could be obtained under other conditions than in the four years of college life. But the inestimable value of our women's colleges lies not so much in their opportunities for actual learning, as in the atmosphere they offer. To live for four years under a *régime* where mental and physical energy are carefully utilized and disciplined, and where the tendency is toward the development of an objective type of mind and the cultivation of a broad intellectual outlook — these are incalculable benefits to woman.

Give to our children, our growing girls, and our young women occupation which, according to their age and capacity, shall develop every faculty of the mind and afford genuine scope for usefulness, and we shall find that the energy which might have been dissipated in unproductive emotions, has been diverted into channels of effective work, and conserved for high and healthful ends.

Chapter 2
THE EVILS OF EDUCATION (1870–1900)

> *All that is distinctly human is man — the field, the ship, the mind, the workshop; all that is truly woman is merely reproductive — the home, the nursery, the schoolroom. There are women, to be sure, who inherit much of male faculty, and some of these prefer to follow male avocations; but in so doing they for the most part unsex themselves; they fail to perform satisfactorily their maternal functions.*
>
> Grant Allen, "Woman's Place in Nature," *Forum* 7 (May 1889): 263.

This and the next two chapters review the debate on women's education in the late nineteenth century. The debate had two distinct, but intertwining strands: one dealing with the alleged physical dangers of advanced education, another concerned with the alleged moral dangers. Early in the nineteenth century, the moral concerns were more prominent, and the founders of female seminaries in the 1830s and 1840s advocated a special program for women to strengthen their sense of maternal duties and provide them with practical training in homemaking. I explore this issue in Chapter 4. In this and the next chapter, I focus on the physical dangers believed to be inherent in advanced education and the confrontation that occurred in the 1870s and 1880s between those who argued that women had a right to an education equal or identical to that offered men and the physicians who warned that a "male" education would render women infertile and sickly.

OVERVIEW OF WOMEN'S EDUCATION IN THE UNITED STATES

The 1870s and 1880s were significant years for women's advanced education in the United States. At the beginning of the period, 11,000 women attended institutions of higher learning, which included a variety of educational establishments, among them colleges, seminaries, and normal schools. (Normal schools were post-secondary schools designed to train students for school teaching.) The goals and standards of these institutions varied widely, and although I allude to some of the differences in my discussion, I am primarily interested in how women's advanced education—that is, any education beyond secondary schooling—responded to and influenced social views concerning women's intellectual capabilities. By 1880, the number of women enrolled in these institutions had increased to 40,000; by 1890, to 56,000.[1] Many women's seminaries and colleges opened during this period: Vassar, 1865; Mills, 1871; Wellesley and Smith, 1875; Bryn Mawr, 1885. Radcliffe, 1879, and Barnard, 1886, appeared as annexes to Harvard and Columbia respectively, and a number of men's colleges began admitting women, including Cornell, 1872, Syracuse, 1871, and many of the midwestern state universities.[2]

These were not the first institutions to offer an education beyond the secondary level to women: Oberlin College in Ohio generally is credited as the first coeducational institution of higher learning, having admitted women into its collegiate department since 1837. In the 1820s and 1830s a number of female seminaries were established. Among the best known were Troy Female Seminary, founded by Emma Willard in 1821; Hartford Seminary, founded by Catharine Beecher in 1823, and Mt. Holyoke Female Seminary, founded by Mary Lyon in 1837.[3] Most of these female institutions catered to the interests of the wealthy, providing instruction to "train" young girls as teachers, wives, mothers, and homemakers. Until the second half of the nineteenth century, the seminaries were virtually the only schools to offer young women an education beyond the elementary level. Eventually, some of the early seminaries became accredited colleges. Today, Mount Holyoke remains the prototype of the female seminary-college, having received collegiate status in 1888.

In the 1840s and 1850s, the first female institutions to call themselves colleges appeared with the explicit goal of providing women with an education similar to that offered by men's colleges: Georgia Female College, 1839; Wesleyan Female College, 1843; Mary Sharp College, 1850; Oxford Female College, 1852; Pennsylvania Female College, 1853; Elmira College, 1855. Many of these institutions attempted to maintain the same standards for admission and graduation as men's colleges, although for financial reasons they often accepted students who lacked sufficient preparation. Consequently, most early collegiate institutions, like most of the women's colleges established later in the nineteenth century, formed preparatory departments, designed to bring newly admitted students up to college level.[4]

The female seminary, with its emphasis on preparing women for marriage and homemaking, remained the dominant institution in women's higher education until 1870, but during the first half of the century several developments occurred that later accelerated the growth in women's higher education. The first was a sharp increase in the number of public elementary schools requiring inexpensive teachers, and there were not enough willing men to fill all available positions. Thus from the start, female seminaries, women's colleges, and normal schools trained many of their students to be teachers. By 1870, there were probably more than 100 normal schools in the United States, whose explicit purpose was to furnish elementary schools with instructors; and their students—the great majority of them women—comprised a large portion, approximately 38% of the 11,000 women attending institutions of higher learning. (See Table 2.3 in Appendix.) Coeducational universities followed the pattern of training women as public school teachers and instituted normal (teacher-training) departments or teachers' colleges that were largely composed of women.

The second development was an increase in the number of public secondary schools. The first high school for girls in the country opened in Worcester in 1824 and was soon followed by coeducational high schools in other cities. After 1865, the number of coeducational public high schools increased rapidly, rising from approximately 160 schools in 1870 to 2,526 in 1890 and 6,005 in 1900.[5] Until the Civil War, most of the institutions of higher learning in the United States were single-sex. But after the war, as the spread in public primary and secondary schools familiarized Americans with coeducation, many of the state universities, particularly in the Midwest, began to admit women. Thus, two important factors—the combination of a need for inexpensive teachers and the presence of an increased number of young women with secondary school educations—contributed to the increase in demand for advanced education for women after 1870.

THE DEBATE OVER WOMEN'S COLLEGES

In the early nineteenth century, proponents of separate schooling for women argued that a carefully designed education for women would benefit the entire community by enabling women to perfect their moral, intellectual, and physical natures and by making them better homemakers and mothers. Emma Willard, in an appeal to New York State legislators for support of a girls' seminary, emphasized that women, as mothers, had a powerful influence on the development of human character and needed formal training to ensure that this influence would be beneficial.[6] In the words of Catharine Beecher, the forerunner of the domestic science movement in the United States:

> And yet what trade or profession of men involves more difficult and complicated duties than that of a housekeeper? Where is skill and science more needed than in the selection, cooking, and economy of food? . . . [or in] the care of infants and young children . . . [or] in woman's post as nurse of the Sick?
> And yet where is the endowment and where is the institution that has for its

aim the *practical training* of woman for any one of these departments of her sacred profession?[7]

The founders of some women's colleges, however, expressed a somewhat different intent, making known their desire to provide women with an education that was "equal" to that available to men. Essentially, they believed that the *purpose* of higher education ought to be the same for both sexes: to train mental faculties and to foster intellectual discipline and creativity of thought. This was a radical stance in the 1870s, given that scientists, academicians, and physicians entertained serious doubts about women's intellectual capabilities, believing that women's brains were too small and underdeveloped to allow them to engage in serious academic study. "Brain work" was thought to be dangerous to women's health. Excessive amounts of intellectual stimulation would impair women's constitutions and reproductive functions. Dr. Edward H. Clarke, a professor of Harvard Medical School, cautioned against subjecting women to a system of education designed for men in his popular work, *Sex in Education; Or, A Fair Chance for the Girls* (1873). Clarke warned that gynecological disorders (ovaritus, collapsed uteri), hysteria, and other nervous diseases could derive from excessive mental stimulation during adolescence, since at this time, the delicate reproductive structures were still developing. In the worst cases, Clarke pointed out, the reproductive organs of college women never fully matured, or these women became sterile or too weak to bear children.[8]

Clarke's views were shared by many male physicians, including Dr. Henry Maudsley, a British physiologist (See "Sex in Mind and in Education," pp. 77–86) and Dr. S. Weir Mitchell, a Philadelphian neurologist famous for the rest cures he prescribed for nervous women.[9] These experts on women's health and physiology carried immense influence with educators and the educated public. Women's colleges took their warnings seriously and instituted precautions to protect the health of their students. Vassar, which became the model for other women's colleges, screened applicants both for intellectual ability and physical fitness, regulated study hours, and required physical exercise of its students. Bryn Mawr strove to protect and improve students' health by "limiting the hours spent in the classroom, by instruction in hygiene, by the supervision of an accomplished physician, by outdoor sports, by the best sanitary conditions, by cheerfulness and joyousness."[10]

Accepting full responsibility for the physical well-being of their students, the women's colleges began to challenge—tentatively at first—the prevailing "scientific" views concerning the physical and intellectual capabilities of women. In 1877, Alice Freeman Palmer (1855–1902), who was later president of Wellesley, declared: "The old notion that low vitality is a matter of course with women, that to be delicate is a mark of superior refinement . . . that sickness is a dispensation of Providence—all these notions meet with no acceptance in college."[11] In 1885, the Association of Collegiate Alumnae, with the assistance of Carroll D. Wright and the Massachusetts Bureau of the Statistics of Labor, published a report on the health of college women, concluding that "the seeking of a college education on the part of women does not in itself necessarily entail a loss of health or serious impairment of the vital forces."[12]

Although women's colleges ultimately disproved the belief that education would physically damage women, educators initially proceeded cautiously and were extremely solicitous for their female students' health. In 1908, M. Carey Thomas, reviewing Bryn Mawr's trials, admitted that women had not known for certain "whether colleges might not produce a crop of . . . invalids. Doctors insisted that they would. . . . Now we have tried it, and tried it for more than a generation, and we know that college women are not only not invalids, but that they are better physically than other women in their own class of life."[13] In recognition of the women's colleges' success in improving the health of their students, S. Weir Mitchell altered his treatment of female nervous disorders from rest cures to increased physical exercise.[14] In little more than a generation, the women's colleges had wrought a small revolution, introducing the educated woman as a model of female health.

While women's colleges desired to give women an education as thorough as that given men, they did not always believe that women ought to be offered an identical education. In its initial 1865–66 catalogue, Vassar stated it wanted "to accomplish for young women what colleges of the first class accomplish for men; that is, to furnish them the means of a thorough, well-proportioned, and liberal education *but one adopted to their wants in life.*"[15] It was precisely because women's "wants" were different from those of men that women's education, as offered by both coeducational and female colleges, frequently was also different. With the exception of Smith College, which claimed that its program was "not intended to fit women for a particular sphere,"[16] coeducational and women's colleges consciously aimed to educate women for the appropriate professions of teaching and homemaking. Although the eastern women's colleges often felt ambivalent about teaching domestic science, they were much less hesitant to offer classes in the arts designed to improve women's grace and attractiveness and to improve their ability to foster a pleasant home life. According to one investigator, Mary Van Kleeck, over half of college women worked for pay upon graduation, but of these, 84% went into teaching—an occupation that was considered excellent preparation for motherhood.[17] Women's colleges never intended education to substitute for marriage. Educators hoped and argued that a college education would enhance women's marriage prospects, although they gladly acknowledged that in the unfortunate cases an education would help those women who could not marry to earn their livings. It was understood that female students who married had an obligation to use their education to promote the happiness and welfare of their families and that this obligation made it impossible for them to continue in another career. A large majority of college women did stop working soon after marriage. For example, Amy Hewes found that of 1,107 single Wellesley women graduating in 1890–1910, 98% did paid work, while less than 2% of the 427 married graduates were employed.[18] A prominent example was set by Alice Freeman, who resigned the presidency of Wellesley in 1887 to marry George Herbert Palmer. In a poem published posthumously by her husband, she wrote:

> ... At a crisis hour
> Of Strength and struggle on the height of life
> He came, and bidding me abandon power,
> Called me to take the quiet name of wife.[19]

THE DEBATE OVER COEDUCATION

The period from 1837 to 1870 was one of experimentation with coeducation. Beginning at Ohio colleges—Oberlin, 1837; Hillsdale, 1844; and Antioch, 1853—coeducation was practiced on a very small scale. But in the 1870s, adoption of coeducation by large influential institutions, most of them in the central and western states, eventually brought general acceptance.[20] These regions of the United States were more receptive initially to coeducation than the east or south, in part because they did not have a long tradition of sex-segregated education to overcome. The Morrill (Land Grant) Act, passed by Congress in 1862, helped establish and strengthen western institutions of higher learning. As it was too impractical and expensive to establish separate colleges for women, the newly built state universities, pressured by taxpayers, slowly admitted women.[21]

The 1890s were years of unprecedented growth for coeducational colleges and universities. The Universities of Minnesota, California, and Wisconsin tripled their student bodies, from approximately 1,000 students each in 1890 to 3,000 each in 1900. Cornell and Michigan increased their number of students by 50%, and two new coeducational institutions, Stanford and Chicago, founded in 1890, became top-ranking universities in enrollment and academic prestige by the end of the century.[22] In 1870, 4,600 women attended coeducational institutions, roughly 40% of all women in college. But by 1900, 61,000—70% of all college women—were enrolled in coeducational institutions.[23]

The arguments against educating women in coeducational institutions included all the arguments advanced against educating women in general: Women's health would suffer; they could not profit from study because their brains were too small; education would coarsen women's femininity, corrupt their moral sensibility, and physically prevent them from bearing children. But there were other arguments applied specifically to teaching men and women in the same classes: Men and women studying side-by-side was "unnatural"; it encouraged sexual indiscretions; it slowed the progress of male students; young men did not like to attend mixed classes; the courses, which were designed for male students, were too strenuous for women's delicate minds and bodies; and coeducational institutions could not provide women with adequate training for their future profession of homemaking. Alice Freeman Palmer, who received her bachelor's degree from Michigan and later became president of Wellesley, compared coeducational institutions and women's colleges:

> [The coeducational college] does not usually provide for what is distinctively feminine. Refining home influences and social oversight are largely lacking; and if they are wanting in the home from which the student comes, it must not be expected that she will show, on graduation, the graces of manner, the niceties of speech and dress, and the shy delicacy which has been encouraged in her more tenderly nurtured sister [attending a women's college]."[24]

The strengths of coeducation, according to Palmer, were "to be found in its tendency to promote independence of judgment, individuality of tastes, commonsense, and foresight in self-guidance, disinclination to claim favor, interest in learning for its own sake; friendly, natural, unromantic [and] non-sentimental relations with men."[25]

Those who supported coeducation argued that it accorded with nature. Since God had united the sexes within the family, the argument went, a coeducational environment was more natural and harmonious, and thus coeducational institutions would be easier to manage than single-sex institutions.[26] Rather than lead to sexual improprieties, supporters of coeducation argued, the presence of women would have a beneficial influence on the behavior of male students. While opponents of women's colleges feared that education would hinder female students from marrying, opponents of coeducational institutions feared the liberal environment would catapult unready women into marrying too soon and unwisely. In 1889, Palmer reassured all concerned: "The early fear that co-education would result in class-room romances has proved exaggerated. These young women do marry; so do others; so do young men. Marriage is not in itself an evil."[27]

At the beginning of the twentieth century, there occurred a reaction against coeducation, a revival of century-old arguments against a practice that by this time had been instituted in the vast majority of elementary and secondary schools and in 70% of the institutions of higher learning. Institutions that had been considered strongholds of coeducation—notably Wisconsin and Chicago—introduced practical changes directed at segregating the sexes.

The specific reasons given for the changes at these institutions were the rapid increase in the numbers of women attending the institution and the flight of men from liberal arts courses, leaving these classes comprised largely of women. At Chicago, the question of separate instruction emerged as a result of a tentative offer of an endowment, provided that women were taught in separate classes. After a great deal of debate and maneuvering on the part of President Harper, the Board of Trustees voted for segregation in October 1902.[28] Articles on Chicago's reaction appeared in newspapers and journals, including one by James Rowland Angell, then a professor at Chicago, that was published in *The Popular Science Monthly*. (See "Some Reflections Upon the Reaction from Coeducation," pp. 87–95.)

For a time, Chicago was the leading proponent of segregation. However, five years after the new plan was introduced, Marion Talbot, dean of women at Chicago, claimed that students in the first two years were separated in only a fifth of their work. A few years later segregation had virtually disappeared at Chicago, except in the sectioning of some large elementary courses.[29]

Although efforts to segregate the sexes proved unsuccessful, most women, at the end of the nineteenth and well into the twentieth century, were still being educated for the traditional feminine professions, teaching and homemaking. Many supporters of women's education initially saw education as a means of expanding women's sphere, but women's colleges and coeducational universities failed to dismantle the established tradition of training women for motherhood and teaching. Nonetheless, by the twentieth century, few people found it possible to argue seriously that women could not compete with men in learning Latin, Greek, or mathematics, or that women's health and reproductive systems could not withstand the rigors of education. However, many Americans continued to believe that higher education made women unfit for marriage, childbearing, and domestic responsibilities. In the early 1900s, Census statistics, speeches by President Theodore Roosevelt, and various college studies refocused public attention on college women's refusal to marry and raise children. The low marriage and birth rates of college women were taken as evidence of the moral degradation brought about by advanced education. I explore this aspect of the woman question—the birth rate question—in the next chapter.

APPENDIX

Table 2.1. Secondary School Graduates From Public and Nonpublic Schools, by Sex and as a Proportion of the 17-Year-Old Population, 1890–1920

	BOYS		GIRLS		
YEAR	TOTAL GRADS	GRADS (%) OF ALL BOYS AGE 17	TOTAL GRADS	GRADS (%) OF ALL GIRLS AGE 17	GIRL GRADS (%) OF ALL GRADUATES
1890	18,549	3	25,182	4	58
1900	38,075	5	56,808	8	60
1910	63,676	7	92,753	10	59
1920	123,684	13	187,582	20	60

Source: National Manpower Council, *Womanpower* (New York: Columbia University Press, 1957), p. 169.

Table 2.2. Women Enrolled in Institutions of Higher Learning, 1870–1920

YEAR	NUMBER OF WOMEN ENROLLED	PERCENTAGE OF ALL WOMEN 18 TO 21	PERCENTAGE OF ALL STUDENTS ENROLLED
1870	11,000	0.7	21.0
1880	40,000	1.9	33.4
1890	56,000	2.2	35.9
1900	85,000	2.8	36.8
1910	141,000	3.8	39.6
1920	283,000	7.6	47.3

Source: Mabel Newcomer, *A Century of Higher Education for American Women* (New York: Harper & Row, 1959), p. 46.

Table 2.3. Women Enrolled in Institutions of Higher Learning According to Type of Institution, 1870–1920

YEAR	TOTAL IN COLLEGE	COEDUCATIONAL INSTITUTIONS	WOMEN'S COLLEGES TOTAL	PUBLIC	PRIVATE	NORMAL
Thousands of Students						
1870	11.1	4.6	6.5	—	2.3	4.2
1880	39.6	23.9	15.7	—	11.3	4.3
1890	56.3	39.5	16.8	—	12.1	4.7
1900	85.4	61.0	24.4	0.2	16.5	7.6
1910	140.6	106.5	34.1	0.1	21.2	12.8
1920	282.9	230.0	52.9	6.4	27.2	19.4
Percentage Distribution						
1870	100	41.1	58.9	—	20.7	38.2
1880	100	60.4	39.6	—	28.7	10.9
1890	100	70.1	29.9	—	21.4	8.5
1900	100	71.4	28.6	0.3	19.3	8.9
1910	100	75.8	24.2	0.1	15.1	9.0
1920	100	81.3	18.7	2.3	9.0	7.5

Source: Mabel Newcomer, *A Century of Higher Education For American Women* (New York: Harper & Row, 1959), p. 49.

Table 2.4. Advanced Degrees Conferred, by Sex, 1870–1920

SCHOOL YEAR ENDING—	ALL DEGREES			BACHELOR'S OR FIRST PROFESSIONAL			MASTER'S OR SECOND PROFESSIONAL			DOCTOR'S OR EQUIVALENT		
	TOTAL	MALE	FEMALE	TOTAL	MALE	FEMALE	TOTAL	MALE	FEMALE	TOTAL	MALE	FEMALE
1870	9,372			9,371	7,993	1,378				1	1	0
1880	13,829			12,896	10,411	2,485	879			54	51	3
1890	16,703			15,539	12,857	2,682	1,015			149	147	2
1900	29,375	23,812	5,563	27,410	22,173	5,237	1,583	1,280	303	382	359	23
1910	39,755	30,716	9,039	37,199	28,762	8,437	2,113	1,555	558	443	399	44
1920	53,516	35,487	18,029	48,622	31,980	16,642	4,279	2,985	1,294	615	522	93

Source: United States Census Bureau Series H 327-338 from *Historical Statistics of the United States: Colonial Times to 1957* (Washington, DC: Government Printing Office, 1961), p. 212.

Table 2.5. Some Colleges Wholly or Partly Open to Women in the Nineteenth Century

Women's Colleges	Date*	Private Coed Colleges	Date
Bryn Mawr (PA)	1885	Antioch (OH)	1852
Mills Seminary (CA)	1871	Boston University (MA)	1869
Mills College (CA)	1885	Cornell University (NY)	1872
Mt Holyoke Seminary (MA)	1837	M.I.T. (MA)	1865
Mt Holyoke College (MA)	1888	Northwestern (IL)	1869
Rockford Seminary (IL)	1849	Oberlin (OH)	1837
Rockford College (IL)	1892	Stanford (CA)	1891
Smith (MA)	1875	Syracuse (NY)	1871
Vassar (NY)	1865	University of Chicago (IL)	1892
Wellesley (MA)	1875	Wesleyan (CT)	1872
Women's College of Baltimore (MD)	1888		

Women's Annexes to Men's Colleges		Foreign Universities	
		Aberdeen (London)	1892
Barnard (Columbia) (NY)	1889	Cambridge University (England)	
College for Women at Western Reserve (OH)	1888	Girton	1872
		Newnham	1873
Harvard Annex (MA)	1879	McGill (Montreal)	1884
Radcliffe (Harvard) (MA)	1894	Oxford University (England)	
Pembroke (Brown) (RI)	1892	Lady Margaret Hall	1879
W. Sophie Newcomb (Tulane)	1886	Somerville Hall	1879
		Queen's College (London)	1853
Coed State Universities		Royal University of Ireland	1882
University of Alabama	1893	University of Dublin Examinations	1896
University of California	1870	University of Edinburgh	1892
University of Illinois	1870	University of Glasgow	1892
University of Indiana	1868	Queen Margaret College	1883
University of Iowa	1856	University of London	1878
University of Kansas	1866	Bedford College	1849
University of Kentucky	1889	King's College	1878
University of Maine	1872	Royal Holloway College	1886
University of Michigan	1870	University College	1878
University of Minnesota	1868	University of Wales	1889
University of Mississippi	1882	University College of Wales	1872
University of Missouri	1870	Victoria (Manchester)	1880
University of Nebraska	1871		
University of North Carolina	1897		
University of Ohio	1873		
University of South Carolina	1894		
University of Texas	1883		
University of West Virginia	1897		
University of Wisconsin	1874		

*Dates throughout are approximate time of admission of women, not the date of origin of the institution.

Sources: Association of Collegiate Alumnae. *Contributions Toward A Bibliography of the Higher Education of Women* (Boston: Public Library, 1897), pp. 25–37 and M. Carey Thomas, "Education of Women," *Monographs on Education in the United States*, ed. Nicholas Murray Butler (New York: J. B. Lyon, 1899).

NOTES

1. Mabel Newcomer, *A Century of Higher Education for American Women* (New York: Harper & Brothers, 1959), p. 46. See Table 2.2 in Appendix for Newcomer's table.
2. Most of these dates of origin are taken from Thomas Woody's classic work, *A History of Women's Education in the United States*, 2 vols. (New York: Science Press, 1929), 2: 149–150. Also see Newcomer, *A Century of Higher Education for American Women* for a discussion of admission of women into state universities, pp. 12–14 and pp. 40–45 for a discussion of women's annexes (coordinate colleges). See Table 2.5 in Appendix for dates of admission of women into some colleges and universities.
3. See Woody, *A History of Women's Education in the United States*, 1: 329–459 for a discussion of the early female seminaries.
4. See Woody, *A History of Women's Education in the United States*, 2: 137–147.
5. W. T. Harris, "Recent Growth of Public High Schools in the United States as Affecting the Attendance of Colleges," *National Educational Association* (1901): 175.
6. See Emma Willard, "A Plan for Improving Female Education," (1819) *Woman and the Higher Education*, ed. Anna C. Brackett (New York: Harper & Brothers, 1893), pp. 1–46.
7. Catharine Beecher, "Woman's Profession Dishonored," *Harper's New Monthly Magazine* 29 (1864): 766; emphasis in original. Also cited in Woody, *A History of Women's Education in the United States*, 1: 400–401.
8. Edward H. Clarke, *Sex in Education; Or, A Fair Chance for the Girls* (Boston: James R. Osgood, 1873), p. 23. For women's responses to Clarke's book, see Anna C. Brackett, "Sex in Education," *The Education of American Girls*, ed. Anna C. Brackett (New York: G. P. Putnam's Sons, 1874), pp. 368–391; Eliza Duffey, *No Sex in Education; Or, An Equal Chance for Both Girls and Boys* (Philadelphia: Stoddart, 1874) and Julia Ward Howe, ed., *Sex and Education: A Reply to Dr. E. H. Clarke's "Sex in Education"* (Boston: Roberts, 1874).
9. See Henry Maudsley, *Body and Mind. An Inquiry into their Connection and Mutual Influence, Specially in Reference to Mental Disorders* (London: Macmillan, 1870) and S. Weir Mitchell, *Wear and Tear, or Hints for the Overworked* (Philadelphia: J. B. Lippincott, 1874; 1st ed., 1871). Clarke, Maudsley, and Mitchell were not the only physicians to speak of the dangers in educating women. See also George M. Beard, *American Nervousness: Its Causes and Consequences* (New York: G. P. Putnam's Sons, 1881) and T. S. Clouston, "Female Education from a Medical Point of View," *The Popular Science Monthly* 24 (December 1883 and January 1884): 214–228; 319–334.
10. *Addresses at the Incorporation of Bryn Mawr College*, 1885, p. 18. Cited in Sheila Rothman, *Woman's Proper Place: A History of Changing Ideals and Practices, 1870 to the Present* (New York: Basic Books, 1978), p. 32.
11. Alice Freeman Palmer, *Why Go to College?* (New York and Boston: T. Y. Crowell, 1897), p. 9. Also cited in Rothman, *Woman's Proper Place*, p. 36.
12. Association of Collegiate Alumnae, *Health Statistics of Women College Graduates* (Boston: Wright & Potter, 1885), p. 77. John Dewey, then instructor of philosophy at the University of Michigan, wrote an extensive review of this study. See John Dewey, "Health and Sex in Higher Education," *The Popular Science Monthly* 28 (March 1886): 606–614.
13. M. Carey Thomas, "Present Tendencies in Women's College and University Education," *Educational Review* 35 (1908): 69.
14. Rothman, *Woman's Proper Place*, pp. 34–36.
15. Cited in Newcomer, *A Century of Higher Education for American Women*, p. 55; emphasis added.
16. Ibid.
17. Mary Van Kleeck, "A Census of College Women," *Journal of the Association of Collegiate Alumnae* 11 (May 1918): 560, 585.
18. Amy Hewes, "Marital and Occupational Statistics of Graduates of Mount Holyoke College,"

Publications of the American Statistical Association 12 (December 1911): 794.
19. Edward T. James, ed., *Notable American Women: A Biographical Dictionary* (Cambridge: Belknap Press, 1971), 3: 6.
20. See Woody, *A History of Women's Education in the United States* 2: 224–303. Dates are from Newcomer, *A Century of Higher Education for American Women*, p. 12. Oberlin was coeducational from its inception in 1833, but 1837 marked the first year women were enrolled in the collegiate department.
21. See Charles R. Van Hise, "Educational Tendencies in State Universities," *Educational Review* 34 (December 1907): 509.
22. These statistics are given by James Rowland Angell, "Some Reflections Upon the Reaction From Coeducation," *The Popular Science Monthly* 62 (November 1902): 6. This article appears on pp. 87–95.
23. These statistics include women attending coeducational classes at coordinate colleges and were compiled by Newcomer, *A Century of Higher Education for American Women*, p. 49. See Table 2.3 in Appendix for a reprint of Newcomer's table. According to Woody, *A History of Women's Education in the United States* 2: 252, there were only 3,044 women attending coeducational institutions in 1875 and 19,959 in 1900.
24. Alice Freeman Palmer, "A Review of the Higher Education of Women," (1889) *Woman and the Higher Education*, ed. Brackett, p. 116.
25. Ibid., p. 115.
26. President Blanchard of Knox College wrote in the *Independent*, January 1870: " . . . shutting the sexes apart to keep them pure is a mistake . . . imagination of the absent sex in the separate schools is worse than its presence in the mixed . . . therefore, by the laws of Nature and the workings of the mind, we ought to expect a college of young men or of young ladies kept separate and apart, would be more disorderly, harder to manage, more unreasonable, and every way worse than where the two are united in an institution sanctified by Christ's presence, and governed by conscience and God's fear." Cited in Woody, *A History of Women's Education in the United States*, 2: 265.
27. Alice Freeman Palmer, "A Review of the Higher Education of Women," (1889) *Woman and the Higher Education*, ed. Brackett, pp. 115–116.
28. See Woody, *A History of Women's Education in the United States*, 2: 282–295.
29. Ibid., p. 289.

ARTICLES ON WOMEN'S EDUCATION APPEARING IN *THE POPULAR SCIENCE MONTHLY*

(Arranged by Date of Publication)

Editor. "The Higher Education of Woman," (April 1874) 4: 748–750.
Henry Maudsley. "Sex in Mind and in Education," (June 1874) 5: 198–215.
"Prof. Huxley on Female Education," (October 1874) 5: 764.
Editor. "Normal Co-Education," (January 1875) 6: 364–366.
Editor. "Who Shall Study the Babies?" (June 1876) 9: 241–243.
A. Hughes Bennett. "Hygiene in the Higher Education of Women," (February 1880) 16: 519–530.
Elizabeth Cumings. "Education as an Aid to the Health of Women," (October 1880) 17: 823–827.
Editor. "The Lessons of the Boston 'Ladies' Deposit,'" (September 1881) 19: 698–703.
"On Science Teaching in the Public Schools," (June 1883) 23: 207–214.
Editor. "The Back-Down of Dr. Dix," (July 1883) 23: 409–411.

T. S. Clouston. "Female Education From A Medical Point of View," (December 1883 and January 1884) 24: 214–228; 319–334.
Mary Putnam-Jacobi. "An Experiment in Primary Education," (August 1885 and September 1885) 27: 468–479; 614–623.
John Dewey. "Health and Sex in Higher Education," (March 1886) 28: 606–614.
Mrs. E. Lynn Linton. "The Higher Education of Woman," (December 1886) 30: 168–180.
Editor. "Higher Education of Women," (June 1887) 31: 269–271.
"Out-Door Play for School-Girls," (April 1888) 32: 856–857.
Mrs. William F. Jenks. "Education for Mothers," (September 1888) 33: 699.
Flora Bridges. "Coeducation in Swiss Universities," (February 1891) 38: 524–530.
Editor. "A Profession for Women," (March 1890) 38: 701–703.
Helene Lange. "Higher Education of Women in Europe," (April 1891) 38: 847–848.
Mary Alling Aber. "An Experiment in Education," (January 1892 and February 1892) 40: 377–392; 517–524.
Editor. "Motherhood," (July 1892) 41: 413–415.
Mary Roberts Smith. "Recent Tendencies in the Education of Women," (November 1895) 48: 27–33.
Sophia Foster Richardson. "Tendencies in Athletics for Women in Colleges and Universities," (February 1897) 50: 517–526.
Caroline W. Latimer. "Scientific Instruction in Girls' Schools," (June 1898) 53: 246–254.
Editor. "Scientific Instruction in Girls' Schools," (July 1898) 53: 413–416.
Alexandra L. B. Ide. "Shall We Teach Our Daughters the Value of Money?" (March 1899) 54: 686–690.
James Rowland Angell. "Some Reflections Upon the Reaction From Coeducation," (November 1902) 62: 5–26.
David Starr Jordan. "The Higher Education of Women," (December 1902) 62: 97–107.
Frances M. Abbott. "Three Decades of College Women," (August 1904) 65: 350–359.
J. McKeen Cattell. "The School and the Family," (January 1909) 74: 84–95.
Mary A. Grupe. "How the Problems of Rural Schools are Being Met," (November 1913) 83: 484–490.

SUGGESTIONS FOR FURTHER READING

Primary Sources

Brackett, Anna C., ed. *The Education of American Girls*. New York: G. P. Putnam's Sons, 1874.
Brackett, Anna C., ed. *Woman and the Higher Education*. New York: Harper & Brothers, 1893.
Catt, Carrie Chapman, "The Home and the Higher Education," *National Educational Association Journal of Proceedings* 41 (July 1902): 100–110.
Clarke, Edward H. *Sex in Education; Or, A Fair Chance for the Girls*. Boston: James R. Osgood, 1873.
Duffey, Eliza. *No Sex in Education; Or, An Equal Chance for Both Girls and Boys*. Philadelphia: Stoddart, 1874.
Palmer, Alice F. "A Review of the Higher Education of Women," *Forum* 12 (September 1889): 28–40.
Palmer, Alice F. *Why Go to College?* New York and Boston: T. Y. Crowell, 1897.
Thomas, M. Carey. "Education of Women," *Monographs on Education in the United States*. Edited by Nicholas Murray Butler. New York: J. B. Lyon, 1899.
Thomas, M. Carey. "Present Tendencies in Women's College and University Education," *Educational Review* 35 (January 1908): 64–85.

Van Hise, Charles R. "Educational Tendencies in State Universities," *Educational Review* 34 (December 1907): 504–520.

For statistical studies of college women, see Chapter 3.

Secondary Sources

Conable, Charlotte Williams. *Women at Cornell: The Myth of Equal Education.* Ithaca: Cornell University Press, 1977.

Frankfort, Roberta. *Collegiate Women: Domesticity and Career in Turn-of-the-Century America.* New York: New York University Press, 1977.

Graham, Patricia Albjerg. "Expansion and Exclusion: A History of Women in American Higher Education," *Signs: Journal of Women in Culture and Society* 3 (Summer 1978): 759–773.

Hofstadter, Richard, and Hardy, C. DeWitt. *The Development and Scope of Higher Education in the United States.* New York: Columbia University Press, 1952.

Newcomer, Mabel. *A Century of Higher Education for Women.* New York: Harper & Brothers, 1959.

Ross, Earle, D. *Democracy's College: The Land-Grant Movement in the Formative Stage.* Ames: Iowa State College Press, 1942.

Rothman, Sheila M. *Woman's Proper Place: A History of Changing Ideals and Practices, 1870 to the Present.* New York: Basic Books, 1978, pp. 26–42, 97–106.

Spender, Dale. *Invisible Women: The Schooling Scandal.* London: Writers and Readers, 1982.

Veysey, Laurence. *The Emergence of the American University.* Chicago: University of Chicago Press, 1965.

Woody, Thomas. *A History of Women's Education in the United States.* 2 vols. New York: Science Press, 1929.

Edward Livingston Youmans (1821–1887), editor of The Popular Science Monthly *during the early 1870s, supported higher education for women in principle, but believed that colleges were not offering courses designed with the severe constitutional limitations of women in mind. Nor did he think that universities were sufficiently preparing women for maternity, childcare, and domestic life. In the following selections, Youmans pointed out the dangers of sending women to coeducational colleges and recommended that the advice of physicians such as Edward Clarke and Henry Maudsley be heeded, lest female students suffer irrevocable harm to their mental and physical health.*

THE HIGHER EDUCATION OF WOMAN

EDITOR'S TABLE

April 1874

OF the great movement of modern culture, one of the most important phases is that now recognized as the "higher education of woman." That woman requires a better education than she has hitherto had, and that it should also be of a higher grade, are undeniable, although the practical questions that arise in the attempt to define and attain it are serious and formidable. The prevalent short-cut solution of the problem—women crave a higher education, therefore open to them the higher institutions—is as far as possible from being an adequate or satisfactory disposition of the case.

It is a constant complaint among the leaders of the woman's movement, that, in consequence of the long subjection of the sex to the domination of men, women have not been allowed or incited to think for themselves. They complain that women's ideas have been moulded by men, in conformity to the state of subordination in which the weaker sex has been held, and that the first thing women have to do is to assert themselves mentally, to develop their own powers in their own way, to form their own opinions, and not be forever dependent upon those who by the radical bias of an opposite constitution are incapable of comprehending woman or of doing justice to her capacities. On this ground it is of course impossible for woman to accept a masculine education. For the existing colleges and universities have not only been originated and developed through centuries exclusively by men, but they have been pervaded by the thoughts and animated by the feelings and tastes, and moulded by the aims and necessities, of men. If women are to free themselves from male control in the matter of one-sided mental influence, it would seem that their first care should be not to *subject* themselves to the action of those institutions the very object of which is to assimilate and determine the intellectual character of students into harmony with their own policy.

We yield to no others in the earnestness of our belief in the higher education of women; but we want to see them take the matter in their own hands, and work out a system of mental cultivation adapted to their own natures and needs. The higher education as embodied in existing institutions cannot meet this requirement. It is, in fact, under indictment for nonadaptation to the present wants of men; and one of the most profound and important of the reforms of our age is that thorough modification of collegiate methods of study that shall bring them up to the demands of modern life. That they now answer to these demands, but very few will maintain. There are many, and the number is increasing, who do not go to college because the education there obtained is thought to be of little use to the possessor, if not indeed a hindrance to him in his future experience with the world. Thousands ignore all considerations of the usefulness of what is to be learned, and go or are sent to college because it is the proper thing, a fashion of society, and has its social benefits; and many undoubtedly go because they have been made to believe that the old education is the perfection of human wisdom for mental discipline, and is, after all, the best thing even for practical life. Yet the distrust of the system is deep, and has already made itself so powerfully felt, that the colleges have been compelled to yield to it, and in many cases to modify their methods of instruction and create supplemental schools devoted to modern knowledge. The higher education of men is thus in a state of conflict and transition; the old education is fading way, and a New Education is rising in its place.

It seems to us that this is the first fact for women to consider in their efforts to attain a higher education. The question that is forced upon men, What shall the higher education be? has even a graver concern for women, for it is not only an open one, but it is an experiment which must be submitted to the test of time, and if mismanaged may be full of peril. It behooves women not to be so carried away by the current clamor about the advantages of education, that they are willing to accept any thing under that name that is dispensed from the schools. Education, like every thing else, may be good or bad, worthless or valuable; but it differs from most other things in this, that, if bad and worthless, it cannot be got rid of. We have yet to realize the important fact that much so-called education is worse than none at all; and that it is better to leave the mind to its spontaneous forces and its self-development under the action of the surrounding influences of Nature and life, rather than to meddle with it inconsiderately, to burden it with worthless knowledge, or to violate its proportions by an extravagant over-culture of some faculties and a total neglect of others. Were the doors of all the colleges of the country to be opened to-morrow to woman, in good faith, and in obedience to a public sentiment that would lead her to avail herself of the opportunity as men do, we believe that the result could not be otherwise than in a high degree disastrous to woman and to society; and this because the education which she would get would be not what she requires, would be put in the place of what she requires, and would indefinitely postpone the attainment of what she requires.

It is well for woman that, in awakening to the necessity of a higher cultivation of her faculties, she is free in the choice of means; but it remains to be seen what she will do with her chance. There is superabounding knowledge, the ripening of all the past — wheat and chaff; there is the world's long experience with education for help or for warning; what, then, will woman do toward constructing a higher education for herself? Will she follow blindly the old traditions, content with any thing, and accept the culture that man has outgrown and is rejecting; or will she be equal to the occasion, and form for herself a curriculum of studies suited to the requirements of her own nature?

That woman has a sphere marked out by her organization, however the notion may be scouted by the reformers, is as true as that the bird and the fish have spheres which are determined by their organic natures. Birds often plunge into the watery deep, and fishes sometimes rise into the air, but one is nevertheless formed for swimming and the other for flight. So women may make transient diversions from the sphere of activity for which they are constituted, but they are nevertheless formed and designed for maternity, the care of children, and the affairs of domestic life. They are the mothers of humankind, the natural educators of childhood, the guardians of the household, and by the deepest ordinance of things they are this, in a sense, and to a degree, that man is not. For woman in these relations, education has hitherto done but little, and humanity has suffered as a consequence. To the mothers of the race, especially, belongs the question of its preservation and improvement. The problem is transcendent, and woman's interest in it more immediate and vital than man's can be. Science has furnished the knowledge that is required, a vast mass of truth that is waiting to be applied for the conduct and ennobling of the domestic sphere. Man has originated it; is it not for woman to use it? And now, when there is so much agitation to give woman larger mental opportunities, and she is pressing for the advantages of a higher education, we have a right to expect that she will consider the subject from her own point of view, and supply the great educational need that has been so long recognized and deplored. The new departure of higher female education should unquestionably be from the results of the medical profession. We believe that physicians have by no means yet taken the share in general education that the interests of society require; but, when the mental cultivation of women is to become systematic and they have their own higher institutions, the agency of physicians will be indispensable. It is not that all women are to be doctors, but that they are to be instructed and become intelligent first of all in the sciences of life, with which also the physician has to deal. If, to get the A. M. of Yale or Harvard, would be worth the struggle for women, as qualifying them for the intelligent fulfillment of their destiny, let the doors be battered down if necessary for their entrance; but, if it would not conduce to this end, and would rather be fatal to it, let the doors remain double-locked. If the present aspiration is to be utilized, the movement must not take a false direction. New institutions are called for, that shall supply a new education on the feminine side. The system of studies may be broad and liberal in the best sense, but what we insist on is that it should be shaped with fundamental reference to the life-needs of female students. From this point of view our existing female colleges are liable to criticism; in so far as they are imitations of the old masculine establishments, they do not meet the wants of the sex, and rather obstruct than aid the true course of feminine cultivation.

NORMAL CO-EDUCATION

EDITOR'S TABLE

January 1875

IN the October MONTHLY we published a letter of Prof. Cochran, from Dr. Clarke's late work, "The Building of a Brain," on the effects of co-education in the Albany Normal School. We have received a suggestive letter from a lady who was connected with the institution at the time, and who says that "the cases of illness or of failing health among the young ladies were sent to me to inquire into and care for;" and she adds that, "to those familiar with the work of the Albany school, at this period, no statistics drawn from its health-roll would count any thing whatever in this discussion." It is mentioned that, "in 1864, when Prof. Cochran left Albany, in a school of over two hundred pupils, there were twenty gentlemen, so that, as far as education is concerned, it would seem that some deference might have been paid to the greatest good of the greatest number." The curriculum is, however, characterized as "oppressive." Our correspondent claims to have investigated the subject, and says that, while "with regard to the facts there is little question, with regard to the causes there is a very important one." Her general view of the case is presented in the following passages from her communication:

> "I have reached the conclusion, from my investigations, that no statistics drawn from mixed schools can prove any thing with regard to co-education or identical education, until the two sexes can be placed in those schools upon equal or similar conditions.
>
> "While there are a hundred outside things that militate against a woman's success in such a school, which find a parallel in the conditions of the male pupils, ill health or want of power among the female pupils can prove nothing. In every quarter woman is unfairly weighted for the race; but especially in our normal schools these conditions have reached their climax. For example:

"In 1856 one of the young ladies, whose failing health warned her of overwork, came to me not more than two months before the time when she should have graduated. Hard as the case seemed, I could only say to her that she must leave school at once. Some facts of her history I obtained from her at that time, but the important points I learned later. She had no home to which she could go. She had been left an orphan at an early age, with a family of brothers and sisters dependent upon her for support. To meet this responsibility she went as a teacher into our district schools. She undertook the hardest positions because they gave a trifle more of pay. She boarded herself, and often went dinnerless to school because the children's bread must not be stinted. She went through mud and snow to her school, with wet feet and scanty clothing, purchasing no rubbers, no warm shawls, because she could not spare the money. She had soon decided that, if she ever lifted those she loved so well from utter poverty, she must fit herself for higher positions, and to this end she began laying aside money that she might attend a normal school.

"So the years passed on, and at last she had saved enough to take her through the two years' course of the normal school at Albany. And now when her classmates were beginning to think of their graduating essays and graduating dresses, her pay-roll was wound up, her summons came, and she turned away from the reward she had sought so tirelessly. The autumn leaves of 1866 fell upon the grave where she found rest for the first time in so many years. This is one of Prof. Cochran's twenty. Another, the same year, was accustomed to take an empty dinner-basket with her to school, and at the hour of lunch to steal away from her companions, that they might not suspect she was too poor to buy a dinner! In my own experience these have been not isolated but representative cases of normal-school invalidism. So familiar have I become with them, that I seem to know beforehand what items I shall obtain in investigating any given case of ill-health. And then the cry arises, 'Co-education does not answer.' It is true we have cases of ill-health among our young ladies which are not to be traced to these causes, but they are so few as hardly to deserve mention. These really make up the bulk of the cases with which we have to deal.

"We have also young gentlemen whose health fails from overwork, but to them the admonition arises in the shape of weak eyes, constant headaches, etc., while with women the more delicately-balanced functions of life are set ajar.

"The young man goes out to teach, and earns sixty dollars per month, while his sister is earning thirty dollars. In half the cases she is the better scholar. The young man goes home to the farm. He is needed in the field, but he is a man—of course he can earn one and a half or two dollars per day; while with his sister they are so glad she has come home to help mother, but it never occurs to any one that she has earned any money. When both return to school, they pay the same price for board, but with him it means that his bed shall be made, his room swept, water brought in, etc., while his washing arrives from the laundry all right every week. But she—the landlady says, 'Of course you will take care of your own room, we always expect our lady-boarders to do that.' She counts over her thirty dollars per month, and says, 'Well, I must do my own washing and ironing if my landlady will allow me.' And the landlady grudgingly consents. Then—'I must make my own calico dresses—I could never afford to pay for that.' To her teacher: 'I wish I could be excused from singing, to-day, I am trying to make a dress.' Or, 'No, I cannot go for a walk, my brother has brought me this whole satchel full of clothes to mend.' In the morning he can easily learn his algebra-lesson while she is arranging on the top of her head the steeple of braids which custom says she must wear. And so the parallel runs on."

From all which, it would seem to be a fair inference that, as the world is at present constituted, co-education is beset with very formidable difficulties. With their inferior strength, their extra burdens, and their more limited pecuniary means, the female students cannot compete with the male students, and in the attempt to do so they break down. The implication is that, in point of fact and practically, there are unequal standards of study to which the two sexes can respectively attain; and, if these standards are to be equalized, either the masculine standard must be lowered, so that the male students will not be pressed to their highest capacity of accomplishment, or the feminine standard must be raised, to the injury of female health. Our correspondent says that there is little question in regard to the facts, but that "woman is unfairly weighted for the race." Whether unfairly or not, she certainly is so seriously weighted that she cannot win in rivalry with her less-weighted competitor. The real question, then, is, whether this difference is accidental and removable, or whether it is radical and permanent, and belongs to the very constitution of the sexes. Upon this point we hope the MONTHLY will soon have something further to say.

WHO SHALL STUDY THE BABIES?

EDITOR'S TABLE

June 1876

THE reader's attention will be arrested by the novelty of our first article, by a distinguished literary Frenchman, giving the result of his observations on the progress of an infant in learning to talk. We confess to some mortification at seeing the name of a man at the head of such a discussion. Not that the dignity of M. Taine is at all compromised, for he never undertook a more important or a more distinguished task than critically noting the steps of mental evolution in a baby. Nevertheless, this would seem to be preëminently the proper work of woman—a work to which we might infer she would be drawn by her feelings, in which she would be interested by her curiosity, and would take up from the temptation of her special opportunities. Yet M. Taine found that it had not been done. He wished to test Max Müller's views in regard to the genesis of language, and wanted a series of observations of infantine mental growth for the purpose. But they had not been made, the facts were wanting, and nothing remained but to make the study himself.

We say this kind of work belongs to woman, and she is perfectly competent to perform it. Why, then, has it not been undertaken, and why has there not grown up a body of carefully-observed and widely-verified facts regarding psychological development in infancy such as would be valuable for arriving at inductive truths for guidance in the rational education of childhood? Undoubtedly, psychology is a backward science, imperfect from the obscurity and complexity of its questions, and its long cultivation by unscientific methods. But the value of observations upon the mental unfolding of infancy is not, by any means, dependent upon the possibility of immediately explaining them. Such observations, if accurately made and intelligently recorded, will have a value of their own independent of the state of psychological science, while they would become a permanent and potent means of its advancement. In most other fields of natural phenomena the facts are far in advance of the theories by which they are organized into science; in the field of mental growth, however, observations are scanty and speculation superabundant.

We are, of course, not to expect that things will come before they are wanted, and, if such observations are not called for, why should they be supplied? But the facts have been long and loudly called for, if not by psychologists, then by practical educators, while woman has had exclusive charge of the education that begins in infancy. She is an educator as a mother, and the culture of childhood has almost universally fallen into her hands as a teacher. We might surely have expected that, with their great excess of opportunity, some few women of ability would have gone carefully and critically and often over the ground which M. Taine has passed over once with such interesting results. But the work that might have been expected, so far as we are aware, has not been done, nor is there any promise of it. The difficulty is, that there has been nothing in woman's education either to interest her in the subject or to qualify her for dealing with it. Observations, to be valuable for scientific purposes, involve an accuracy of perception and an intellectual discrimination which are not to be had except by patient and methodical training of the observing powers. This is the one thing that has not been included in female education. Neither languages, nor mathematics, nor history, nor mental philosophy, nor music, nor general literature, affords any exercise whatever of the observing faculties. A student may become proficient in all these branches, while the intellectual interest in the phenomena of daily experience, and the objects of common life, remains as dormant as it is in the savages. Nay, more, absorption in these modes of mental activity, which involve chiefly the memory and reflective powers, is fatally unfavorable to observation, as it brings the mind under the control of mental habits that exclude it. No woman can make valuable observations on mental progress in infancy that has not had a culture fitted for it, first, by a long practice, such as she gives to music, in independent observation in some branch of objective science, as botany, for example; and, secondly, by a thorough knowledge of the constitution of the child, especially the functions of its nervous mechanism. With their heads filled with history, aesthetics, algebra, French, and German, they will never attain to these qualifications for studying the character of children. The seminaries do not prepare them for it; the high-schools and the normal schools do not confer it. Nor is this all, nor the worst. There is no appreciation of it or aspiration for it. The so-called woman's movement, which professes to aim at her higher improvement and the enlargement of her activities, is not in this direction. It looks to public, professional, and political life, as woman's future and better sphere of action. In the new colleges for women that are springing up in all directions with munificent endowments, the supreme consideration seems to be to ignore sex, and frame the feminine curriculum of study on the old masculine models, and keep it up to the masculine standards. The spirit of these schools is that of a slavish imitation. They are organized with no reference to the urgent and living needs of society, but they go in for the traditional trumperies of the old colleges; and, instead of studying science in its personal, domestic, and social bearings, the women demand Latin and Greek, and as much of it as the masculine intellect has proved capable of surviving. Children are imitators. Savages are imitators. What else are the women in their demands for new and ampler opportunities of culture? They will study classics, and let the men study the babies; but, if they are incompetent, of course the men *must* do it. For this business of studying the science of infancy must be pursued by somebody, thoroughly and exhaustively. It is nothing less than a transcendent problem of human character lying at the foundation of the social state; for only as the human being is understood in its deeper organic laws, prenatal and infantine, as well as in its subsequent unfolding, can we arrive at settled and scientific views regarding the rights, claims, duties, and true interests of the individual in society. If not a new research, it is at least a new impulse and stage of research, and we say again that we should think intelligent and ambitious women would be glad to have a share in it, and would have wisdom enough to include it in their extended schemes of female education.

Henry Maudsley (1835–1918), a renowned British physician and professor of Medical Jurisprudence at University College in London, was greatly influenced by Edward H. Clarke's book, Sex in Education *(1873), which warned against subjecting women to a higher education that was designed for men. The following article first appeared in a British magazine,* Fortnightly Review *(April 1874) and then in* The Popular Science Monthly *two months later. Maudsley, like Clarke before him, criticized educational institutions for not adequately addressing the needs of female students and for ignoring the important biological differences between men and women. Maudsley argued that because women labored under an "inferiority of constitution" that "for one quarter of each month during the best years of life is more or less sick and unfit for hard work," it was imprudent to model women's education after that of men. Maudsley's article, like Clarke's book, fostered public debate; see Elizabeth Garret Anderson, "Sex in Mind and Education: A Reply,"* Fortnightly Review *15 (May 1874): 582–594.*

SEX IN MIND AND IN EDUCATION

HENRY MAUDSLEY, M.D.

June 1874

IT is quite evident that many of those who are foremost in their zeal for raising the education and social status of woman, have not given proper consideration to the nature of her organization, and to the demands which its special functions make upon its strength. These are matters which it is not easy to discuss out of a medical journal; but, in view of the importance of the subject at the present stage of the question of female education, it becomes a duty to use plainer language than would otherwise be fitting in a literary journal. The gravity of the subject can hardly be exaggerated. Before sanctioning the proposal to subject woman to a system of mental training which has been framed and adapted for men, and under which they have become what they are, it is needful to consider whether this can be done without serious injury to her health and strength. It is not enough to point to exceptional instances of women who have undergone such a training, and have proved their capacities when tried by the same standard as men; without doubt there are women who can, and will, so distinguish themselves, if stimulus be applied and opportunity given; the question is, whether they may not do it at a cost which is too large a demand upon the resources of their nature. Is it well for them to contend on equal terms with men for the goal of man's ambition?

Let it be considered that the period of the real educational strain will commence about the time when, by the development of the sexual system, a great revolution takes place in the body and mind, and an extraordinary expenditure of vital energy is made, and will continue through those years after puberty when, by the establishment of periodical functions, a regularly recurring demand is made upon the resources of a constitution that is going through the final stages of its growth and development. The energy of a human body being a definite and not inexhaustible quantity, can it bear, without injury, an excessive mental drain as well as the natural physical drain which is so great at that time? Or, will the profit of the one be to the detriment of the other? It is a familiar experience that a day of hard physical work renders a man incapable of hard mental work, his available energy having been exhausted. Nor does it matter greatly by what channel the energy be expended; if it be used in one way it is not available for use in another. When Nature spends in one direction, she must economize in another direction. That the development of puberty does draw heavily upon the vital resources of the female constitution, needs not to be pointed out to those who know the nature of the important physiological changes which then take place. In persons of delicate constitution who have inherited a tendency to disease, and who have little vitality to spare, the disease is apt to break out at that time; the new drain established having deprived the constitution of the vital energy necessary to withstand the enemy that was lurking in it. The time of puberty and the years following it are therefore justly acknowledged to be a critical time for the female organization. The real meaning of the physiological changes which constitute puberty is, that the woman is thereby fitted to conceive and bear children, and undergoes the bodily and mental changes that are connected with the development of the reproductive system. At each recurring period there are all the preparations for conception, and nothing is more necessary to the preservation of female health than that these changes should take place regularly and completely. It is true that many of them are destined to be fruitless so far as their essential purpose is concerned, but it would be a great mistake to suppose that on that account they might be omitted or accomplished incompletely, without harm to the general health. They are the expressions of the full physiological activity of the organism. Hence it is that the outbreak of disease is so often heralded, or accompanied, or followed by suppression or irregularity of these functions. In all cases they make a great demand upon the physiological energy of the body: they are sensitive to its sufferings, however these be caused; and, when disordered, they aggravate the mischief that is going on.

When we thus look the matter honestly in the face, it would seem plain that women are marked out by Nature for very different offices in life from those of men, and that the healthy performance of her special functions renders it improbable she will succeed, and unwise for her to persevere, in running over the same course at the same pace with him. For such a race she is certainly weighted unfairly. Nor is it a sufficient reply to this argument to allege, as is sometimes done, that there are many women who have not the opportunity of getting married, or who do not aspire to bear children; for whether they care to be mothers or not, they cannot dispense with those physiological functions of their nature that have reference to that aim, however much they might wish it, and they cannot disregard them in the labor of life without injury to their health. They cannot choose but to be women: cannot rebel successfully against the tyranny of their organization, the complete development of function whereof must take place after its kind. This is not the expression of prejudice nor of false sentiment; it is the plain statement of a physiological fact. Surely, then, it is unwise to pass it by; first or last it must have its due weight in the determination of the problem of woman's education and mission; it is best to recognize it plainly, however we may conclude finally to deal with it.

It is sometimes said, however, that sexual difference ought not to have any place in the culture of the mind, and one hears it affirmed with an air of triumphant satisfaction that there is no sex in mental culture. This is a rash statement, which argues want of thought or insincerity of thought in those who make it. There is sex in mind as distinctly as there is sex in body; and, if the mind is to receive the best culture of which its nature is capable, regard must be had to the mental qualities which correlate differences of sex. To aim, by means of education and pursuits of life, to assimilate the female to the male mind, might well be pronounced as unwise and fruitless a labor as it would be to strive to assimilate the female to the male body by means of the same kind of physical training and by the adoption of the same pursuits. Without doubt there have been some striking instances of extraordinary women who have shown great mental power, and these may fairly be quoted as evidence in support of the right of women to the best mental culture; but it is another matter when they are adduced in support of the assertion that there is no sex in mind, and that a system of female education should be laid down on the same lines, follow the same method, and have the same ends in view, as a system of education for men.

Let me pause here to reflect briefly upon the influence of sex upon mind. In its physiological sense, with which we are concerned here, mind is the sum of those functions of the brain which are commonly known as thought, feeling, and will. Now, the brain is one among a number of organs in the commonwealth of the body; with these organs it is in the closest physiological sympathy by definite paths of nervous communication, has special correspondence with them by internuncial nerve-fibres; so that its functions habitually feel and declare the influence of the different organs. There is an intimate consensus of functions. Though it is the highest organ of the body, the coördinating centre to which impressions go and from which responses are sent, the nature and functions of the inferior organs with which it lives in unity affect essentially its nature as the organ of mental functions. It is not merely that disorder of a particular organ hinders or oppresses these functions, but it affects them in a particular way; and we have good reason to believe that this special pathological effect is a consequence of the specific physiological effect which each organ exerts naturally upon the constitution and function of mind. A disordered liver gives rise to gloomy feelings; a diseased heart, to feelings of fear and apprehension; morbid irritation of the reproductive organs, to feelings of a still more special kind—these are familiar facts; but what we have to realize is, that each particular organ has, when not disordered, its specific and essential influence in the production of certain passions or feelings. From of old the influence has been recognized, as we see in the doctrine by which the different passions were located in particular organs of the body, the heart, for example, being made the seat of courage, the liver the seat of jealousy, the bowels the seat of compassion; and although we do not now hold that a passion is aroused anywhere else than in the brain, we believe nevertheless that the organs are represented in the primitive passions, and that, when the passion is aroused into violent action by some outward cause, it will discharge itself upon the organ and throw its functions into commotion. In fact, as the uniformity of thought among men is due to the uniform operation of the external senses, as they think alike because they have the same number and kind of senses, so the uniformity of their fundamental passions is due probably to the uniform operation of the internal organs of the body upon the brain; they feel alike because they have the same number and kind of internal organs. If this be so, these organs come to be essential constituents of our mental life.

The most striking illustration of the kind of organic action which I am endeavoring to indicate, is yielded by the influence of the reproductive organs upon the mind; a complete mental revolution being made when they come into activity. As great a change takes place in the feelings and ideas, the desires and will, as it is possible to imagine, and takes place in virtue of the development of their functions. Let it be noted, then, that this great and important mental change is different in the two sexes, and reflects the difference of their respective organs and functions. Before experience has opened their eyes, the dreams of a young man and maiden differ. If we give attention to the physiology of the matter, we see that it cannot be otherwise, and if we look to the facts of pathology, which would not fitly be in place here, they are found to furnish the fullest confirmation of what might have been predicted. To attribute to the influence of education the mental differences of sex which declare themselves so distinctly at puberty, would be hardly less absurd than to attribute to education the bodily differences which then declare themselves. The comb of a cock, the antlers of a stag, the mane of a lion, the beard of a man, are growths in relation to the reproductive organs which correlate mental differences of sex as marked almost as these physical differences. In the first years of life, girls and boys are much alike in mental and bodily character, the differences which are developed afterward being hardly more than intimated, although some have thought the girl's passion for her doll evinces even at that time a forefeeling of her future functions; during the period of reproductive activity, the mental and bodily differences are declared most distinctly; and when that period is past, and man and woman decline into second childhood, they come to resemble one another more again. Furthermore, the bodily form, the voice, and the mental qualities of mutilated men approach those of women; while women whose reproductive organs remain from some cause in a state of arrested development, approach the mental and bodily habits of men. . . .

If the foregoing reflections be well grounded, it is plain we ought to recognize sex in education, and to provide that the method and aim of mental culture should have regard to the specialties of woman's physical and mental nature. . . . There is sex in mind, and there should be sex in education.

Let us consider, then, what an adapted education must have regard to. In the first place, a proper regard to the physical nature of women means attention given, in their training, to their peculiar functions and to their foreordained work as mothers and nurses of children. Whatever aspirations of an intellectual kind they may have, they cannot be relieved from the performance of those offices so long as it is thought necessary that mankind should continue on earth. Even if these be looked upon as somewhat mean and unworthy offices in comparison with the nobler functions of giving birth to and developing ideas; if, agreeing with Goethe, we are disposed to hold—"Es wäre doch immer hübscher wenn man die Kinder von den Baumen schüttelte;" it must still be confessed that for the great majority of women they must remain the most important offices of the best period of their lives. Moreover, they are work which, like all work, may be well or ill done, and which, in order to be done well, cannot be done in a perfunctory manner, as a thing by the way. It will have to be considered whether women can scorn delights, and live laborious days of intellectual exercise and production, without injury to their functions as the conceivers, mothers, and nurses of children. For, it would be an ill thing, if it should so happen that we got the advantages of a quantity of female intellectual work at the price of a puny, enfeebled, and sickly race. In this relation, it must be allowed that women do not and cannot stand on the same level as men.

In the second place, a proper regard to the mental nature of woman means attention given to those qualities of mind which correlate the physical differences of her sex. Men are manifestly not so fitted mentally as women to be the educators of children during the early years of their infancy and childhood; they would be almost as much out of place in going systematically to work to nurse babies as they would be in attempting to suckle them. On the other hand, women are manifestly endowed with qualities of mind which specially fit them to stimulate and foster the first growths of intelligence in children, while the intimate and special sympathies which a mother has with her child as a being which, though individually separate, is still almost a part of her nature, give her an influence and responsibilities which are specially her own. The earliest dawn of an infant's intelligence is its recognition of its mother as the supplier of its wants, as the person whose near presence is associated with the relief of sensations of discomfort, and with the production of feelings of comfort; while the relief and pleasure which she herself feels in yielding it warmth and nourishment strengthen, if they were not originally the foundation of, that strong love of offspring which with unwearied patience surrounds its wayward youth with a thousand ministering attentions. It can hardly be doubted that, if the nursing of babies were given over to men for a generation or two, they would abandon the task in despair or in disgust, and conclude it to be not worth while that mankind should continue on earth. But "can a woman forget her sucking child, that she should not have compassion on the son of her womb?" Those can hardly be in earnest who question that woman's sex is represented in mind, and that the mental qualities which spring from it qualify her especially to be the successful nurse and educator of infants and young children.

Furthermore, the female qualities of mind which correlate her sexual character adapt her, as her sex does, to be the helpmate and companion of man. It was an Eastern idea, which Plato has expressed allegorically, that a complete being had in primeval times been divided into two halves, which have ever since been seeking to unite together and to reconstitute the divided unity. It will hardly be denied that there is a great measure of truth in the fable. Man and woman do complement one another's being. This is no less true of mind than it is of body; is true of mind indeed as a consequence of its being true of body. Some may be disposed to argue that the qualities of mind which characterize women now, and have characterized them hitherto, in their relations with men, are in great measure, mainly if not entirely, the artificial results of the position of subjection and dependence which she has always occupied; but those who take this view do not appear to have considered the matter as deeply as they should; they have attributed to circumstances much of what unquestionably lies deeper than circumstances, being inherent in the fundamental character of sex. It would be a delusive hope to expect, and a mistaken labor to attempt, to eradicate by change of circumstances the qualities which distinguish the female character, and fit woman to be the helpmate and companion of man in mental and bodily union.

So much may be fairly said on general physiological grounds. We may now go on to inquire whether any ill effects have been observed from subjecting women to the same kind of training as men. The facts of experience in this country are not such as warrant a full and definite answer to the inquiry, the movement for revolutionizing the education of women being of a recent date. But in America the same method of training for the sexes in mixed classes has been largely applied; girls have gone with boys through the same curriculum of study, from primary to grammar schools, from schools to graduation in colleges, working early under the stimulus of competition, and disdaining any privilege of sex. With what results? With one result certainly— that, while those who are advocates of the mixed system bear favorable witness to the results upon both sexes, American physicians are beginning to raise their voices in earnest warnings and protests. It is not that girls have not ambition, nor that they fail generally to run the intellectual race which is set before them, but it is asserted that they do it at a cost to their strength and health which entails life-long suffering, and even incapacitates them for the adequate performance of the natural functions of their sex. Without pretending to indorse these assertions, which it would be wrong to do in the absence of sufficient experience, it is right to call attention to them, and to claim serious consideration for them; they proceed from physicians of high professional standing, who speak from their own experience, and they agree, moreover, with what perhaps might have been feared or predicted on physiological grounds. It may fairly be presumed that the stimulus of competition will act more powerfully on girls than on boys; not only because they are more susceptible by nature, but because it will produce more effect upon their constitutions when it is at all in excess. Their nerve-centres being in a state of greater instability, by reason of the development of their reproductive functions, they will be the more easily and the more seriously deranged. . . .

A small volume, entitled "Sex in Education," which has been published recently by Dr. Edward Clarke, of Boston, formerly a professor in Harvard College, contains a somewhat startling description of the baneful effects upon female hea 1 which have been produced by an excessive educational strain. It is asserted that the number of female graduates of schools and colleges who have been permanently disabled to a greater or less degree by improper methods of study, and by a disregard of the reproductive apparatus and its functions, is so great as to excite the gravest alarm, and to demand the serious attention of the community. . . . The course of events is something in this wise: The girl enters upon the hard work of school or college at the age of fifteen years or thereabouts, when the function of her sex has perhaps been fairly established; ambitious to stand high in class, she pursues her studies with diligence, perseverance, constancy, allowing herself no days of relaxation or rest out of the school-days, paying no attention to the periodical tides of her organization, unheeding a drain "that would make the stroke oar of the university crew falter." For a time all seems to go well with her studies; she triumphs over male and female competitors, gains the front rank, and is stimulated to continued exertions in order to hold it. But in the long-run Nature, which cannot be ignored or defied with impunity, asserts its power; excessive losses occur; health fails, she becomes the victim of aches and pains, is unable to go on with her work, and compelled to seek medical advice. Restored to health by rest from work, a holiday at the sea-side, and suitable treatment, she goes back to her studies, to begin again the same course of unheeding work, until she has completed the curriculum, and leaves college a good scholar but a delicate and ailing woman, whose future life is one of more or less suffering. For she does not easily regain the vital energy which was recklessly sacrificed in the acquirement of learning; the special functions which have relation to her future offices as woman, and the full and perfect accomplishment of which is essential to sexual completeness, have been deranged at a critical time; if she is subsequently married, she is unfit for the best discharge of maternal functions, and is apt to suffer from a variety of troublesome and serious disorders in connection with them. In some cases the brain and the nervous system testify to the exhaustive efforts of undue labor, nervous and even mental disorders declaring themselves.

. . . In addition to the ill effects upon the bodily health which are produced directly by an excessive mental application, and a consequent development of the nervous system at the expense of the nutritive functions, it is alleged that remoter effects of an injurious character are produced upon the entire nature, mental and bodily. The arrest of development of the reproductive system discovers itself in the physical form and in the mental character. There is an imperfect development of the structure which Nature has provided in the female for nursing her offspring.

> "Formerly," writes another American physician, Dr. N. Allen, "such an organization was generally possessed by American women, and they found but little difficulty in nursing their infants. It was only occasionally in case of some defect in the organization, or where sickness of some kind had overtaken the mother, that it became necessary to resort to the wet-nurse, or to feeding by hand. And the English, the Scotch, the German, the Canadian, the French, and the Irish women who are living in this country, generally nurse their children; the exceptions are rare. But how is it with our American women who become mothers? It has been supposed by some that all, or nearly all of them, could nurse their offspring just as well as not; that the disposition only was wanting, and that they did not care about having the trouble or confinement necessarily attending it. But this is a great mistake. This very indifference or aversion shows something wrong in the organization, as well as in the disposition; if the physical system were all right, the mind and natural instincts would generally be right also. While there may be here and there cases of this kind, such an indisposition is not always found. It is a fact that large numbers of our women are anxious to nurse their offspring, and make the attempt; they persevere for a while — perhaps for weeks or months — and then fail. . . . There is still another class that cannot nurse at all, having neither the organs nor nourishment necessary to make a beginning."

Why should there be such a difference between American women and those of foreign origin residing in the same locality, or between them and their grandmothers? Dr. Allen goes on to ask. The answer he finds in the undue demands made upon the brain and nervous system, to the detriment of the organs of nutrition and secretion:

> "In consequence of the great neglect of physical exercise, and the continuous application to study, together with various other influences, large numbers of our American women have altogether an undue predominance of the nervous temperament. If only here and there an individual were found with such an organization, not much harm comparatively would result; but when a majority, or nearly a majority have it, the evil becomes one of no small magnitude."

To the same effect writes Dr. Weir Mitchell, an eminent American physiologist:

> "Worst of all, to my mind, most destructive in every way, is the American view of female education. The time taken for the more serious instruction of girls extends to the age of eighteen, and rarely over this. During these years they are undergoing such organic development as renders them remarkably sensitive. . . . To-day the American woman is, to speak plainly, physically unfit for her duties as woman, and is, perhaps, of all civilized females, the least qualified to undertake those weightier tasks which tax so heavily the nervous system of man.She is not fairly up to what Nature asks from her as wife and mother. How will she sustain herself under the pressure of those yet more exacting duties which nowadays she is eager to share with man?"

Here, then, is no uncertain testimony as to the effects of the American system of female education: some women who are without the instinct or desire to nurse their offspring, some who have the desire but not the capacity, and others who have neither the instinct nor the capacity. The facts will hardly be disputed, whatever may finally be the accepted interpretation of them. . . .

The foregoing considerations go to show that the main reason of woman's position lies in her nature. That she has not competed with men in the active work of life was probably because, not having had the power, she had not the desire to do so, and because, having the capacity of functions which man has not, she has found her pleasure in performing them. It is not simply that man, being stronger in body than she is, has held her in subjection, and debarred her from careers of action which he was resolved to keep for himself; her maternal functions must always have rendered, and must continue to render, most of her activity domestic. . . .

Setting physiological considerations aside, it is not possible to suppose that the whole explanation of woman's position and character is that man, having in the beginning found her pleasing in his eyes and necessary to his enjoyment, took forcible possession of her, and has ever since kept her in bondage, without any other justification than the right of the strongest. Superiority of muscular strength, without superiority of any other kind, would not have done that, any more than superiority of muscular strength has availed to give the lion or the elephant possession of the earth. If it were not that woman's organization and functions found their fitting home in a position different from, if not subordinate to, that of men, she would not so long have kept that position. If she is to be judged by the same standard as men, and to make their aims her aims, we are certainly bound to say that she labors under an inferiority of constitution by a dispensation which there is no gainsaying. This is a matter of physiology, not a matter of sentiment; it is not a mere question of larger or smaller muscles, but of the energy and power of endurance of the nerve-force which drives the intellectual and muscular machinery; not a question of two bodies and minds that are in equal physical conditions, but of one body and mind capable of sustained and regular hard labor, and of another body and mind which for one quarter of each month during the best years of life is more or less sick and unfit for hard work. It is in these considerations that we find the true explanation of what has been from the beginning until now, and what must doubtless continue to be, though it be in a modified form. It may be a pity for woman that she has been created woman, but, being such, it is as ridiculous to consider herself inferior to man because she is not man, as it would be for man to consider himself inferior to her because he cannot perform her functions. There is one glory of the man, another glory of the woman, and the glory of the one differeth from that of the other. . . .

James Rowland Angell (1869–1949) was born into an academic family. His grandfather was president of Brown University; his father president of the Universities of Vermont and Michigan. Angell, in turn, earned a bachelor's degree from the University of Michigan (1890), a master's in psychology from Harvard (1892), and joined the psychology department at the University of Chicago in 1894 (at the age of 25), becoming head of the department in 1905, a senior dean in 1908, dean of the faculties in 1911, and acting president of the university in 1918–1919. In 1921, Angell became president of Yale University, a position he held until his retirement in 1937. His first major work, Psychology (1904), was used extensively as a textbook in many American colleges.

The following excerpt is from an article Angell wrote while teaching at Chicago. Since its origin in 1890, the University of Chicago had one of the most liberal policies of any American institution with regard to coeducation. But in 1900, the question of segregation arose in response to an offer of an endowment to be given on condition that women were taught in separate classes. The Board of Trustees voted in favor of separate instruction in October 1902. A month later, Angell, a lifelong supporter of coeducation, reviewed the recent social and historical events that had led to Chicago's change in policy.

SOME REFLECTIONS UPON THE REACTION FROM COEDUCATION

JAMES ROWLAND ANGELL

November 1902

THE authorities of many of our great coeducational universities have been of late much perplexed and depressed at the astonishing number of young women who insist on patronizing these institutions. Taken in moderation the coeducational young woman has succeeded in approving herself to a considerable majority of her instructors. But she has recently shown a disposition to outnumber the young men in her classes, and this is resented by certain of her mentors as an obvious impropriety. The occasion has been seized upon by reactionaries here and there to magnify the drawbacks of coeducation, and there can be no question that many members of the faculties of institutions committed to this system are restive under the extant conditions, and apprehensive for the future. A few, especially certain of those educated in eastern non-coeducational institutions or in foreign universities, are severely, not to say bitterly, critical in their attitude, and eager for anything so it be a change. . . .

In order to see the present situation in its just perspective one must bear in mind the remarkable educational development which has occurred in recent years in those parts of the country where coeducation is indigenous. The decade from 1890 to 1900 witnessed a growth wholly unprecedented in most of the strong coeducational colleges and universities of the central and western states. Academic standards were raised, equipments were lavishly provided in accordance with modern demands, faculties were enlarged and admirably trained specialists were secured in every department. In many institutions graduate courses of high merit were developed. The increase in the number of students was equally remarkable. The University of Minnesota leaped from 1,183 to over 3,000. The University of California from 763 to 3,024. The University of Wisconsin rose from 966 to 2,619. Cornell had 1,390 students in 1890, and 2,458 in 1900. At the University of Michigan the figures for the same period were 2,420 and 3,482. Moreover, in this same decade two coeducational universities were founded, Leland Stanford, Jr., University and the University of Chicago, which at once took rank with the foremost institutions of the country. In the year 1900 the former reported 1,389 students, the latter 3,520. Many other universities might be cited, such as the State Universities of Iowa, Ohio, Missouri, Kansas, Nebraska and Illinois, but they all tell the same story of the tropical development of higher education throughout this central and western region.

. . . In all the other important universities the percentage of women has materially increased, and in some instances passed the fifty-per-cent. mark. Thus in 1900 the course in literature, arts and general science showed at the University of California 55 per cent. women; at Minnesota 53 per cent.; at Chicago 47 per cent.; at Michigan 47 per cent.; and at Northwestern 44 per cent.

If professional schools are included in the computations, the figures take on a very different complexion. Thus at Michigan, for instance, in 1900 the law school with nearly 900 male students and the medical departments with over 600 served, together with the engineering school of 350, to keep the proportion of women to men in the whole institution within 3 per cent. of what it was in 1890, the figures being 16 per cent. and 19 per cent. for the beginning and end of the decade respectively. In almost every case the increase in the percentage of women occurred in the face of an unparalleled increase in the number of male students. Indeed, a recent writer has presented statistics to prove that the increase in the number of men in coeducational institutions has for some years past been relatively more rapid than the increase in the colleges for men alone. This, too, is a hard saying for those who maintain that women are driving men out of the coeducational institutions. But it does not mean, as we shall presently see, that women and men are increasing with equal rapidity in the same courses of study.

One finds his natural anticipations fulfilled in the almost entire absence of women from the courses in law and technology. Medicine attracts a small quota, but to all intents and purposes the serious embarrassments of the present situation may be treated as if the women were all segregated in the department of liberal arts. . . . Thus it will be found that the classes in English literature have been largely appropriated by women. History, the modern languages and classical studies show in many institutions a large invasion of women, while mathematics, geology and biology, together with the philosophical and social sciences, furnish a transition to the exact sciences and to studies of an advanced character immediately preparatory for law, medicine and technology, in which the men have things almost to themselves. . . .

In those parts of the country where collegiate coeducation is the prevailing system, the great mass of the young men are expecting immediately after graduation to enter upon a business or professional career, and this intention frequently leads them early in their college course to desert the humanities and the more purely cultural studies, so called, in favor of what they, or the faculties of the professional schools, consider the branches of immediately practical value. Literature and the classics rapidly surrender their claims upon these young men to economics, political science, constitutional history, physics, chemistry, biology, etc. In every large undergraduate body there is naturally always a considerable group of men who conceive of their educational opportunities in a more liberal manner than this, and another group cherishing more or less definite intention of graduate specialization in some of the departments of collegiate work other than those allied with the professional schools. Taken together these groups supply a considerable masculine leaven to what might otherwise in many of the courses in the humanities be a hopelessly feminine lump.

The women taken in mass are also widely controlled by professional considerations. However it may be in the colleges exclusively for women, there can be no reasonable question that up to the present time, at least, large numbers of the women in coeducational institutions have been looking forward to self-support. Or, at all events, if this were not an explicit purpose in their collegiate course, it was a very comforting possible consequence, exercising an indirect influence over their selection of studies. By common consent medicine and teaching are the two professions really open to women. Consequently the average college woman who intends to live by her brains feels that she must choose between the two. The opportunities open to women in a business career have hitherto been too precarious or too purely clerical to attract extensive attention from college-bred women. It is not necessary to point out the reasons which lead the vast majority of college women, who are looking to self-support, to choose the profession of teaching, nor yet to dwell upon the further reasons which practically confine the larger number of them to work in the secondary schools, whose fate is already largely in their hands. Admitting the facts, we at once have a reason why large numbers of women elect the courses they do in college. It is the old familiar process of demand and supply. In no small measure, then, the educational phenomenon from which we started, *i.e.*, the segregation of the sexes in pursuit of different subjects, is attributable to social and economic causes. . . .

When one comes to speak of the influence of native abilities, native tastes and intellectual interests in the election of courses, the available data are altogether less tangible. The prevalent doctrines concerning the mental differences of men and women are matters of dogma readily susceptible of neither proof nor disproof. In polite letters, as in society, woman has long figured as the adorable parent of men, not of ideas; as the repository of delicate sentiment rather than of accurate knowledge and in general, as the residuary legatee of all those interests which men do not care to cultivate. The evidence adducible in support of the correctness of this view is often proclaimed as conclusive; and ranged behind it is the authority of sundry notable physiologists, psychologists, sociologists and gentlemen of fashion. The contrary view in accordance with which women are allowed to possess ideas, some of them even original, is supported by evidence almost as intelligible and by partisans quite as eminent and quite as confident. The fact seems to be that it is extremely difficult to demonstrate how much of a woman's intellectual bent is due to the sexual bias of her mind, and how much to the influences which surround her from cradle to grave. It may be, for example, that literature is intrinsically feminine in its character, and that exact science is dominantly masculine. In the meantime it must certainly be granted that the conditions environing many girls are from childhood on such as tend to cultivate a mild but definite variety of sentimentalism. . . .

It seems improbable, however, that the relatively small number of men in the courses in belles-lettres in coeducational institutions is due in any considerable degree to this asserted fact regarding native masculine tastes, nor in any indisposition on the part of the men to sit in classes with women. For, as regards the second point, it must be remembered that there are many courses in which both men and women are well represented, and in which the ratio of the sexes to one another has changed but little throughout considerable periods. This is true, for example, of certain courses in history. The first point gains an interesting side light from the observation that in several important eastern institutions for men, where the modes of exposition and instruction in literature do not differ absolutely from those in vogue in western coeducational colleges, the attendance upon such courses shows in recent years no extreme shrinkage, and, as regards certain courses in English, even exhibits a marked development. The most obvious explanation of this difference between the east and the west (over and above the influence of the stimulating personality of certain successful instructors) is unquestionably to be found in the social conditions of which we have already spoken. The appreciation of the educational value of literature is necessarily more circumscribed in new and less wealthy communities than in those which have been long established, and the pressure toward the obviously practical is inevitably far greater. The astonishing development of technological schools in western universities affords striking confirmation of this last named tendency—a tendency by the way which finds a counterpart in the rapid growth of the so-called scientific schools in great eastern universities.

The final explanation which some misogynists would offer for the predominance of women in English courses, *i.e.*, that English literature is ordinarily the easiest of college courses unhappily proves too much. Such statistics as are available tend to show that were this generally conceded, women would rarely preponderate in such classes. On the whole, therefore, it seems probable that although the taste for literature is more largely developed in the women of our coeducational colleges, while the taste for exact science is largely the property of the men, the factor which has produced the most extreme and anomalous conditions of sex segregation, is, especially as regards the men, professional, economic and social in character. Moreover, that which passes as native taste is itself often a mere expression of social and domestic pressure, emphasizing with relentless insistence certain interests as sexually appropriate or practically valuable. . . .

A warning hand must be held out at this point against the amiable and ubiquitous inconsequentialist who insists on confusing the problem of coeducation with the problem of the higher education of women. The two are indeed connected, but by no means identical. Collegiate coeducation as a system assumes as a premise that women are to receive collegiate education, if they so desire, and a study of the curricula of women's colleges indicates that women generally wish to pursue those branches offered in the colleges for men. The coeducational problem is not, however, fundamentally one of curricula. It is a problem of determining the conditions under which men and women shall study in any curriculum whatsoever.

The criticisms of the present day upon coeducation dealing in part with strictly educational questions, and in part with general social considerations, differ in a somewhat suggestive manner as concerns one particular from those which were most commonly encountered in the early days of coeducation. The most frequent probably of all criticisms was the hygienic one. Although it was a matter of prehistoric knowledge that women could work all day in the field, many learned persons predicted a speedy decline for the audacious young female who attempted to follow the same collegiate course as her brother. The young person referred to has, however, both in coeducational colleges and in colleges for women, generally insisted on the retention of oppressively good health. And she has done even worse things to discredit the general calling of prophet by discovering numbers of educated men who were willing and eager to attempt matrimony with her assistance. Worst of all, when she has married, she has had a normal number of vigorous children. The irreconcilables on these points generally deny themselves the luxury of the available statistics. This is by no means to call in question the possibility held out to an injudicious girl of ruining her health by social and mental dissipation at a coeducational college. She can in this way undoubtedly emulate some of her sisters at women's colleges and certain of her brothers at men's universities.

But the intelligent contemporary opponent of coeducation has largely lost interest in the health of college women, and he has of late more often turned his attention to the baleful influence of the sex on social and intellectual standards. It is maintained, for instance, by an occasional instructor that women lower the level of scholarship in his classes. He finds it impossible to make such rigorous exactions of them as he would of men, and in consequence the whole tone of his class in contaminated. There seems reason to believe that this opinion is largely subjective in its basis, and it is suggestively rare on the lips of instructors educated in coeducational institutions. The instructor may have allowed himself to secure from his classes what he deemed the best work by the aid of a class room manner which he properly considers incompatible with the presence of ladies. In such cases one may fairly question the propriety of the pedagogical method. He may cherish a purely sentimental attitude toward women. This is a not infrequent circumstance in the case of young men brought up in men's colleges, and exposed for the first time to the ravages of coeducation. In this case time or matrimony or both are likely to cure his complaint. Certainly there are plenty of instructors who have taught in men's colleges without detecting any such decline of standard upon transferring the scene of their labors to coeducational institutions. Indeed, a contrary opinion has been not infrequently expressed, and one even hears the antithetical argument soberly advanced that women inevitably outstrip the men in mere class room exercises, and that the latter should consequently be spared such depressing competition. A few eminent Harvard professors have even gone so far as to rank their students at Radcliffe higher than the men in the corresponding Harvard classes. This may of course be the exception which proves the rule. One may safely surmise that there are no wider differences in point of scholarship between coeducational institutions as a class and men's colleges, than there are among men's colleges themselves. Indeed, in elective courses, in which the men and women are represented in anything like equal numbers, a common verdict of instructors concerning their relative merits is that the women are on the average the better students. They seldom attain the eminence of the ablest men, but the ablest men are excessively rare. There is certainly no palpable proof and not even good circumstantial evidence to convict women of lowering the undergraduate standards of scholarship.

A subtler form of this same criticism aimed at American education in general, but especially applicable to coeducational institutions, is the assertion that women exercise a repressive influence upon the spirit of research for which they have as a sex neither capacity nor appreciation. Inasmuch as a real university must get its highest inspiration from the spirit of investigation it is obviously a matter of paramount importance to prevent women from securing any considerable influence in university life. This argument can be made rhetorically effective, but it begs the whole question and will carry no conviction to one not already convinced. . . .

Coeducation is often charged with an insidious suppression of freedom of expression in the class room. *Academische Freiheit* is endangered, therefore, by other enemies than college presidents and boards of trustees. Some subjects are evidently ill qualified for discussion in mixed classes. But it may be doubted whether the restrictions emanating from this source have been unmixed evils. The meritricious and obscene jests of literature, which some eminent scholars in men's colleges delight to dwell upon, can perhaps be spared. Surely there is no serious reason to fear that the male youth of the land will fail to secure outside the class room all the really indispensable development on this side of his appreciation for humor. In the fields of biology, where embarrassments might be supposed most inevitable, the difficulties have by no means proved so serious as anticipated. Nevertheless it must be admitted without scruple that women cause some restraint upon freedom of speech as charged in the indictment. . . .

The asserted violation of reasonable social proprieties constitutes one of the most frequent and important sources of annoyance to the advocates of coeducation. To behold the campus dotted with couples, billing and cooing their way to an A.B., is a thing, it is said, to rejoice Venus or Pan rather than Minerva, and were it the frequent or necessary outcome of coeducation, the future of the system would certainly be in jeopardy. No university can safely become a matrimonial bureau, nor yet a clearing house for flirtations. With the entire absence of supervision, which characterizes the attitude of many coeducational institutions toward the social life of students, it is not to be wondered at that an occasional silly boy and an occasional silly girl should occasionally do some extremely silly thing. It only remains to remember that the same boy and girl will, with remarkably few exceptions, do equally silly things whatever educational surroundings may be given them. Furthermore, institutions which have attempted to control these matters by fixed rules of deportment seem on the whole to have succeeded in producing rather more risqué escapades than those which eschew restrictions altogether. Public opinion has generally proved a safe guide in this direction. It would be folly to pretend that no social transgressions have occurred, but on the whole judged by any standard reasonably applicable to the situation, the relations of the men and women in the majority of such colleges seem to have been wholesome and unobjectionable. . . .

Even though actual flirtation is avoided, many critics insist that boys' interests are stimulated in other boys' sisters at a time when it would be quite as well if they could be diverted into entirely different channels, even football. Girls, it is said, are unduly excited by masculine attention at a critical time in their physiological development, when they might better be engaged in storing their minds with useful learning. On the other side of this account it is to be observed that the shock of the class room goes far to shatter the traditional masculine idol in the feminine mind. This destructive process is in the case of most young women in such coeducational colleges begun in the primary schools and carried without interruption up to the academic level. By means of this anti-romantic treatment girls are unquestionably spared much painful disillusionizing, and they are brought through a difficult period with probably a minimum of silliness and mawkish sentimentality. Moreover, they are often spared certain highly morbid experiences familiar to the authorities of girls' colleges. In the case of the boy there is abundant evidence to warrant the opinion that the grosser forms of vice to which he falls an occasional prey are rendered distinctly less alluring by daily contact with women of refinement and intelligence. That he ordinarily becomes effeminized by such contact is a fantastic theory of some critics which finds absolutely no tangible evidence upon which to rest. . . .

The complementary criticism that women necessarily suffer a loss of refinement and feminine nicety seems equally difficult of proof. It is certainly hard to point out any unavoidable features of coeducation which should inevitably preclude the development or retention of good manners and fine feeling. If the psychologists are correct, the acquirement of what we call manners belongs largely to a period antedating college life, and although a coarsening of the moral and esthetic fiber during this period would unquestionably appear in less refined conduct, it is not obvious that any such change commonly occurs, much less must occur under coeducational conditions. The fact seems rather to be that the college must inevitably expect to reflect in large measure the manners with which its students come already supplied, and these which are often admirable will be determined by the social standards prevailing in the families and communities from which they come. . . .

One may deplore the fact sincerely, as every lover of established order does, but he can not gainsay that the development of American social and economic life is rapidly carrying increasing numbers of women out of the beaten path of the domestic treadmill with its everlasting insistence upon the incident of sex into fields where social service is gauged by other standards than those of child-bearing, house-keeping and adorning pink teas. Efficiency is certain to be the touchstone by which women are tried in these new fields, and they will go or stay in proportion as they do better or less well than men. The picture which one learned professor has recently drawn of an uprising of men to force by violence a return of women to their proper sphere, is the product of an inflamed imagination attempting to portray an oriental Utopia. In reality men are chiefly responsible for the changes now going forward, but they are neither the doctrinaires of academic dignity nor yet the leaders of cotillons; they are the seekers after commercial, industrial and professional efficiency. So long as the economic situation remains what it is as regards the principles and motives that control in it, no amount of merely hysterical criticism and opposition is likely seriously to modify the case. And so long, therefore, as many women prefer self-support to marriage on the terms they find the latter offered to them, women will remain primary items in the economic situation, and they cannot be treated in this realm from the merely sexual point of view.

Coeducation is a reflection, often unconscious, of the tendencies which have produced this condition. It represents historically, as well as intrinsically, the democratic disposition to offer equal educational opportunities so far as possible to every human being. The touchstone by which it tests worthiness for such opportunities is social service. So long as women show themselves worthy by this standard to receive the highest forms of education, they will be given opportunity to obtain it, and moreover they will probably, in western institutions at least, obtain it under coeducational auspicies. Justly or unjustly the western mind is suspicious of a fallacy lurking in the proclaimed equality of instruction in women's colleges and annexes, with that given in men's colleges. Then, too, if there were no other considerations, the economic waste involved in supporting separate institutions for men and women would tell heavily in favor of coeducation in many western communities. 'Equal but different' is not in educational matters a generally palatable doctrine away from the Atlantic seaboard. If the male is intellectually an altogether superior individual, the female ought on democratic principles to be given a chance for improvement by contact with him. . . .

David Starr Jordan (1851–1931) was educated at Gainesville (New York) Female Seminary and Cornell University, where he received a master's degree (1872) before earning a medical degree from Indiana Medical School (1875). During the next ten years, Jordan taught natural sciences at North Western Christian University (later renamed Butler University) in Indianapolis (1875–1879) and at the University of Indiana (1879–1885). In 1885, he became president of Indiana University, leaving in 1891 to become the first president of Stanford University. Under Jordan's leadership, Stanford made serious efforts to attract women to its faculty and programs. The following selection was part of an address that President Jordan gave to the Los Angeles Federation of Women's Clubs in May 1902. The Popular Science Monthly printed the address in December during the debate over Chicago's reaction to coeducation. (See previous selection by Angell.) Jordan was an important proponent of women's education, and his arguments are particularly interesting for what they reveal about liberal attitudes toward this issue.

THE HIGHER EDUCATION OF WOMEN

DAVID STARR JORDAN

December 1902

THE subject of the higher training of young women may resolve itself into three questions:

1. *Shall a girl receive a college education?*
2. *Shall she receive the same kind of a college education as a boy?*
3. *Shall she be educated in the same college?*

As to the first question: It must depend on the character of the girl. Precisely so with the boy. What we should do with either depends on his or her possibilities. No parent should let either boy or girl enter life with any less preparation than the best he can give. It is true that many college graduates, boys and girls alike, do not amount to much after the schools have done all they can. It is true also that higher education is not a question alone of preparing great men for great things. It must prepare even little men for greater things than they would otherwise have found possible. And so it is with the education of women. The needs of the time are imperative. The highest product of social evolution is the growth of the civilized home, the home that only a wise, cultivated and high-minded woman can make. To furnish such women is one of the worthiest functions of higher education. No young women capable of becoming such should be condemned to anything lower. Even with those who are in appearance too dull or too vacillating to reach any high ideal of wisdom, this may be said—it does no harm to try. A few hundred dollars is not much to spend on an experiment of such moment. Four of the best years of one's life spent in the company of noble thoughts and high ideals cannot fail to leave their impress. To be wise, and at the same time womanly, is to wield a tremendous influence, which may be felt for good in the lives of generations to come. It is not forms of government by which men are made and unmade. It is the character and influence of their mothers and their wives. The higher education of women means more for the future than all conceivable legislative reforms. And its influence does not stop with the home. It means higher standards of manhood, greater thoroughness of training, and the coming of better men. Therefore let us educate our girls as well as our boys. A generous education should be the birthright of every daughter of the republic as well as of every son. . . .

1. SHALL WE GIVE OUR GIRLS THE SAME EDUCATION AS OUR BOYS?

Yes, and no. If we mean by the *same*, an equal degree of breadth and thoroughness, an equal fitness for high thinking and wise acting, yes, let it be the same. If we mean this: Shall we reach this end by exactly the *same* course of studies? then the answer must be, No. For the same course of study will not yield the same results with different persons. The ordinary 'college course' which has been handed down from generation to generation is purely conventional. It is a result of a series of compromises in trying to fit the traditional education of clergymen and gentlemen to the needs of a different social era. The old college course met the needs of nobody, and therefore was adapted to all alike. The great educational awakening of the last twenty years in America has lain in breaking the bonds of this old system. The essence of the new education is constructive individualism. Its purpose is to give to each young man that training which will make a man of *him*. Not the training which a century or two ago helped to civilize the mass of boys of that time, but that which will civilize this particular boy. . . . [A] young woman . . . is an individual as well as he, and her work gains as much as his by relating it to her life. But an institution which meets the varied needs of varied men can also meet the varied needs of the varied women. The intellectual needs of the two classes are not very different in many important respects. In so far as these are different the elective system gives full play for the expression of such differences. It is true that most men in college look forward to professional training and that very few women do so. But the college training is not in itself a part of any profession, and it is broad enough in its range of choice to point to men and women alike the way to any profession which may be chosen. Those who have to do with the higher education of women know that the severest demands can be met by them as well as by men. There is no demand for easy or 'goody-goody' courses of study for women except as this demand has been encouraged by men. In this matter the supply has always preceded the demand.

There are, of course, certain average differences between men and women as students. Women have often greater sympathy or greater readiness of memory or apprehension, greater fondness for technique. In the languages and literature, often in mathematics and history, they are found to excel. They lack, on the whole, originality. They are not attracted by unsolved problems and in the inductive or 'inexact' sciences they seldom take the lead. The 'motor' side of their minds and natures is not strongly developed. They do not work for results as much as for the pleasure of study. In the traditional courses of study—traditional for men—they are often very successful. Not that these courses have a fitness for women, but that women are more docile and less critical as to the purposes of education. And to all these statements there are many exceptions. In this, however, those who have taught both men and women must agree; the training of women is just as serious and just as important as the training of men, and no training is adequate for either which falls short of the best.

2. SHALL WOMEN BE TAUGHT IN THE SAME CLASSES AS MEN?

This is partly a matter of taste or personal preference. It does no harm whatever to either men or women to meet those of the other sex in the same class rooms. But if they prefer not to do so, let them do otherwise. No harm is done in either case, nor has the matter more than secondary importance. Much has been said for and against the union in one institution of technical schools and schools of liberal arts. The technical quality is emphasized by its separation from general culture. But I believe that better men are made when the two are brought more closely together. The culture studies and their students gain from the feeling of reality and utility cultivated by technical work. The technical students gain from association with men and influences of which the aggregate tendency is toward greater breadth of sympathy and a higher point of view.

A woman's college is more or less distinctly a technical school. In most cases, its purpose is distinctly stated to be such. It is a school of training for the profession of womanhood. It encourages womanliness of thought as more or less different from the plain thinking which is called manly. The brightest work in woman's colleges is often accompanied by a nervous strain, as though its doer were fearful of falling short of some outside standard. The best work of men is natural, is unconscious, the normal result of the contact of the mind with the problem in question.

In this direction, I think, lies the strongest argument for coeducation. This argument is especially cogent in institutions in which the individuality of the student is recognized and respected. In such schools each man, by his relation to action and realities, becomes a teacher of women in these regards, as, in other ways, each cultivated woman is a teacher of men.

In woman's education, as planned for women alone, the tendency is toward the study of beauty and order. Literature and language take precedence over science. Expression is valued more highly than action. In carrying this to an extreme the necessary relation of thought to action becomes obscured. The scholarship developed is not effective, because it is not related to success. The educated woman is likely to master technique, rather than art; method, rather than substance. She may know a good deal, but she can do nothing. Often her views of life must undergo painful changes before she can find her place in the world.

In schools for men alone, the reverse condition often obtains. The sense of reality obscures the elements of beauty and fitness. It is of great advantage to both men and women to meet on a plane of equality in education. Women are brought into contact with men who can do things — men in whom the sense of reality is strong, and who have definite views of life. This influence affects them for good. It turns them away from sentimentalism. It gives tone to their religious thoughts and impulses. Above all, it tends to encourage action as governed by ideals, as opposed to that resting on caprice. It gives them better standards of what is possible and impossible when the responsibility for action is thrown upon them.

In like manner, the association with wise, sane and healthy women has its value for young men. This value has never been fully realized, even by the strongest advocates of coeducation. It raises their ideal of womanhood, and the highest manhood must be associated with such an ideal. This fact shows itself in many ways; but to point out its existence must suffice for the present paper.

At the present time the demand for the higher education of women is met in three different ways:

1. In separate colleges for women, with courses of study more or less parallel with those given in colleges for men. In some of these the teachers are all women, in some mostly men, and in others a more or less equal division obtains. In nearly all these institutions, those old traditions of education and discipline are more prevalent than in colleges for men, and nearly all retain some trace of religious or denominational control. . . .

2. In annexes for women to colleges for men. In these, part of the instruction to the men is repeated for the women, though in different classes or rooms, and there is more or less opportunity to use the same libraries and museums. In some other institutions, the relations are closer, the privileges of study being similar, the difference being mainly in the rules of conduct

by which the young women are hedged in, the young men making their own.

It seems to me that the annex system cannot be a permanent one. The annex student does not get the best of the institution, and the best is none too good for her. Sooner or later she will demand it, or go where the best is to be had. The best students will cease to go to the annex. The institution must then admit women on equal terms, or not admit them at all. There is certainly no educational reason why a woman should prefer the annex of one institution when another equally good throws its doors wide open to her.

3. The third system is that of coeducation. In this system young men and young women are admitted to the same classes, subjected to the same requirements, and governed by the same rules. This system is now fully established in the State institutions of the North and West, and in most other colleges in the same region. Its effectiveness has long since passed beyond question among those familiar with its operation. Other things being equal, the young men are more earnest, better in manners and morals, and in all ways more civilized than under monastic conditions. The women do more work in a more natural way, with better perspective and with saner incentives than when isolated from the influence of society of men. There is less of silliness and folly where a man is not a novelty. In coeducational institutions of high standards, frivolous conduct or scandals of any form are rarely known. The responsibility for decorum is thrown from the school to the woman, and the woman rises to the responsibility. Many professors have entered western colleges with strong prejudices against coeducation. These prejudices have not often endured the test of experience with men who have made an honest effort to form just opinions.

It is not true that the character of the college work has been in any way lowered by coeducation. The reverse is decidedly the case. It is true that untimely zeal of one sort or another has filled the West with a host of so-called colleges. It is true that most of these are weak and doing poor work in poor ways. It is true that most of these are coeducational. It is also true that the great majority of their students are not of college grade at all. In such schools low standards rule, both as to scholarship and as to manners. The student fresh from the country, with no preparatory training, will bring the manners of his home. These are not always good manners, as manners are judged. But none of these defects is derived from coeducation; nor are any of these conditions made worse by it.

Very lately it is urged against coeducation that its social demands cause too much strain both on young men and young women. College men and college women, being mutually attractive, there are developed too many receptions, dances and other functions in which they enjoy each other's company. But this is a matter easily regulated. Furthermore, at the most the average young woman is college spends in social matters less than one tenth the time she would spend at home. With the young man the whole matter represents the difference between high-class and low-class associates and associations. When college men stand in normal relation with college women, meeting them in society as well as in the class room, there is distinctly less of drunkenness, rowdyism and vice than obtains under other conditions. And no harm comes to the young woman through the good influence she exerts. To meet freely the best young men she will ever know, the wisest, cleanest and strongest, can surely do no harm to a young woman. Nor will the association with the brightest and sanest young women of the land work any harm to the young men. This we must always recognize. The best young men and the best young women, all things considered, are in our colleges. And this has been and will always be the case. . . .

With all this it is necessary for us to recognize actual facts. There is no question that a reaction has set in against coeducation. The number of those who proclaim their unquestioning faith is relatively fewer than would have been the case ten years ago. This change in sentiment is not universal. It will be nowhere revolutionary. Young women will not be excluded from any institution where they are now welcomed, nor will the almost universal rule of coeducation in state institutions be in any way reversed. The reaction shows itself in a little less civility of boys towards their sisters and the sisters of other boys; in a little more hedging on the part of the professors; in a little less pointing with pride on the part of college executive officials. There is nothing tangible in all this. Its existence may be denied or referred to ignorance or prejudice.

But such as it is, we may for a moment inquire into its causes. First as to those least worthy. Here we may place the dislike of the idle boy to have his failures witnessed by women who can do better. I have heard of such feelings, but I have no evidence that they play much actual part in the question at issue. Inferior women do better work than inferior men because they are more docile and have much less to distract their minds. But there exists a strong feeling among rowdyish young men that the preference of women interferes with rowdyish practices. This interference is resented by them, and this resentment shows itself in the use of the offensive term 'coed' and of more offensive words in vogue in more rowdyish places. I have not often heard the term 'coed' used by gentlemen, at least without quotation marks. Where it is prevalent, it is a sign that true coeducation—that is, education in terms of generous and welcome equality—does not exist. I have rarely found opposition to coeducation on the part of really serious students. The majority are strongly in favor of it but the minority in this as in many other cases makes the most noise. The rise of a student movement against coeducation almost always accompanies a general recrudesence of academic vulgarity.

A little more worthy of respect as well as a little more potent is the influence of the athletic spirit. In athletic matters, the young women give very little assistance. They cannot play on the teams, they can not yell, and they are rarely generous with their money in helping those who can. A college of a thousand students, half women, counts for no more athletically than one of five hundred, all men. It is vainly imagined that colleges are ranked by their athletic prowess, and that every woman admitted keeps out a man, and this man a potential punter or sprinter. There is not much truth in all of this, and if there were, it is of no consequence. College athletics is in its essence by-play, most worthy and valuable for many reasons, but nevertheless only an adjunct to the real work of the college, which is education. If a phase of education otherwise desirable interferes with athletics, so much the worse for athletics.

Of like grade is the feeling that men count for more than women, because they are more likely to be heard from in after life. Therefore, their education is of more importance, and the presence of women impedes it.

A certain adverse influence comes from the fact that the oldest and wealthiest of our institutions are for men alone or for women alone. These send out a body of alumni who know nothing of coeducation, and who judge it with the positiveness of ignorance. Most men filled with the time-honored traditions of Harvard and Yale, of which the most permeating is that of Harvard's and Yale's infallibility, are against coeducation on general principles. Similar influences in favor of the separate education of women go out from the sister institutions of the East. The methods of the experimenting, irreverent, idol-breaking West find no favor in their eyes.

The only serious new argument against coeducation is that derived from the fear of the adoption of universities of woman's standards of art and science rather than those of men, the fear that amateurism would take the place of specialization in our higher education. Women take up higher education because they enjoy it; men because their careers depend upon it. Only men; broadly speaking, are capable of objective studies. Only men can learn to face fact without flinching, unswayed by feeling or preference. The reality with woman is the way in which the fact affects her. Original investigation, creative art, the 'resolute facing of the world as it is'—all belong to man's world, not at all to that of the average woman. That women in college do as good work as the men is beyond question. In the university they do not, for this difference exists, the rare exceptions only proving the rule, that women excel in technique, men in actual achievement. If instruction through investigation is the real work of the real university, then in the real university the work of the most gifted women may be only by play.

It has been feared that the admission of women to the university would vitiate the masculinity of its standards, that neatness of technique would replace boldness of conception, and delicacy of taste replace soundness of results.

It is claimed that the preponderance of high-school-educated women in ordinary society is showing some such effects in matters of current opinion.

For example, it is claimed that the university extension course is no longer of university nature. It is a lyceum-course designed to please women who enjoy a little poetry, play and music, read the novels of the day, dabble in theosophy, Christian science, or physic psychology, who cultivate their astral bodies and think there is something in palmistry, and are edified by a candy coated ethics of self-realization. There is nothing ruggedly true, nothing masculine left in it. Current literature and history are affected by the same influences. Women pay clever actors to teach them—not Shakespeare or Goethe, but how one ought to feel on reading King Lear or Faust or Saul. If the women of society do not read a book it will scarcely pay to publish it. Science is popularized in the same fashion by ceasing to be science and becoming mere sentiment or pleasing information. This is shown by the number of books on how to study a bird, a flower, a tree, or a star, through an opera glass, and without knowing anything about it. Such studies may be good for the feelings or even for the moral nature, but they have no elements of that 'fanaticism for veracity,' which is the highest attribute of the educated man.

These results of the education of many women and a few men, by which the half-educated woman becomes a controlling social factor have been lately set in strong light by Dr. Münsterberg. But they are used by him, not as an argument against coeducation, but for the purpose of urging the better education of more men. They form likewise an argument for the better education of more women. The remedy for feminine dilettantism is found in more severe training. Current literature as shown in profitable editions reflects the taste of the leisure class. The women with leisure who read and discuss vapid books are not representative of woman's higher education. Most of them have never been educated at all. In any event this gives no argument against coeducation. It is thorough training, not separate training, which is indicated as the need of the times. Where this training is taken is a secondary matter, though I believe, with the fullness of certainty that better results can be obtained, mental, moral and physical in coeducation, than in any monastic form of instruction.

A final question: Does not coeducation lead to marriage? Most certainly it does; and this fact can not be and need not be denied. The wonder is rather that there are not more of such marriages. It is a constant surprise that so many college men turn from their college associates and marry some earlier or later acquaintance of inferior ability, inferior training and often inferior personal charm. The marriages which result from college association are not often premature—college men and college women marry later than other men and women—and it is certainly true that no better marriages can be made than those founded on common interests and intellectual friendships.

A college man who has known college women, as a rule, is not drawn to those of lower ideals and inferior training. His choice is likely to be led toward the best he has known. A college woman is not led by mere propinquity to accept the attentions of inferior men.

Where college men have chosen friends in all cases both men and women are thoroughly satisfied with the outcome of coeducation. It is part of the legitimate function of higher education to prepare women, as well as men, for happy and successful lives.

An Eastern professor, lately visiting a Western State university, asked one of the seniors what he thought of the question of coeducation.

> 'I beg your pardon,' said the student, 'what question do you mean?'
> 'Why coeducation,' said the professor, 'the education of women in colleges for men.'
> 'Oh,' said the student, 'coeducation is not a question here.'

And he was right. Coeducation is never a question where it has been fairly tried.

Chapter 3
THE BIRTH RATE QUESTION (1890-1905)

When woman first claimed admission to the privileges of higher education, men pointed out that a female who studied in botany that plants had sex-organs, would be unfit to associate with their [sic] respectable sisters. When she knocked at the gates of medicine, men declared that a woman who could listen to a lecture in anatomy was unworthy of honorable wifehood. When she asked for chloroform to assuage the pangs of childbirth, men quickly informed her that if women bear their children without pain, they will be unable to love them. When the married woman demanded the right to own property, men swore that such a radical step would totally annihilate woman's influence, explode a volcano under the foundations of family union, and destroy the true felicity of wedded life, and they assured us they opposed the change, not because they loved justice less, but because they loved woman more. During the many years that woman fought for citizenship, men gathered in gambling-dives and barrooms and sadly commiserated [with] each other on the fact that woman was breaking up the home. Now woman demands the control of her own body, and there are men who reply that if women learn how to prevent pregnancy, they will abolish maternity. It seems there are always some men who are haunted by the fear that women are planning the extinction of the race. To attempt to reason with such men is folly, and we can only hope that a general knowledge of contraceptive methods, judiciously applied, will eliminate this type.

Victor Robinson, *Pioneers of Birth Control in England and America* (New York: Voluntary Parenthood League, 1919), pp. 90-92.

The United States Census of 1900 brought into focus a disturbing demographic fact: American women were having many fewer children than they had several generations before. For 1,000 women, age 15 to 49, there were 474 children under 5 years of age in 1900, compared with 634 such children in 1860. Moreover, white native-born women had many fewer children than white foreign-born women—462 children under 5 per 1,000 native-born women, age 15 to 44, compared with 710 children for foreign-born women.[1] And the better educated, married middle-class woman had the fewest children of all—on average about two, compared to a national average of 3.56.[2] President Theodore Roosevelt called national attention to the declining birth rates of native-born Americans with his widely publicized comments on "race suicide."[3] For many Americans in the late nineteenth century, who viewed the family as the quintessential institution in social life, a drop in fertility, particularly among the upper classes, seemed to threaten the basis of civilized society.

Rising divorce rates were another focus of concern. Divorce contributed to the decline in fertility by disrupting marriages that might have produced more children. Moreover, the increase in divorce symbolized the general dissolution of family ties. In 1870, for every 1,000 marriages there were roughly 1.5 divorces; in 1900 this number increased to 4.0.[4] Although the absolute levels may seem to us insignificant, given the much higher rates to which we have grown accustomed, the increase deeply troubled Americans at the turn of the century. Together, the statistics on the decline in fertility and the increase in divorce generated a national crisis, not because the family as an institution was rapidly dissolving, but because many people believed it was.

It must be noted from the outset that the perceptions of Americans in the late nineteenth and early twentieth centuries concerning the falling birth rate were not entirely accurate. Their fundamental observation that women in the late nineteenth century bore fewer children was correct. They were also correct to assume that the decline was deliberate, as birth control was increasingly practiced, especially among the better educated middle class. But a popular belief that the sharp decline in fertility *began* around 1860 was mistaken. Government statistics from 1790 onwards revealed that the "demographic transition"—the development of a much smaller nuclear family—began at least as early as the late eighteenth century and did not occur abruptly. The transition from large to small families took place gradually over a period of more than one hundred years.[5]

Although the United States did not have an organized, national system to register births until 1915, Americans had access to Census statistics that enabled them to trace the trend of births from 1800 onwards. From 1800 to 1820 the Census Bureau counted the number of white children under the age of ten. From 1830 to 1920 it counted the number of white children under the age of five. In 1830 and 1840 the Census Bureau counted the number of black children under the age of ten, and from 1850 onwards it counted the number of black children under the age of five. From these records demographers could and did estimate the birth rate of Americans in several ways. The simplest way was to take the number of surviving children under a given age, usually five years old, and divide that number by the total population, arriving at what demographers call the "crude birth rate"—the number of children per 1,000 population. Another method was to divide the number of surviving children by the number of women of childbearing age (usually age 15 to 44 or 49). Most nineteenth-century demographers understood the difficulties inherent in these procedures. First, both methods ignore the problem of infant mortality, since the statistics do not include the number of children born in any given year who die before the age of five. Second, the crude birth rate, which uses the total population as the divisor, may give a distorted view of fertility because it is affected by changes in the proportion of men and women of non-childbearing age in the population. If in one year many more older people die than in the previous year, the birth rate will rise sharply, even if the same number of women have the same number of children as in the earlier year. This problem does not occur if the second method is used, since only women of childbearing age are counted in the divisor.

As early as 1843, George Tucker, a professor of moral philosophy and political economy at the University of Virginia, used Census enumeration of white children under ten to demonstrate that there had been a steady decline in the fertility of white women from 1800 to 1840. But Tucker's work was largely forgotten by Americans a half-century later when John S. Billings, the director of the 1890 Census, using the crude birth rate, focused on the decline occurring from 1880 to 1890. This fact becomes the more remarkable given that Billings knew of Tucker's work and referred to it in his article, "The Diminishing Birth-Rate in the United States," published in *Forum* (1893). Billings' emphasis on the decline occurring in the late nineteenth century prevailed until 1905, when Walter F. Willcox, director of the 1900 and 1910 Census, demonstrated in two reports, the first published in 1905, the second in 1911, that the decline could be traced as far back as 1810. Table 3.1 is a composite of Tucker's, Billings', and Willcox's statistics.

Although Americans knew (or could have known) about the declining birth rate as early as the 1840s, the "birth rate question" did not emerge as a focus of public concern until the late nineteenth century after two important social developments had occurred. The first was massive immigration; the second was a salient increase in the number of women with college educations. Around the turn of the twentieth century, Americans frequently referred to these two developments to explain the decline in fertility.

Francis A. Walker, director of the Census in 1870 and 1880, was among the first to observe the disparity between the birth rates of native-born and foreign-born Americans. After the 1890 Census showed a continuation of this discrepancy, Walker pointed out that the decline in the rate of increase of the American population (and so the decline in the birth rate) began with the rapid influx of immigrants:

> Now, this correspondence might be accounted for in three different ways: (1) It might be said that it was a mere coincidence, no relation of cause and effect existing between the two phenomena. (2) It might be said that the foreigners came because the native population was relatively declining, that is, failing to

Table 3.1. Birth Rates of American Women, 1800–1900

	TUCKER (1843)	BILLINGS (1893)	WILLCOX (1905)	WILLCOX (1911)
	CHILDREN UNDER 10 PER 1,000 WOMEN	CHILDREN UNDER 1 PER 1,000 POPULATION	CHILDREN UNDER 5 PER 1,000 WOMEN AGE 15–49	CHILDREN UNDER 5 PER 1,000 WOMEN AGE 16–44
1800	709			976
1810	701			976
1820	678			928
1830	662			877
1840	646			835
1850			626	699
1860			634	714
1870			572	649
1880		30.95	559	635
1890		26.68	486	554
1900			474	541

Sources: George Tucker, *Progress of the United States in Population and Wealth in Fifty Years* (New York: Press of Hunt's Merchants' Magazine, 1843), p. 90; John S. Billings, "The Diminishing Birth-Rate in the United States," *Forum* 15 (June 1893): 467; Walter F. Willcox, "Proportion of Children in the United States," *Bureau of the Census Bulletin 22* (Washington: Government Printing Office, 1905), p. 11; and Walter F. Willcox, "The Change in the Proportion of Children in the United States and in the Birth Rate in France During the Nineteenth Century," *Publications of the American Statistical Association* 12 (March 1911): 495.

keep up its pristine rate of increase. (3) It might be said that the growth of the native population was checked by the incoming of the foreign elements in such large numbers. . . .

The true explanation of the remarkable fact we are considering, I believe to be the last of the three suggested. The access of foreigners, at the time and under the circumstances, constituted a shock to the principle of population among the native element. . . . And it is to be noted, in passing, that not only did the decline in the native element as a whole, take place in singular correspondence with the excess of foreign arrivals, but it occurred chiefly in just those regions to which the newcomers most freely resorted. . . .

. . . [F]oreign immigration into this country has, from the time it first assumed large proportions, amounted, not to a re-enforcement of our population, but to a replacement of native by foreign stock. That if the foreigners had not come, the native element would long have filled the places the foreigners usurped, I entertain not a doubt.[6]

Walker believed that the decline in the birth rate of native-born Americans was caused by immigration, or more precisely, by the adverse social and economic conditions for which immigrants were held responsible. Robert

Hunter, paraphrasing Walker, summed up the argument in *Poverty* (1904) thus:

> The appearance of vast numbers of men, foreign in birth and often in language, with a poorer standard of living, with habits repellent to our native people, of an industrial grade suited only to the lowest kind of manual labor, was exactly such a cause as by any student of population would be expected to affect profoundly the growth of the native population. Americans shrank alike from the social contact and the economic competition thus created. They became increasingly unwilling to bring forth sons and daughters who should be obliged to compete in the market for labor and in the walks of life with those whom they did not recognize as of their own grade and condition.[7]

The problem, as Walker, Hunter, and others saw it, was that immigrants worked for low wages and continued to have large families because they were accustomed to a low standard of living. But native-born Americans, forced to accept comparably low wages, restricted the number of their offspring to avoid lowering their standard of living. There can be no doubt that the thrust of the anti-immigration arguments was racist. Immigrants, particularly those from southern Europe, were considered by many to be inferior to native-born Americans. To save Anglo-Saxon America from further degeneration, immigration would have to be restricted. Robert Hunter put the issue bluntly:

> It [immigration] is a question of babies and birth-rates, and whatever decision is made regarding immigration, it is perforce a decision concerning the kind of children that shall be born. The decision for Congress to make consciously and deliberately is simply whether or not it is better for the world that the children of native parents should be born instead of the children of foreign parents. The making of the decision cannot be avoided. It is made now, although unconsciously, and it is a decision against the children of native parents. . . . This is the race-suicide, the annihilation of our native stock, which unlimited immigration forces upon us, none the less powerfully because it is gradually and stealthily done.[8]

Immigration seemed to pose such a dangerous threat to the progress of American (i.e., white Anglo-Saxon) civilization that some people even argued that the sacred, unalterable law of natural selection should be altered. For although immigrants were clearly not the "fittest," somehow they managed to outsurvive the far superior Anglo-Saxons. Many considered this a horrible anomaly, but I have found no one who used it to challenge the assumptions of social-Darwinian theory. However, Frederick Bushee suggested that something be done to change the course of natural selection:

> It [immigration] is a race question, and the birth rate shows the racial group that is to survive. If however, it is found that the status of society which has the highest development tends to be blotted out by the increase of the lower strata, the *cause of progress will demand that the course of natural selection be interfered with* by removing the continual external pressure on the native stock.[9]

The height of the furor on immigration and the birth rate occurred from 1901–1910.[10] During this period, the educated classes came under attack for not doing their share to people the nation. In 1902, President Charles Eliot of Harvard University discussed the marriage and birth rates of alumni from the all male classes of 1872–1877 in a brief report that stirred public interest. Eliot found that more than 25 years after graduation, 28% of these alumni remained unmarried, and those that had married had had an average of two children. These numbers appeared even more alarming when viewed in the perspective that close to 90% of American men and women married eventually.[11] The following year, 1903, a number of additional studies were published confirming Eliot's findings for Harvard and demonstrating that similar statistics characterized the graduates of other male colleges. In May 1903, Edward Thorndike of Teachers' College, Columbia University found that the married alumni of Middlebury, Wesleyan, and New York University, who had graduated in the 1870s, had fewer than three children, while married alumni from earlier classes (1800–1810) had approximately five children. In June 1903, George J. Engelmann published statistics he had gathered for the alumni of Princeton, Brown, Yale, Bowdoin, and Harvard. He stated that Harvard's record was somewhat atypical, finding that roughly 80% of the alumni graduating from other colleges in the 1870s had married by 1900. But among married graduates the average number of offspring was two. And in September 1903, G. Stanley Hall (professor of psychology at Clark University) and Theodate L. Smith together published a detailed study on the marriage and birth rates of the male graduates of Harvard, Yale, Amherst, Brown, Bowdoin, Dartmouth, Middlebury, and Williams. Table 3.2 presents a composite of some of the statistics presented by Eliot, Thorndike, Engelmann, and Hall and Smith.

Eliot looked upon his findings with regret and attributed the excessively low birth rates of Harvard alumni to the postponement of marriage. Thorndike was also worried by the statistics and attributed the decline to a "conscious restriction" due to a variety of causes:

> Greater prudence, higher ideals of education for children, more interest in the health of women, interests of women in affairs outside the home, the increased knowledge of certain fields of physiology and medicine, a decline in the religious sense of the impiety of interference with things in general, the longing for freedom from household cares—any or all of these may be assigned as the motive for restriction.[12]

Engelmann defended educated men, arguing that after infant mortality rates were taken into consideration, college men did better than the general population at replacing themselves. Engelmann dismissed the implications of higher education for the fertility of men with the explanation that men needed their intellectual training to "win in the struggle for existence," but he could not make the same allowance for women: "The educated female is in a different class; the fecundity of the female college graduate in this country is lower than that of any other native group."[13]

Hall and Smith suggested that the decline in fertility for the educated class was not really as large as it seemed, if it was remembered that in the eighteenth century it frequently took several women to give birth to such large families.[14] They seemed much less concerned than the other male investigators about the declining fertility of the educated class, believing it "absurd" and "illogical" that such a term as "race suicide" be used to describe the "decrease of one class, and that a class relatively small in relation to the entire population."[15] Having also studied the fertility of college women (see Table 3.3), Hall and Smith found the disparity in the marriage rates of college men and women "far less than we supposed." In sum, they came to a very different conclusion than Engelmann:

> Indeed, considering the facts that in our social system man makes the advances and that woman is by nature more prone than man to domesticity and parenthood, it is not impossible that men's colleges do more to unfit for these than do those for women.[16]

Despite the fact that much of the statistical data available in the late nineteenth century implied that the declining birth rate was a widespread social phenomenon, characteristic of all groups of native-born whites, and blacks as well,[17] there were virulent attacks against women, particularly highly educated women, who were criticized severely for not recognizing that "the greatest thing for any woman is to be a good wife and mother."[18]

The attacks against colleges for creating a class of antisocial, irresponsible—that is, childless—women must be seen within the larger tradition of arguments against educating women. In the 1870s, with the admittance of large numbers of women into coeducational and women's colleges, physicians issued dire warnings that advanced education endangered women's health and reproductive systems. The hostility of the medical community toward educating women persisted into the twentieth century, despite the appearance of studies documenting that college education in and of itself did not negatively affect women's health. (See Chapter 2.)

In the 1890s and early 1900s, as the birth rate attracted national attention, those opposed to women's higher education blamed colleges for causing the decline in fertility. They supported their accusations by pointing to a number of studies published from 1900 to 1918 indicating that fewer college women married and that college women had on average fewer children than American women as a whole. The low birth rate of college women was attributed in part to the relatively late age at which these women married. Thus, many Americans held colleges accountable for instilling reprehensible values in their

Table 3.2. Birth Rates of College Men, 1700–1900

HARVARD (ELIOT, 1903)

Year of Graduation	Married Alumni Total	Married Alumni % of Class	No. Children Per Married Alumnus
1872	82	72	2.01
1873	96	73	1.88
1874	124	75	1.99
1875	90	64	1.90
1876	106	75	2.00
1877	136	72	2.10

ENGELMANN (1903)

College	Year of Graduation	Number in Class	Percent Married	Number of Surviving Children To Each Married Graduate	Number of Surviving Children To Each Member of Class Married and Single
Princeton	1876	118	80.4	2.7	2.3
Brown	1872	53	88.7	2.26	2.—
Yale	1860–1879	1,105	78.4	2.28	1.79
Yale	1879	118	81.3	2.05	1.66
Yale	1873	113	82.3	1.98	1.57
Bowdoin	1875 and 1877	107	86.9	1.88	1.56
Harvard	1872–1880	1,401	71.4	1.86	1.34
	1870–1880	3,015	75.7	2.10	1.49

Yale, Princeton, Brown and Bowdoin

Y. P. Br. Bo.	1860–1880	1,614	79.4	2.28	1.81
Harvard	1872–1880	1,401	71.4	1.86	1.34

This table is arranged according to rate of reproduction.

Table 3.2. (continued)

MIDDLEBURY (THORNDIKE, 1903)		
1802–1809	64	5.6
1810–1819	161	4.8
1820–1829	163	4.1
1830–1839	189	3.9
1840–1849	83	3.4
1850–1859	90	2.9
1860–1869	114	2.8
1870–1874	50	2.3
1875–1879	32	1.8

YALE (HALL & SMITH, 1903)

Surviving Children

Year of Graduation	No. of Graduates	Total	Per Alumnus	Per Married Alumnus	% of Class Married
1710	2	19	9.50		
1720	10	46	4.60		
1730	18	75	4.16		
1740	21	89	4.23		
1750	17	64	3.76		
1760	33	110	3.33		
1872	129	220	1.70	2.20	78
1880	123	152	1.23	1.63	75
1885	118	48	.40	.94	43
1890*	146	110	.75	1.22	62
1898*	301	16	.05	.36	5

*Statistics incomplete because they were compiled in 1903 when alumni were still in their twenties and thirties.

Sources: Charles Eliot, "President's Report for 1901–1902," *Annual Reports of the President and Treasurer of Harvard College, 1901–1902* (Cambridge: Harvard University Press, 1903), pp. 31–32. Edward L. Thorndike, "The Decrease in the Size of American Families," *The Popular Science Monthly* 63 (May 1903): 64. George J. Engelmann, "Education Not the Cause of Race Decline," *The Popular Science Monthly* 63 (June 1903): 173, 183. G. Stanley Hall and Theodate L. Smith, "Marriage and Fecundity of College Men and Women," *The Pedagogical Seminary* 10 (September 1903): 285–288.

Table 3.3. Birth Rates of College Women, 1870–1920

Year of Graduation	Number of Women Studied	Number of Women Married	% Married	Avg Age At Marriage	Approx Age*	Avg Number of Children Born to Each Alumna	Avg Number of Children Born to Each Married Alumna	Avg Number of Children Born to Each Mother	Number of Childless Women Number	Number of Childless Women % Married
VASSAR COLLEGE (HALL & SMITH, 1903)[a]
1865–1876	323	179	55		48–59	1.13	2.03	3.09	61	34
1877–1886	378	192	51		38–47	.77	1.53	2.57	78	41
1887–1896**	603	169	28		28–37	.22	.79	1.58	85	50
SMITH COLLEGE (HALL & SMITH, 1903)[a]
1877–1888	370	158	43		36–47	.85	1.99	2.08	6	4
1889–1898**	1130	331	28		26–35	.14	.77	1.22	199	60
WELLESLEY COLLEGE (HALL & SMITH, 1903)[a]
1879–1888	436	203	47		36–45	.71	1.81	2.37	71	35
1889–1898**	1162	296	25		26–35	.15	1.04	1.67	192	65
MT. HOLYOKE SEMINARY (HEWES, 1911)[b]
1842–1849	41	35	85	(31)***	81–88	2.10	2.77	3.30	5	16
1850–1859	163	123	75	(83) 25.9	71–80	1.72	3.38	3.84	10	12
1860–1869	271	165	61	(131) 27.5	61–70	1.28	2.64	3.26	25	19
1870–1879	234	139	60	(104) 27.7	51–60	1.22	2.75	3.62	25	25
1880–1889	297	171	57	(143) 30.6	41–50	1.22	2.54	3.05	24	17
1890–1892	70	35	50	(33) 27.6	38–40	.90	1.91	2.33	6	18
Total	1076	668	63	(525) 29.3		1.42	2.66	3.46	95	18
				27.2						

MT. HOLYOKE COLLEGE (HEWES, 1911)[b]

1890–1899	377	158	(156)	42	28.1	.76	1.85	2.44	38	24
1900–1909**	1206	285	(283)	24	26.0	.23	.92	1.49	109	38
Total**	1583	443	(439)	28	26.8	.35	1.25	1.39	147	33

COMPOSITE OF 16,739 WOMEN GRADUATING FROM COLLEGES, 1865–1915[c] (VAN KLEECK, 1918)

Prior to 1880	189	109		57	57+	1.33	2.32	2.88	21	19
1880–1890	821	435		53	47–56	1.16	2.18	2.82	97	22
1890–1900	3178	1594		50	37–46	1.00	1.99	2.55	346	22
1900–1910	7531	3511		47	27–36	.61	1.32	1.84	997	28
1910–1915**	5019	895		18	22–26	.09	.50	1.17	512	47

*Approximate age of women at the time of study assuming they graduated from college at age 22 (or from the seminary at age 20).

**These figures are incomplete as a significant portion of the women included had not reached the end of their childbearing years at the time of study.

***Numbers in parentheses represent the number of married women upon which Hewes calculates birth rates, the average age at time of marriage, and the percentage of childless marriages. For further explanation see note b.

[a]Hall and Smith collected their data from class secretaries at the women's colleges; when secretaries failed to respond they used official records of the college. For Smith College (1889–1898) the number of children born is based on six of the ten classes reporting; for Wellesley College (1879–1888) the number of children born is based on eight of the ten classes reporting; for Wellesley (1889–1898) the number of children born is based on five of the ten classes reporting.

[b]Hewes selected these responses from a total of 2,827 responses received to more than twice that number of questionnaires sent out in 1910. All data was taken from living persons. For reasons not given, Hewes decided to exclude women who had married after the age of 40 from her calculations of average age at time of marriage and from her number of children born.

[c]Van Kleeck compiled these statistics from a total of 24,503 graduates of nine colleges.

Sources: G. Stanley Hall and Theodate L. Smith, "Marriage and Fecundity of College Men and Women," *The Pedagogical Seminary* 10 (September 1903): 301, 302, 304; Amy Hewes, "Marital and Occupational Statistics of Graduates of Mount Holyoke College," *Publications of the American Statistical Association* 12 (December 1911): 4–19; Mary Van Kleeck, "A Census of College Women," *Journal of the Association of Collegiate Alumnae* 11 (May 1918): 578, 581.

unmarried graduates and for delaying the marriages of those women graduates who did marry. Table 3.3 presents birth rates for the college women studied by Hall and Smith, Amy Hewes, and Mary Van Kleeck.

The statistics presented in these studies were remarkably consistent: Only half the women graduating from college in the 1880s and 1890s (who would have ranged in age from 32 to 52 at the time the studies were done) had married. The average number of children for these college educated mothers was around 2.5, and since approximately 20% of these married college women remained childless, the average number of children per married woman fell to about two. These statistics generated much concern, and as a result college women frequently were singled out for the reproach that they were not fulfilling their moral and social obligations to marry and have children.

Mary Roberts Smith, associate professor of sociology at Stanford, was among the first to suggest that the low birth rates of college women might be due not to their college background, but to their general upbringing. Arguing that it was unjust to compare college women to "that vague standard, the 'average woman of the Census,'" Smith selected 343 married college women and compared them with their sisters, cousins, and friends who had not gone to college. Although Smith's study could not be taken as representative of all married college women, her conclusions nonetheless attracted attention. Smith found that college education delayed marriage on average only two years: The college women in her study married at an average age of 26; the non-college women at 24. Moreover, while only 8% of Smith's college women had married under the age of 23, compared to 21% for non-college women, this difference was made up by the age of 30, by which time, 87% of both groups had married. As the average age of women marrying in Massachusetts from 1875 to 1895 was 25, Smith concluded that "the marriage of educated women is seen by comparison not to have been postponed so much beyond that of the average population as might have been expected." After excluding those married less than two years at the time of the study, Smith also found that 19% of the college women were childless, compared to 15% for the non-college women, and that the average number of children per college woman was 1.65 (in 9.6 years of marriage) compared to 1.87 (for 11.6 years of marriage) for the noncollege women.[19]

The declining birth rate among college women has been interpreted by writers in the nineteenth century and historians today as a measure of an increase in feminist* awareness. As early as 1867, an anti-reconstructionist observed:

---

*Feminist and feminism are problematic terms because they mean different things to different people. Here I use "feminist" to denote a conscious awareness among women of their individual autonomy, coupled with an insistent, purposeful demand for equal rights.

The anti-offspring practice has been carried in New England and wherever New England ideas prevail. It is esoteric, the interior doctrine of the woman's rights movement. These female reformers see that if they are to act the part of men in the world, they must not be burdened with the care of young children . . . in proportion as women's rights ideas prevail, are parents becoming ashamed of large families. It is not that the New England women are unable to bear as many children as formerly, but that they will not.[20]

While most participants in the woman movement in the nineteenth century never repudiated the traditional structure and function of the family, an increasing number of women supported "voluntary motherhood"—the right of women to refuse husbands' sexual demands in order to control reproduction. Voluntary motherhood was an ideology intended to protect women from too frequent sexual demands and to preserve their sexual and moral purity by advocating that they abstain from sexual intercourse rather than use less desirable means of birth control. Even the most staunch supporters of voluntary motherhood rarely denied that women's foremost social duty was to have and raise children.[21]

Some historians today (prominent among them are Daniel Scott Smith and Carl N. Degler) argue that the decline in the birth rate testifies to an increasing sense of autonomy among nineteenth-century women, a consciousness of self that permitted women to see themselves as individuals with interests distinct from those of men.[22] Sometimes this expanding consciousness of self is associated with the broadening experiences that colleges provided for women. Yet, while colleges contributed to women's sense of autonomy by providing their students with the skills and encouragement they needed to work outside the home, it must be remembered that most women college graduates left paid employment soon after marrying. (See p. 59.) Moreover, although many college women desired to use their intellectual training outside the home, it does not follow that this aspiration derived solely from their college experiences. Career aspirations were likely to exist in women before they left home to go to college, and as Mary Roberts Smith argued, the desire to limit the size of families was as much related to economic class as it was to college education. In sum, the declining birth rate evolved out of wide-reaching social and economic changes: the development of an urban, industrial society, increased immigration, and wider educational and vocational opportunities for women. To say simply that educating women produced a sense of individuality that in turn led to the decline in the birth rate is not sufficient and may in fact confound cause and effect. For it is also true that smaller families facilitated the development of this consciousness of self by providing women with the leisure to devote to their education. Historian Linda Gordon points out the impossibility of selecting one of these formulations to the exclusion of the other, declaring simply that "smaller families were both cause and effect of feminism."[23]

Another important point to keep in mind is that men played a significant role in bringing about the reduction of fertility. Given the birth control methods of the day (abstention and withdrawal were among the most widely used), it would have been impossible for women to have limited family size so successfully without the cooperation of their husbands. Comments from physicians, women, and men themselves support the point that men were concerned and actively participated in restricting the size of their families.[24]

Thus, the birth rates of college women, which gave weight to the vehement attacks against women's education, must be considered in the larger context to which they belong. Industrialization and urbanization had altered the economic conditions of life. There was less incentive for people living in the cities at the end of the century to have large families than there had been for Americans living on farms earlier in the century. As industrialization and urbanization progressed, the economic value of children declined, particularly in middle-class families where the birth rate was the lowest. In cities, especially, fewer children meant fewer mouths to feed; smaller families alleviated the economic burden for men and women supporting families.

In the following selections, the writers overlooked men's role in fertility and, as is true for much of the nineteenth-century literature, discussed the fertility of men and women as if they had children independently of one another. In any case, it is a distortion of the statistical evidence—evidence available to these writers—to attribute the declining birth rate solely to the perversity and irresponsibility of college-educated women. In fact, many college women and other people at the time attributed their desire to limit the number of their children to an increasingly deep interest in children's welfare, since by limiting offspring, parents could devote more of their attention and resources to each child. (See "Alumna's Children" pp. 137–143.) Commenting on this point, Degler writes, "as the number of children in a family declined, the emotional investment and the affection for them were encouraged to rise, just as greater concern for children encouraged women, as rearers of children, to have fewer."[25] In other words, the declining birth rate attested to the development of a new attitude toward children and childrearing, an attitude that, as I explore in the next chapter, also found expression in the domestic science movement.

NOTES

1. Walter F. Willcox, "The Proportion of Children in the United States," *Bureau of the Census Bulletin 22* (Washington, D.C.: Government Printing Office), pp. 23–24. *The Popular Science Monthly* extensively discussed the decline in fertility occurring in the late nineteenth and early twentieth centuries; Willcox's figures were reported in a brief article, "The Proportion of Children in the United States," (December 1905) 67: 762.
2. Bureau of the Census statistics for the birth rates of women by level of education do not exist, but many private studies found the average number of children per married college women to be under two. The figure 3.56 is given by Daniel Scott Smith, "Family Limitation, Sexual

Control and Domestic Feminism in Victorian America," *A Heritage of Her Own: Toward a New Social History of American Women,* eds. Nancy F. Cott and Elizabeth H. Pleck (New York: Simon and Schuster, 1979), p. 226. and by Carl N. Degler, *At Odds: Women and the Family in America From the Revolution to the Present* (New York: Oxford University Press, 1980), p. 181.

3. For the fullest statement of Roosevelt's position on the birth rate question, see Theodore Roosevelt, "Birth Reform, From The Positive, Not the Negative, Side," *The Foes of Our Own Household* (New York: George H. Doran, 1917), pp. 250–272. See also Roosevelt's famous letter to Mrs. Bessie Van Vorst (October 18, 1902) in *Presidential Addresses and State Papers of Theodore Roosevelt* (New York: P. F. Collier, n.d.; New York: Kraus Reprint, 1970), 1–2: 508–510.

4. These statistics are from Paul Jacobson's classic work, *American Marriage and Divorce* (New York: Rinehart, 1959), p. 90. Although many Americans alluded to the increase in the divorce rate in their discussions of declining fertility, they were far more concerned with families' decisions to restrict their number of children. Lack of space prevents me from treating nineteenth-century attitudes toward divorce. For an excellent discussion of this topic, see William L. O'Neill, *Divorce in the Progressive Era* (New Haven: Yale University Press, 1967). For a shortened version, see O'Neill, "Divorce in the Progressive Era," *American Quarterly* 17 (1965): 203–217. Also reprinted in Michael Gordon, ed., *The American Family in Social-Historical Perspective* (New York: St. Martin's Press, 1973), pp. 251–266.

5. There was also much controversy during the late nineteenth century over whether the fertility rate of blacks was increasing or decreasing. Census statistics revealed that the birth rate of blacks was declining, and several reports appeared commenting on this trend. See, for example, Willcox, "The Proportion of Children," pp. 5, 7, 8, 19 and John S. Billings, "The Diminishing Birth-Rate in the United States," *Forum* 15 (June 1893): 470.

6. Francis A. Walker, "Immigration and Degradation," reprinted in Francis A. Walker, *Discussions in Economics and Statistics,* ed. Davis R. Dewey (New York: Henry Holt, 1899), 2: 422–425. "Immigration and Degradation" originally appeared in *Forum* 11 (August 1891): 634–644. Ten years later, a Washington economist, R. R. Kuczynski, revived Americans' concern with the discrepancy between the birth rates of immigrant and native-born Americans with two influential articles, both entitled "The Fecundity of the Native and Foreign Born Population in Massachusetts," *Quarterly Journal of Economics* 16 (1901–1902): 1–36; 141–186.

7. Robert Hunter, *Poverty* (London: Macmillan, 1904), p. 308. Compare with Francis A. Walker, "Immigration and Degradation," *Discussions in Economics and Statistics,* 2: 424.

8. Hunter, *Poverty,* pp. 313–314.

9. Frederick A. Bushee, "The Declining Birth Rate and Its Cause," *The Popular Science Monthly* 63 (August 1903): 361.

10. Another movement occurring conjointly with the rise of anti-immigration and racist sentiment was the eugenics movement. Eugenics—the biological science pertaining to the improvement of the human species—was founded by a British biologist, Sir Francis Galton, during the latter half of the nineteenth century and was named by him in 1885. The term *eugenics* derived from the Greek word, *eugenia,* meaning "well-born," and Galton defined the science as the study of the agencies under social control that could improve or impair the racial qualities of future generations. In the early twentieth century, arguments taken from eugenics were often used to castigate educated or wealthy men and women who did not have children. *The Popular Science Monthly* gave extensive coverage to eugenics; for example, see Thorndike, "Professor Pearson on the Distribution of Fertility," 63: 84; Webb, "Physical Degeneracy or Race Suicide," 69: 512–529; Pearson, "The Scope and Importance to the State of The Science of National Eugenics," 71: 385–412, especially pp. 409–411; Kassel, "Fertility and Genius," 71: 452–454; Michaud, "Shall We Improve Our Race?" 72: 75–78; Ridgeway, "The Application of Zoological Laws to Man," 73: 500–522; Allemann, "Immigration and the Future

American Race," 75: 586–596; Nearing, "'Race Suicide' Vs. Over Population," 78: 81–83; Thorndike, "Eugenics: With Special Reference to Intellect and Character," 83: 125–138; and Dublin, "The Trend of American Vitality," 86: 313–319.
11. See Jacobson, *American Marriage and Divorce*, pp. 21, 35, 159. The statistics given by Jacobson suggest that marriage rates for men and women over the age of 15 increased gradually but not steadily from 1870 to 1920, with 1890 representing the low point for both groups.
12. Edward Thorndike, "The Decrease in the Size of American Families," *The Popular Science Monthly* 63 (May 1903): 65.
13. George J. Engelmann, "Education Not the Cause of Race Decline," *The Popular Science Monthly* 63 (June 1903): 179. In this article, however, Engelmann recognized that it was unfair and misleading to place the entire responsibility for declining fecundity on the wife, claiming "in truth it is the husband to an equal and even a greater extent [who is responsible for the voluntary prevention of childbearing]," p. 180.
14. G. Stanley Hall and Theodate L. Smith, "Marriage and Fecundity of College Men and Women," *The Pedagogical Seminary* 10 (September 1903): 278.
15. Ibid., p. 282–283.
16. Ibid., p. 314.
17. The fertility rates for white and black women from 1800 to 1920 are given by the Bureau of the Census as follows:

Number of Children Under 5 per 1,000 Women, Aged 20–44

YEAR	WHITES	BLACKS
1800	1,342	—
1810	1,358	—
1820	1,295	—
1830	1,145	—
1840	1,085	—
1850	892	1,087
1860	905	1,072
1870	814	997
1880	780	1,090
1890	685	930
1900	666	845
1910	631	736
1920	604	608

Source: Bureau of the Census, *Historical Statistics of the United States, Colonial Times to 1957* (Washington, D.C.: Government Printing Office, 1961), p. 24.

18. Theodore Roosevelt, "Letter to Mrs. Bessie Van Vorst October 18, 1902," *Presidential Addresses and State Papers of Theodore Roosevelt* (New York: P. F. Collier, n.d.; New York: Kraus Reprint, 1970), 1–2: 510.
19. Mary R. Smith, "Statistics of College and Non-College Women," *Publications of the American Statistical Association* 7 (March/June 1900): 11.
20. *Universal Suffrage: Female Suffrage* (Philadelphia: J. B. Lippincott, 1867). Cited in Linda

Gordon, *Woman's Body, Woman's Right: A Social History of Birth Control in America* (New York: Grossman, 1976), p. 149.
21. See Gordon, *Woman's Body, Woman's Right*, pp. 95–115 and Linda Gordon, "Voluntary Motherhood: The Beginnings of Feminist Birth Control Ideas in the United States," *Clio's Consciousness Raised: New Perspectives on the History of Women*, eds. Mary S. Hartman and Lois Banner (New York: Harper & Row, 1974), pp. 54–71.
22. Daniel Scott Smith uses the term *domestic feminism* in reference to the link between women's increase in autonomy within the family and the decline in marital fertility; see Smith, "Family Limitation," *A Heritage of Her Own*, eds. Cott and Pleck, pp. 231–240. Also see Degler, *At Odds*, pp. 182–192.
23. Gordon, *Woman's Body, Woman's Right*, p. 151.
24. Lydia K. Commander, a journalist who interviewed physicians about "the causes of the prevalence and popularity of the small family," found a common response was the financial burden of raising and educating children. See Commander, *The American Idea?* (New York: A. S. Barnes, 1907); Commander, "Has the Small Family Become an American Ideal?" *Independent* 56 (April 14, 1904): 836–840; Commander, "Why Do Americans Prefer Small Families?" *Independent* 57 (October 13, 1904): 847–850. See also Winnifred Harper Cooley, *The New Womanhood* (New York: Broadway Publishing Company, 1904), pp. 84–93 and Engelmann, "Education Not the Cause of Race Decline," in Note 13.
25. Degler, *At Odds*, p. 226.

ARTICLES ON THE BIRTH RATE QUESTION APPEARING IN *THE POPULAR SCIENCE MONTHLY*

(Arranged by Date of Publication)

Nathan Allen. "The Law of Human Increase," (November 1882) 22: 39–48.
Stewart Gordon. "Our Marriage and Divorce Laws," (June 1883) 23: 224–237.
Nathan Allen. "Changes in New England Population," (August 1883) 23: 433–444.
E. T. Merrick. "Our Marriage and Divorce Laws," (September 1883) 23: 663–668.
O. E. Randall. "Birth-Rate in A New Hampshire Town," (February 1884) 24: 555.
Lucy M. Hall. "Higher Education of Women and the Family," (March 1887) 30: 612–618.
S. H. Mead. "Population and the Food-Supply," (October 1888) 33: 843.
Grant Allen. "Plain Words on the Woman Question," (December 1889) 36: 170–181.
E. F. Andrews. "Grant Allen on the Woman Question," (February 1890) 36: 552–553.
Mrs. M. F. Armstrong. "The Mission of Educated Women," (March 1890) 36: 601–608.
A. A. M. "A Defense of 'Advanced' Women," (March 1890) 36: 699–700.
Alice B. Tweedy. "Is Education Opposed to Motherhood?" (April 1890) 36: 751–760.
Carroll D. Wright. "The Native and Foreign-Born Population: VIII—Lessons from the Census," (October 1892) 41: 756–762.
Sydney G. Fisher. "Has Immigration Increased Population?" (December 1895) 48:244–255.
Sydney G. Fisher. "The Drift of Population in France," (January 1896) 48: 412.
Sydney G. Fisher. "Immigration and Crime," (September 1897) 49: 625–630.
H. S. Pritchett. "The Population of the United States During the Next Ten Centuries," (November 1900) 58: 49–53.
Edward L. Thorndike. "Marriage Among Eminent Men," (August 1902) 61: 328–329.
"The Question of the Birth-Rate," (April 1903) 62: 567–569.
Edward L. Thorndike. "The Decrease in the Size of American Families," (May 1903) 63: 64–70.
Edward L. Thorndike. "Professor Pearson on the Distribution of Fertility," (May 1903) 63: 84.

George J. Engelmann. "Education Not the Cause of Race Decline," (June 1903) 63: 172–184.
C. E. Smith. "The Question of Racial Decline," (July 1903) 63: 275–276.
Editor. "The Question of Racial Decline," (July 1903) 63: 276–277.
Frederick A. Bushee. "The Declining Birth Rate and its Cause," (August 1903) 63: 355–361.
C. "The Birth Rate in Fiction," (August 1903) 63: 373.
"The Size of Families of College Graduates," (April 1904) 64: 570–571.
An Alumna. "Alumna's Children," (May 1904) 65: 45–51.
Another Alumna. "Alumna's Children Again," (July 1904) 65: 279–281.
Frances M. Abbott. "Three Decades of College Women," (August 1904) 65: 350–359.
A. Lapthorn Smith. "Higher Education of Women and Race Suicide," (March 1905) 66: 466–473.
Olivia R. Fernow. "Does Higher Education Unfit Women for Motherhood?" (April 1905) 66: 573–575.
"The Birth Rate Again," (April 1905) 66: 576–577.
"Proportion of Children in the United States," (December 1905) 67: 762–763.
Sidney Webb. "Physical Degeneracy or Race Suicide?" (December 1906) 69: 512–529.
George B. Mangold. "The Waste of Children," (June 1907) 70: 549–556.
Charles Kassel. "Fertility and Genius," (November 1907) 71: 452–454.
Karl Pearson. "The Scope and Importance to the State of the Science of National Eugenics," (November 1907) 71: 385–412; especially pp. 409–411.
Gustav Michaud. "Shall We Improve Our Race?" (January 1908) 72: 75–78.
William Ridgeway. "The Application of Zoological Laws to Man," (December 1908) 73: 500–522.
J. McKeen Cattell. "The School and the Family," (January 1909) 74: 84–95.
Albert Perry Brigham. "Capacity of the United States for Population," (September 1909) 75: 209–220.
"The Possible Population of the United States," (October 1909) 75: 414–416.
Albert Allemann. "Immigration and the Future American Race," (December 1909) 75: 586–596.
James S. Stevens. "The Population of the United States," (April 1910) 76: 382–388.
Charles Franklin Emerick. "Is the Diminishing Birth Rate Volitional?" (January 1911) 78: 71–80.
Scott Nearing. "'Race Suicide' Vs. Over Population," (January 1911) 78: 81–83.
"The Population of the United States in 1910," (April 1911) 78: 409–410.
"Vital Statistics and the Decreasing Death Rate," (December 1912) 81: 617–618.
"The Population of the United States in 1910," (August 1912) 81: 204–205.
"Vital Statistics and the Marriage Rate," (July 1913) 83: 103–104.
Edward L. Thorndike. "Eugenics: With Special Reference to Intellect and Character," (August 1913) 83: 125–138.
"Legal Limitations of Marriage," (November 1913) 83: 518–519.
Louis I. Dublin. "The Trend of American Vitality," (April 1915) 86: 313–319.
Earl Barnes. "The Celibate Women of Today," (June 1915) 86: 550–556.
Elaine Goodale Eastman. "The Waste of Life," (August 1915) 87: 187–194.

SUGGESTIONS FOR FURTHER READING

Primary Sources

Abbott, Frances M. "College Women and Matrimony Again," *Century* 51 (March 1896): 796–798.
"Are Women to Blame?" *North American Review* 148 (May 1889): 622–642.
Anthony, Susan B. "Reply to President Roosevelt's Race Suicide Theory," *Socialist Woman* 2 (September 1908): 6.
Billings, John S. "The Diminishing Birth-Rate in the United States," *Forum* 15 (June 1893): 467–477.

Bureau of the Census. *A Century of Population Growth: From the First Census of the United States to the Twelfth, 1790–1900.* Washington, DC: Government Printing Office, 1909.

Bureau of the Census. *Historical Statistics of the United States, Colonial Times to 1957.* Washington, DC: Government Printing Office, 1961.

Calhoun, Arthur W. *A Social History of the American Family From Colonial Times to the Present.* Vol. 3. Cleveland: Arthur H. Clark, 1919.

Carlton, Frank. "Broad Aspects of Race Suicide," *Arena* 36 (December 1906): 607–612.

Commander, Lydia K. *The American Idea?* New York: A. S. Barnes, 1907.

Commander, Lydia K. "Has the Small Family Become an American Ideal?" *Independent* 56 (April 14, 1904): 836–40.

Commander, Lydia K. "Why do Americans Prefer Small Families?" *Independent* 57 (October 13, 1904): 847–50.

Emerick, Charles Franklin. "College Women and Race Suicide," *Political Science Quarterly* 24 (June 1909): 269–283.

Kuczynski, R. R. "The Fecundity of the Native and Foreign Born Population in Massachusetts," *Quarterly Journal of Economics* 16 (1901–1902): 1–36; 141–186.

Roosevelt, Theodore. "Birth Reform, From the Positive, Not the Negative, Side," *The Foes of Our Own Household.* New York: George H. Doran, 1917, pp. 250–272.

Ross, E. A. "Western Civilization and the Birth-Rate," *American Journal of Sociology* 12 (March 1907): 607–632.

Shinn, Milicent Washburn. "The Marriage Rate of College Women," *Century* 50 (October 1895): 946–948.

Smith, Theodate L. "Dr. Engelmann on Race Decline," *Nation* 76 (1903): 494–495.

Walker, Francis A. "Immigration and Degradation," *Forum* 11 (August 1891): 634–644. Reprinted in Francis A. Walker. *Discussions in Economics and Statistics.* Vol. 2 edited by Davis R. Dewey. New York: Henry Holt, 1899, pp. 417–426.

"Why I Have No Family" by "A Childless Wife" *Independent* 58 (March 23, 1905): 654–659.

Statistical Studies of College Women

Association of Collegiate Alumnae. *Health Statistics of Women College Graduates.* Boston: Wright & Potter, 1885.

Hall, G. Stanley, and Smith, Theodate L. "Marriage and Fecundity of College Men and Women," *The Pedagogical Seminary* 10 (September 1903): 275–314.

Hewes, Amy. "Marital and Occupational Statistics of Graduates of Mount Holyoke College," *Publications of the American Statistical Association* 12 (December 1911): 771–796.

Nearing, Nellie Seeds. "Education and Fecundity," *Publications of the American Statistical Association* 14 (June 1914): 156–175. Reprinted in *Education and Fecundity.* Chautauqua, New York: Chautauqua Print Shop, 1917.

Smith, Mary R. "Statistics of College and Non-College Women," *Publications of the American Statistical Association* 7 (March/June 1900): 1–26.

Van Kleeck, Mary. "A Census of College Women," *Journal of the Association of Collegiate Alumnae* 11 (May 1918): 557–591.

Secondary Sources

Marriage and Birth Rates

Coale, Ansley J., and Zelnik, Melvin. *New Estimates of Fertility and Population in the United States.* Princeton: Princeton University Press, 1963.

Grabill, Wilson H.; Kiser, Clyde V.; and Whelpton, Pascal K. *The Fertility of American Women.* New York: John Wiley & Sons, 1958.

Jacobson, Paul H. *American Marriage and Divorce.* New York: Rinehart, 1959.
May, Elaine Tyler. *Great Expectations: Marriage and Divorce in Post-Victorian America.* Chicago: University of Chicago Press, 1980.
Okun, Bernard. *Trends in Birth Rates in the United States Since 1870.* Baltimore: Johns Hopkins Press, 1958.
O'Neill, William L. *Divorce in the Progressive Era.* New Haven: Yale University Press, 1967.
Spengler, Joseph J. *Population and America's Future.* San Francisco: W. H. Freeman, 1975.
Westoff, Charles F. *College Women and Fertility Values.* Princeton: Princeton University Press, 1967.
Yasuba, Yasukichi. *Birth Rates of the White Population in the United States, 1800–1860: An Economic Study.* Baltimore: Johns Hopkins Press, 1962.

Fertility and Birth Control

Degler, Carl N. *At Odds: Women and the Family in America from the Revolution to the Present.* New York: Oxford University Press, 1980.
Gordon, Linda. *Woman's Body, Woman's Right: A Social History of Birth Control in America.* New York: Grossman, 1976.
Kennedy, David M. *Birth Control in America: The Career of Margaret Sanger.* New Haven: Yale University Press, 1970.
Smith, Daniel Scott. "Family Limitation, Sexual Control and Domestic Feminism in Victorian America," *Feminist Studies* 1 (Winter/Spring 1973): 40–57. Reprinted in Cott, Nancy F., and Pleck, Elizabeth H., eds. *A Heritage of Her Own: Toward a New Social History of American Women.* New York: Simon and Schuster, 1979, pp. 222–245. Also reprinted in Hartman, Mary S., and Banner, Lois, eds. *Clio's Consciousness Raised: New Perspectives on the History of Women.* New York: Harper & Row, 1974, pp. 119–136.
Wilson, Margaret Gibbons. *The American Woman in Transition: The Urban Influence 1870–1920.* Westport: Greenwood Press,1979, pp. 41–75.

For readings on women's sexuality, see Chapter 1.

Grant Allen (1848–1899), born in Canada, attended schools in the United States, France, and England, took a degree from Oxford University (1870), and became principal of Queen's College in Jamaica before settling in London. Allen was a prolific writer, publishing articles and books on many social and scientific topics. Finding it difficult to earn his living from scientific writings, however, Allen began to write fiction and discovered, as the London Times put it, "that his worst fiction was more profitable than his best science." His most popular novel, The Woman Who Did (1895), tells the story of Herminia Barton, a "very advanced" woman, who refuses to marry her lover, Alan Merrick, because she believes that marriage feeds men's monopolistic instincts and requires women to relinquish their freedom and individuality. Having persuaded Alan to become her lover, Herminia gives birth to their daughter Dolores after his unexpected death. At the end of the book, Dolores, ashamed of her birthright, rejects her mother, who commits suicide in order to leave Dolores free to marry the man she loves.

Allen, an avowed agnostic, disciple of Darwin, and advocate of free love (all unpopular stances in his day), sympathized with his heroine, but nonetheless could foresee no life for her but one of suffering and martyrdom. Reviewers condemned the book for its immoral views. It outsold all of Allen's other works. Despite the book's seemingly immoral outlook, Allen vigorously upheld conventional beliefs about women's natural duty to have children. In describing Herminia, he carefully distinguished her from "unnatural" women who remained childless. "Herminia was far removed indeed from the blatant and decadent sect of 'advanced women' who talk as though motherhood were a disgrace and a burden.... She knew that to be a mother is the best privilege of her sex, a privilege of which unholy manmade institutions [women's colleges] now conspire to deprive half the finest and noblest women in our civilized communities. Widowed as she was, she still pitied the unhappy beings doomed to the cramped life and dwarfed heart of the old maid; pitied them as sincerely, as she despised those unhealthy souls who would make of celibacy, wedded or unwedded, a sort of anti-natural religion for women." Allen made the same argument in the following selection, and as usual, his work elicited many replies. One of them, "Is Education Opposed to Motherhood?" by Alice B. Tweedy, follows Allen's article.

PLAIN WORDS ON THE WOMAN QUESTION

GRANT ALLEN

December 1889

IF any species or race desires a continued existence, then above all things it is necessary that that species or race should go on reproducing itself. This, I am aware, is an obvious platitude; but I think it was John Stuart Mill who once said there were such things in the world as luminous platitudes.

Some truths are so often taken for granted in silence, that we are in danger at times of quite losing sight of them. And as some good friends of mine have lately been accusing me of "barren paradoxes," I am anxious in this paper to avoid all appearance of paradox, barren or fertile, and to confine myself strictly to the merest truisms. Though the truisms, to be sure, are of a particular sort too much overlooked in controversy nowadays by a certain type of modern lady writers.

Let us look then briefly at the needful conditions under which alone the human race can go on reproducing itself. . . .

Well, suppose, now, every man and every woman in a given community were to marry; and suppose they were in each case to produce two children, a boy and a girl; and suppose those children were in every case to attain maturity; why, then, the next generation would exactly reproduce the last, each father being represented by his son, and each mother by her daughter, *ad infinitum.* (I purposely omit, for simplicity's sake, the complicating factor of the length and succession of generations, which by good luck in the case of the human species practically cancels itself.) But, as a matter of fact, all the children do not attain maturity: on the contrary, nearly half of them die before reaching the age of manhood—in some conditions of life, indeed, and in some countries, more than half. Roughly speaking, therefore (for I don't wish to become a statistical bore), it may be said that in order that two children may attain maturity and be capable of marriage, even under the most favorable circumstances, four must be born. The other two must be provided to cover risks of infant or adolescent mortality, and to insure against infertility or incapacity for marriage in later life. They are wanted to make up the categories of soldiers, sailors, imbeciles, cripples, and incapables generally. So that even if every possible person married, and if every married pair had four children, we should only just keep up the number of our population from one age to another. . . .

Now, I have the greatest sympathy with the modern woman's demand for emancipation. I am an enthusiast on the Woman Question. Indeed, so far am I from wishing to keep her in subjection to man, that I should like to see her a great deal more emancipated than she herself as yet at all desires. Only, her emancipation must not be of a sort that interferes in any way with this prime natural necessity. . . .

For let us look again for a moment at what this all but universal necessity of maternity implies. Almost every woman must bear four or five children. In doing so she must on the average use up the ten or twelve best years of her life—the ten or twelve years that immediately succeed her attainment of complete womanhood. For note, by the way, that these women must also for the most part marry young: as Mr. Galton has shown, you can quietly and effectually wipe out a race by merely making its women all marry at twenty-eight: married beyond that age, they don't produce children enough to replenish the population. Again, during these ten or twelve years of child-bearing at the very least, the women can't conveniently earn their own livelihood; they must be provided for by the labor of the men — under existing circumstances (in favor of which I have no Philistine prejudice) by their own husbands. . . .

Seeing, then, that these necessities are laid by the very nature of our organization upon women, it would appear as though two duties were clearly imposed upon the women themselves, and upon all those men who sympathize in their welfare: First, to see that their training and education should fit them above everything else for this their main function in life; and, second, that in consideration of the special burden they have to bear in connection with reproduction, all the rest of life should be made as light and easy and free for them as possible. We ought frankly to recognize that most women must be wives and mothers; that most women should therefore be trained, physically, morally, socially, and mentally, in the way best fitting them to be wives and mothers; and that all such women have a right to the fullest and most generous support in carrying out their functions as wives and mothers.

And here it is that we seem to come in conflict for a moment with most of the modern Woman-Question agitators. I say for a moment only, for I am not going to admit, even for that brief space of time, that the doctrine I wish to set forth here is one whit less advanced, one whit less radical, or one whit less emancipatory than the doctrine laid down by the most emancipated women. On the contrary, I feel sure that while women are crying for emancipation they really want to be left in slavery: and that it is only a few exceptional men, here and there in the world, who wish to see them fully and wholly enfranchised. And those men are not the ones who take the lead in so-called Woman's Rights movements.

For what is the ideal that most of these modern women agitators set before them? Is it not clearly the ideal of an unsexed woman? Are they not always talking to us as though it were not the fact that most women must be wives and mothers? Do they not treat any reference to that fact as something ungenerous, ungentlemanly, and almost brutal? Do they not talk about our "casting their sex in their teeth"? — as though any man ever resented the imputation of manliness. Nay, have we not even, many times lately, heard those women who insist upon the essential womanliness of women described as "traitors to the cause of their sex"? Now, we men are (rightly) very jealous of our virility. We hold it a slight not to be borne that any one should impugn our essential manhood. And we do well to be angry; for virility is the key-note to all that is best and most forcible in the masculine character. Women ought equally to glory in their femininity. A woman ought to be ashamed to say she has no desire to become a wife and mother. Many such women there are, no doubt — it is to be feared, with our existing training, far too many; but, instead of boasting their sexlessness as a matter of pride, they ought to keep it dark, and to be ashamed of it — as ashamed as a man in a like predicament would be of his impotence. They ought to feel they have fallen short of the healthy instincts of their kind, instead of posing as in some sense the cream of the universe, on the strength of what is really a functional aberration.

Unfortunately, however, just at the present moment, a considerable number of the ablest women have been misled into taking this unfeminine side, and becoming real "traitors to their sex" in so far as they endeavor to assimilate women to men in everything, and to put upon their shoulders, as a glory and privilege, the burden of their own support. Unfortunately, too, they have erected into an ideal what is really an unhappy necessity of the passing phase. They have set before them as an aim what ought to be regarded as a *pis-aller*. And the reasons why they have done so are abundantly evident to anybody who takes a wide and extended view of the present crisis—for a crisis it undoubtedly is—in the position of women.

In the first place, the movement for the higher education of women, in itself an excellent and most praiseworthy movement, has at first, almost of necessity, taken a wrong direction, which has entailed in the end much of the present uneasiness. Of course, nothing could well be worse than the so-called education of women forty or fifty years ago. Of course, nothing could be narrower than the view of their sex then prevalent, as eternally predestined to suckle foods and chronicle small beer. But when the need for some change was first felt, instead of reform taking a rational direction—instead of women being educated to suckle strong and intelligent children, and to order well a wholesome, beautiful, reasonable household—the mistake was made of educating them like men—giving a like training for totally unlike functions. The result was that many women became unsexed in the process, and many others acquired a distaste, an unnatural distaste, for the functions which Nature intended them to perform. At the present moment a great majority of the ablest women are wholly dissatisfied with their own position as women, and with the position imposed by the facts of the case upon women generally; and this as the direct result of their false education. They have no real plan to propose for the future of women as a sex; but in a vague and formless way they protest inarticulately against the whole feminine function in women, often even going the length of talking as though the world could get along permanently without wives and mothers.

In the second place, a certain real lack of men to marry, here and now, in certain classes of society, and those the classes that lead thought, has made an exceptional number of able women at present husbandless, and thus has added strength to the feeling that women must and ought to earn their own living. How small and local this cause is I shall hereafter try to show: but there can be no doubt that it has much to do with the present discontents among women. . . .

[I]t is a fact, that both in England and America the marriageable men of the middle and upper classes are not to the fore, and that accordingly in these classes — the discussing, thinking, agitating classes — an undue proportion of women remains unmarried. The causes of this class disparity are not far to seek. In America the young man has gone West. In England he is in the army, in the navy, in the Indian Civil Service, in the Cape Mounted Rifles. He is sheep-farming in New Zealand, ranching in Colorado, growing tea in Assam, planting coffee in Ceylon; he is a cowboy in Montana, or a wheat-farmer in Manitoba, or a diamond-digger at Kimberley, or a merchant at Melbourne: in short, he is anywhere and everywhere except where he ought to be, making love to the pretty girls in England. For, being a man, I, of course, take it for granted that the first business of a girl is to be pretty.

Owing to these causes, it has unfortunately happened that a period of great upheaval in the female mind has coincided with a period when the number of unmarried women in the cultivated classes was abnormally large. The upheaval would undoubtedly have taken place in our time, even without the co-operation of this last exacerbating cause. The position of women was not a position which could bear the test of nineteenth-century scrutiny. Their education was inadequate; their social status was humiliating; their political power was *nil*; their practical and personal grievances were innumerable: above all, their relation to the family — to their husbands, their children, their friends, their property — was simply insupportable. A real Woman Question there was, and is, and must be. The pity of it is that the coincidence of its recognition with the dearth of marriageable men in the middle and upper classes has largely deflected the consequent movement into wrong and essentially impracticable channels.

For the result has been that, instead of subordinating the claims of the unmarried women to the claims of the wives and mothers, the movement has subordinated the claims of the wives and mothers to the claims of the unmarried women. Almost all the Woman's Rights women have constantly spoken, thought, and written as though it were possible and desirable for the mass of women to support themselves, and to remain unmarried forever. The point of view they all tacitly take is the point of view of the self-supporting spinster. Now, the self-supporting spinster is undoubtedly a fact — a deplorable accident of the passing moment. Probably, however, even the most rabid of the Woman's Rights people would admit, if hard pressed, that in the best-ordered community almost every woman should marry at twenty or thereabouts. We ought, of course, frankly to recognize the existence of the deplorable accident; we ought for the moment to make things as easy and smooth as possible for her; we ought to remove all professional barriers, to break down the absurd jealousies and prejudices of men, to give her fair play, and if possible a little more than fair play, in the struggle for existence. So much our very chivalry ought to make obligatory upon us. That we should try to handicap her heavily in the race for life is a shame to our manhood. But we ought at the same time fully to realize that she is an abnormity, not the woman of the future. We ought not to erect into an ideal what is in reality a painful necessity of the present transitional age. We ought always clearly to bear in mind—men and women alike—that to all time the vast majority of women must be wives and mothers; that on those women who become wives and mothers depends the future of the race; and that, if either class must be sacrificed to the other, it is the spinsters whose type perishes with them that should be sacrificed to the matrons who carry on the life and qualities of the species.

For this reason a scheme of female education ought to be mainly a scheme for the education of wives and mothers. And if women realized how noble and important a task it is that falls upon mothers, they would ask no other. If they realized how magnificent a nation might be molded by mothers who devoted themselves faithfully and earnestly to their great privilege, they would be proud to carry out the duties of their maternity. Instead of that, the scheme of female education now in vogue is a scheme for the production of literary women, schoolmistresses, hospital nurses, and lecturers on cookery. All these things are good in themselves, to be sure—I have not a word to say against them; but they are not of the center. They are side-lines off the main stream of feminine life, which must always consist of the maternal element. "But we can't know beforehand," say the advocates of the mannish training, "which women are going to be married, and which to be spinsters." Exactly so; and therefore you sacrifice the many to the few, the potential wives to the possible lady lecturers. You sacrifice the race to a handful of barren experimenters. What is thus true of the blind groping after female education is true throughout of almost all the Woman Movement. It gives precedence to the wrong element in the problem. What is essential and eternal it neglects in favor of what is accidental and temporary. What is feminine in women it neglects in favor of what is masculine. It attempts to override the natural distinction of the sexes, and to make women men—in all but virility.

The exact opposite, I believe, is the true line of progress. We are of two sexes: and in healthy diversity of sex, pushed to its utmost, lies the greatest strength of all of us. Make your men virile: make your women womanly. Don't cramp their intelligence: don't compress their waists: don't try to turn them into dolls or dancing girls: but freely and equally develop their feminine idiosyncrasy, physical, moral, intellectual. Let them be healthy in body: let them be sound in mind: if possible (but here I know even the most advanced among them will object), try to preserve them from the tyranny of their own chosen goddess and model, Mrs. Grundy. In one word, emancipate woman (if woman will let you, which is more than doubtful), but leave her woman still, not a dulled and spiritless epicene automaton.

That last, it is to be feared, is the one existing practical result of the higher education of women, up to date. Both in England and America, the women of the cultivated classes are becoming unfit to be wives or mothers. Their sexuality (which lies at the basis of everything) is enfeebled or destroyed. In some cases they eschew marriage altogether—openly refuse and despise it, which surely shows a lamentable weakening of wholesome feminine instincts. In other cases, they marry, though obviously ill adapted to bear the strain of maternity; and in such instances they frequently break down with the birth of their first or second infant. This evil, of course, is destined by natural means to cure itself with time: the families in question will not be represented at all in the second generation, or will be represented only by feeble and futile descendants. In a hundred years, things will have righted themselves; but meanwhile there is a danger that many of the most cultivated and able families of the English-speaking race will have become extinct, through the prime error of supposing that an education which is good for men must necessarily also be good for women....

If the "advanced" women will meet us on this platform, I believe the majority of "advanced" men will gladly hold out to them the right hand of fellowship. As a body we are, I think, prepared to reconsider, and to reconsider fundamentally, without prejudice or preconception, the entire question of the relations between the sexes—which is a great deal more than the women are prepared to do. We are ready to make any modifications in those relations which will satisfy the woman's just aspiration for personal independence, for intellectual and moral development, for physical culture, for political activity, and for a voice in the arrangement of her own affairs, both domestic and national. As a matter of fact, few women will go as far in their desire to emancipate woman as many men will go. It was Ibsen, not Mrs. Ibsen, who wrote the "Doll's House." It was women, not men, who ostracized George Eliot. The slavishness begotten in women by the *régime* of man is what we have most to fight against, not the slave-driving instinct of the men—now happily becoming obsolete, or even changing into a sincere desire to do equal justice. But what we must absolutely insist upon is full and free recognition of the fact that, in spite of everything, the race and the nation must go on reproducing themselves. Whatever modifications we make must not interfere with that prime necessity. We will not aid or abet women as a sex in rebelling against maternity, or in quarreling with the constitution of the solar system. Whether we have wives or not—and that is a minor point about which I, for one, am supremely unprejudiced—we must at least have mothers. And it would be well, if possible, to bring up those mothers as strong, as wise, as free, as sane, as healthy, as earnest, and as efficient as we can make them. If this is barren paradox, I am content to be paradoxical; if this is rank Toryism, I am content for once to be reckoned among the Tories.

IS EDUCATION OPPOSED TO MOTHERHOOD?

ALICE B. TWEEDY
April 1890

. . . .

IN "Plain Words on the Woman Question," Mr. Grant Allen discusses a topic which, he thinks, is "too much overlooked by modern lady writers." It is the continuation of the human race. Intrenched behind population and marriage statistics, he opens fire upon feminine reformers from a quarter where they have made little defense. His text is—if a race will continue, it must reproduce itself. His argument may be briefly given as follows:

I. Marriages are decreasing in England and America.
II. Women of the cultivated classes are becoming unfitted for motherhood.
III. The movement which demands the independence and higher education of women is responsible for this—it creates a "spiritless epicene automaton" and the "self-supporting spinster."
IV. The emancipation of women, especially from Mrs. Grundy, is desirable; but it must not conflict with the existence of mothers who are necessary to the race.

. . . Granting, then, that these conditions are truthfully represented, what can be said to the argument which Mr. Allen founds upon them?

I. Marriages are decreasing in many civilized countries. There are local causes for this tendency among men, but the principal and prevailing one with women is that they are passing from the rule of force to a state of freedom, and use their newly found liberty to reject what seems to hamper and handicap them. They are emerging from the condition in which marriage is consequent upon physical or social constraint, and they have not generally arrived at the point where it is for them the result of deliberate choice or response to natural instinct. In China, India, Persia, and Arabia, where marriages are still controlled by force, the number would be diminished at once, without the influence of higher education or industrial training, were the women allowed simple freedom of choice. This decrease would not indicate "the dulling of feminine instinct," but the vitality of it—since marriages are made there in defiance of natural selection, and represent the worst condition of servitude. In more civilized states the popularity of marriage does not depend wholly upon the way in which women regard it, but upon the way in which it is treated by men. The laws of some countries render it easier for a man to live illegitimately with a woman than to marry her; true marriage is discouraged by social usage and dishonored by false philosophy. Few thoughtful minds will deny that the customs which render it difficult for a young man to marry, which send him hither and thither to gain a fortune, succeed in a profession, or dissipate his strength, when he should be choosing his sweetheart, are harmful, and divert men "from the true problem of their sex to fix it on side issues of comparative unimportance."

Boys, as well as girls, should be taught that the full meaning of human life is missed unless they deserve and find a fitting mate. Authors who represent wifehood and maternity as onerous and unattractive, however necessary, or who surround illicit and incomplete love with superficial glamour, are open to the charge of depreciating marriage. Guests are not tempted to a banquet by fear of starvation, nor are men attracted toward matrimony for the interests of the race. Instead of showing that marriage offers the greatest possibility of happiness, it is often described by men as an unintellectual, slavish, and pitiable condition. . . .

II. Mr. Allen seems to regard as evidence that women "are becoming unfitted for motherhood" the fact that they do not glory in their femininity, and charges also that women reformers speak and write "as though it were desirable that the mass of women should remain unmarried forever." Worse even than this, he asserts: "At the present moment a great majority of the ablest women are wholly dissatisfied with their own position as women, and with the position imposed by the facts of the case upon women generally; this as the direct result of their false education." Here are two ideas badly entangled for want of definition—the natural and the artificial position of women. Mr. Allen gives us on the following page his opinion of "the position" (artificial) of women in language strong enough for the most blatant reformer. "The position of women was not a position which could bear the test of nineteenth-century scrutiny; . . . their relation to the family, to their husbands, their children, their friends, their property, was simply insupportable." (Does he demand of these ablest women that they should be satisfied with a position he calls "insupportable"?) But, let him not be distressed because woman does not openly "glory" in her natural position of womanhood. There is no failure of healthy instinct here, but a natural feminine divergence from the masculine feeling. The differentiation of the sexes is a subject upon which we have no adequate data. We might as well try to surmise the habits of the wild cat from the domesticated pussy, as to speculate upon the essential qualities of free womanhood. But, so far as woman's physical constitution indicates anything, it points toward greater reserve on her part than is exhibited by man. This corresponds with the almost universal inclination of women to be more modest than men. Therefore, though a woman may prefer her own sex and be proud of her privileges as woman, she will not voluntarily go about "glorifying" her womanhood. If Mr. Allen should meet a young woman who announced herself a candidate for motherhood, it is doubtful whether he would approve of her, although she embodied his theory. . . .

Though men have greatly outgrown tyranny of thought and action, there is still alive much masculine arrogance. With many it is entirely unconscious; it probably is so with Mr. Allen when he calls a literary or scientific education "mannish." I do not know of any purely mannish training except that received by the monks of La Trappe, and that which fits men to be soldiers, sailors, blacksmiths, or workmen whose physical force is a necessity to their calling. A college or university education, although in past years given exclusively to men, was never supposed to fit them for any essentially masculine occupation, not even to become the fathers of the future race. It was preliminary to a professional or literary career, and intended to develop the powers of mind. And mind—emphasize as we will the physical differences of the body that goes with it—has no discoverable gender. The lavish way in which the epithet of "masculine" or "feminine" has been applied to particular minds is utterly destructive of precision of thought. Vigorous minds are called masculine and those of the namby-pamby, sentimental sort are dubbed feminine. This classification may be historically justifiable by the slight appearance women have made in the literary and scientific world, but there have been clear-headed women enough in all antiquity, and there are too many well-developed minds among them to-day, not to make them resent further tricking out with masculine trappings. The wise old Greeks saw fit to personify mind in a woman; the moderns seem to be afraid of such a result.

If education must be specialized, and women should be fitted to become wise mothers, then, in all fairness, men should be trained to become intelligent fathers. Their lack in this respect is as palpable to any just mind as the failure of women in motherhood. That there should be fathers, and good fathers, is no less important, from a utilitarian standpoint, than that there should be good mothers. Indeed, it may be questioned whether there are not annually more children lost to the world through the wickedness and ignorance of male parents than would be gained by the conversion of all "self-supporting spinsters" into model matrons. It is not necessary to enter into detail here, but appalling statistics are easily obtainable. Until no foundling hospital, no abandoned family exists, it is ungenerous to reproach woman with evading or "shirking" her natural duties. Postponement of marriage by men results in another not inconsiderable evil, false marriage [marriage under coercion] of many young women. Nature often revenges herself here by a lack of mothers. The wiser plan would be to follow the teaching of Nature and not dissociate the sexes, particularly during impressionable years. In study, work, or society, do not bar them from each other; then they will not form the erroneous notions that taint maturity. Let them be "human, instead of half-human."

III. The most evident good of education to woman, aside from the discipline of mind and development of power, is in its teaching observation of nature and the intelligent use rather than the repression of any instinct or force. Those who assert that these influences "unsex" woman, render her "unwomanly," should explain what is meant. She may lose some of the characteristics that have distinguished her in the past, but while analytic or radical minds call these characteristics local and temporary, conservatives cling to them as part of essential womanhood. It may be observed that, although Mr. Allen holds fast to the term of "radical," he agrees with our dear old great-grandmothers in this apprehension that education and independence unfit women to become mothers. To these timid souls may be recommended a greater trust in Nature, that she will be able to maintain the differences necessary to a continuance of the race. Clothes and customs vary with time and place; sex is stable and not injured by anything but physical condition. The traditional idea that womanhood can be modified in some occult way by occupation, training, or environment, is wholly unscientific and *baneful*; for it undoubtedly serves to nurse in many a woman that "slavishness of soul" which Mr. Allen, as a true well-wisher of woman, deplores. Physical condition, then, is the constant coefficient in the problem. Anything injurious to the health of either man or woman incapacitates each just so much for the fullest requirement of life. . . .

But, it may be asked, "Are doubting physicians not justified at all—are there no women students who break down or die?" There are such cases of overstrain or feeble constitution which find their parallel among men, but the percentage among women is so small that it leaves the health average still above the generality of women.

But, argues Mr. Allen, there is "the self-supporting spinster"; "almost every woman should marry"; and she is "a deplorable accident." Now, it is possible that while I may deem her admirable, another may consider her "deplorable" —it is a matter of taste merely. But, that she is not an "accident," rather an eternal verity, stands confessed in Mr. Allen's "almost." Unless, indeed, the entire community should be paired off—which is not desirable for economic reasons—spinsters and bachelors will continue to exist. It does not materially affect the issues of the race whether they are dependent or independent, and we may fearlessly praise in them the qualities which please us most. If the condition of "self-supporting spinsterhood" is more attractive than the condition of wifehood, there is menace for the future. It would be alarming, if we could believe with Mr. Allen in anything so unflattering to masculine endeavor; but, unfortunately, there are no statistics to prove whether this is due to dulled feminine instinct, or to the failure of man to make love at the right time. In the interim, from collateral evidence, the latter cause appears more trustworthy.

IV. If freedom from Mrs. Grundy is desirable, it is patent that education and independence are gradually liberating woman. The counter-charge is often made that the educated woman is too regardless of that favored deity.

From a biological point of view, Mr. Allen endows four years of college training with enormous potentiality. In this he evidently follows the eminent leader, Mr. Herbert Spencer, who asserts that the infertility of "upper-class girls" in England is due to "overtaxing of their brains"! Whether the majority of English "upper-class girls" are educated to that extreme point, and whether the question is not begged in the use of the word "overtaxation," may be left to the reader. It is strange that powerful heredity and palpable causes of race deterioration should be ignored by physiologists in order to throw the *onus* of this accusation upon mental culture. Insurance tables are made out more scientifically than this forecast of a girl's future. If in education, or in the industrial independence of women, there existed any tendency toward infertility, it would be barely discoverable in our generation, little more so in the next, and possibly in the third generation something might be ascertained from careful statistics following Mr. Galton's method. Nature does not retrograde so rapidly. . . .

Women can no longer be coerced into marriage, nor will they marry from a sense of duty to humanity. But for these reasons there need be no fear that the race will perish. There is as much prospect that roses will refuse to bloom in June as that women will ever become invincible to love. This force, and this alone, can make of them light-hearted mothers in place of the weary wrecks whose perverted motherhood has been anything but a boon to humanity. As long as it is taught that motherhood oppresses woman physically and restricts her intellectually, so long the average woman may dread or rebel against it. When she studies it in all its conditioning, she finds it does not impose such a fate upon her. She learns to discriminate between the ordering of Nature and the blunders of mankind, and recognizes that normal physical development can not be antagonistic to mental growth. . . .

"Alumna's Children" and "Alumna's Children Again" were written by two college-educated women, who related their personal experiences and attitudes concerning their former classmates' decisions to have children. The anonymity of the authors and the first writer's admission that she was rushing in where angels feared to tread testify to the controversiality of the birth rate question and to the great reluctance of many women to speak openly on this subject. The first writer concurred with physicians' conclusions that college-educated women were frequently too weak to bear children, while the second writer disagreed and discussed the healthy mothers she knew. More significant than the different experiences of these two women is their shared insistence that college-educated women, along with all other groups of American women, considered motherhood to be "life's greatest good" and were content to devote themselves exclusively to rearing children.

ALUMNA'S CHILDREN

AN ALUMNA

May 1904

THE latest publication of vital statistics in Massachusetts has again called attention to a subject often discussed in this magazine and elsewhere—the decreasing number of children in native American families. According to the majority of opinions given, this decrease is due mostly to 'social ambition.' This means that the women who should be, in a real sense, the pillars of our society prefer other things to bringing up their own children. If this is true, it seems a very serious indictment of the American woman.

But is the case settled yet? While social ambition may be operative in many cases, perhaps peculiarly among those coming to the notice of a specialist in medicine, may there not be some data that the statistician can not collect — some pertinent facts which in the nature of the case are not within reach of the investigators?

Among all the talk by learned men and high officials, it is strange that no member of the class under discussion has spoken to the question. On further thought the reason is obvious; the case is necessarily of great delicacy and incapable of proof. But because the charge seems to me in many cases so peculiarly unjust, hereby do I rush in where angels have feared to tread.

Dr. Engelmann in his especially interesting article[1] spoke particularly of the college graduates, that 'group having a lower birth rate than any other.' There may be no need to separate alumna from the rest of her racial group

---

[1] George J. Engelmann, "Education Not the Cause of Race Decline," *The Popular Science Monthly* 63 (June 1903): 172-184. (Ed.)

for consideration, for the body of college women now is made up of nearly all the elements of what may be called the middle class. But because narrowing a subject makes it easier to view; because the birth rate of the alumnae is the very lowest; and, especially, because I happen to know more of the conditions among college girls, I confine myself to that group.

There is no need to question the figures—that 1.8 children is the average family of an alumna wife; but let us consider in the beginning just what that 1.8 children mean. Incidentally, we may think a moment of the marriage rate among college women. Both these relatively low numbers are inspiring in one respect—in the thought of the elements which have been eliminated. If less than 50 per cent. of college women marry, yet of that number few take husbands 'for a home' or because they have nothing else to do. Perhaps there are as many happy marriages of companionship among a hundred college women as among a hundred women selected from any class, and does the state lose by the elimination of all others?

Alumna's marriage, then, means that a mature, independent, trained woman deliberately chooses to give the direction of her life to a man, because she loves him well enough to find in so doing her greatest happiness. Of such a mating are alumna's children born—of a 'selected' father, of a mother who has at least had an opportunity for knowledge—born to a heritage of intelligent love and care. So they ought to be a power for good, even though they are few. But just because they are of such a quality, society wants more of them; and it behooves the state to determine why their numbers are so few.

Yesterday I received some evidence on this question which seemed to me pertinent. I spent the day with a member of this group 'having a lower birth rate than any other.' She had recently buried her only child, hardly a month old. As I was on my way to her, my mind went over her past year; her hope that she might at last be strong enough to bear a well child, the months of illness, the forty-eight hours of agony, the supreme joy so soon followed by anxiety, and the awful loss. And when I saw her face I could not speak. But she spoke, and with a smile: 'Don't pity me so. It paid! it pays!' During the hours I spent with her she showed me two books of letters, mostly from college friends of ours. One collection was received when her baby came, the other when he went. 'I am so happy to know that your life has been made complete'—this thought was expressed over and over again in the letters of congratulation. Mothers or childless, all these women seemed to know that any woman's life is incomplete until she has known motherhood. Of those notes that came at the little one's death from childless women, married or single, all said, in one phrase or another, 'how much sadder than yourself am I, who have no child to die.' These letters inevitably suggest the question, why are so few of these women mothers, when all of them speak of motherhood as life's greatest good? It seemed to me a very solemn question, and I went over the list of those whom I know best and found what seems to me a suggestive unanimity.

There is A, the brightest girl in our class, kept from the really brilliant literary career of which she is capable by her physical weakness. She loves a man who is her ideal mate and he returns her love, but they live their lives apart. A short time ago I said something to her about her being married. "Be married!" she said. "What right have I to be married? My physician tells me that I am no more a woman physically than is a twelve-year-old girl. What right have I to give to any man an invalid wife and take away from him his hope of children? I shall never be married!"

B has just adopted a baby, 'because I may never hope to have one of my own,' as she wrote me. C, apparently well during college days, came to decennial the mother of three children, but such an invalid that she only with difficulty sat up during class dinner. D had one child who died at birth, and no other has ever come to her. E, an especially close friend of mine, has one child and longs for more, but her physician husband is unwilling that she should again take the risk, saying she was 'never meant to bear children.' F's case is almost the same; a woman of magnificent physique, she refused to heed her doctor. Her first baby lived, but she barely escaped herself; her second child was sacrificed to save its mother's life; 'and I can never hope for another,' she said to me, her eyes full of tears. G also would not believe her physician, but her hope was finally justified. Though three times she was disappointed, her fourth suffering gave her a son, who, she says, much more than pays for all. H has two strong, beautiful children. 'I wish we had six,' said her husband, a college dean, by the way, 'but the two that we have cost their mother so much that we shall never have any more.'

These women are all among my classmates, but the conditions are not peculiar to my class or my college. I could cite as many instances among other college friends, but they are so nearly identical that they would seem merely a repetition. Two friends of mine now are fighting hard for the lives which have been threatened ever since their first babies came, in each case over a year ago. The example of greatest courage, perhaps, is not a college woman, though decidedly a schooled woman. Five times she went to the very gates of death for her great hope, but only once did she see the face of a living child of hers, and he died at six months.

In connection with a woman's ruling passion, I always think of that gracious lady preeminent as scholar and citizen, who recently left this world so much the poorer, especially for those who enjoyed the distinction of her friendship. I once heard a woman ask her whether she had any children. "Do you suppose," she replied, "that if I had any children, I should be running around the country talking?" And her tone said 'since all that my life seemed meant for, fails,' though all other honors were hers, save only motherhood.

Throughout my acquaintance, among not only my college friends but also my husband's college friends, I find, it has seemed to me, a large proportion of childless homes. And wherever a word has been dropped in my hearing as to the feeling of the wife in the matter, it has always been referred to as a great sorrow. I have been considering the question for some years and have tried to receive any light that appeared.

In many homes that I know there is an only child. It may seem that here are mothers who can have children but do not want them. The only child does not mean this, but that the one came so near costing its mother her life that he to whom she is dearer than even his hope of children can not bear to let her undergo the ordeal again. Dr. Engelmann has referred to the fact that men more often than their wives wish to limit the number of their children. I shall never forget the pathos of the day when K, a boy who had graduated from college and married hardly a year before, came to tell me of the birth of his son. For twenty-four hours his wife had striven between life and death; as soon as she was out of immediate danger, he came to me, her long-time friend, and broke down. 'If this is what babies cost,' he said, 'there will never be any more at our house.' The son born that day is seven years old, but he is still an only child. I know many instances where children are few because the one or two who have come 'have cost their mother too much.'

These women are not cowards. Undoubtedly the first six hundred women who cross the campus to-morrow morning would make a Balaklava charge without a desertion. But how many men would go one by one into an advance in which they had been told there was no hope of winning and every chance of being left burdensome cripples for life? I have known many, many college women who have said, 'I *will* have my baby!' Some of them died for the faith that was in them, some of them are happy mothers; a good many are invalids, of higher or lower degree.

Occasionally we find an alumna who, not strong before maternity, is well thereafter. The stunted system develops, and she becomes the woman that she was meant to be. And what beautiful families these women have! I recall two; in one there are four children under six, in the other, five under twelve, and all hearty, beautifully brought up children.

That it is alumna's misfortune, not her fault, that her children are so few I do not expect to *prove*. The testimony for the defendant can not, in the nature of the case, be brought into court. Even were I made an accredited observer, the examples quoted could give no scientific proof, since so small a per cent. of the class has been examined. But naturally they influence my opinion because they are 100 per cent. of the cases which I happen to know all about. But I can not even say 'name and address given on application,' as the patent medicine advertisements promise.

The theory which attributes childlessness to physical weakness is by no means a new one. It has been consciously or unconsciously suggested over and over again by students of vital statistics. Dr. Holmes touches upon it in a medical address given in 1867; the declining birth rate was attracting attention even then. And again and again in discussions of the subject by students who are advancing various theories this element in the problem appears. Among later utterances, Dr. Engelmann said, 'Race decline is not due to education, not of the educated man at least. The educated woman is in a different class.' Professor Thorndike concludes that 'the condition is due to a decrease in fertility in the racial group to which college men *and their wives* belong.' In passing we might quote another sentence of his: "The opinion of metropolitan physicians may here be as wide of the mark as the common belief that unwillingness is the cause of the failure of the women of the better class to nurse their own children."[2]

If you grant me for a time that the cause of the 1.8 — it seems like the judgment of Solomon to speak of tenths of a child! — be physical inability, what is the cause of this inability among the, let us say, schooled American women, with the rate the lowest among those who have been longest in school? What is the cause of the extirpation of that function which one would think would be of all others promoted by natural selection? Is our system of education an element in this result? These questions are surely vital in more senses than one.

Thus far I have been sure of my ground, even if I could not make it clear. Now the way is more obscure, for undoubtedly different influences operate in different classes to undermine the health of our girls. If this weakness of function appears especially among college girls, is then the college course at fault? The birth rate is only a little lower among the alumnae, and we may find that their disability is due to conditions not directly a part of the college course, which each college woman undergoes and only nearly all other women. Observation has almost universally brought the report that the average girl improves in health during her college course.

Is then the responsibility in the high school, where the greater part of our girls do their preparatory work? Very many girls break down here, we know. Frequently a high school teacher attributes a high school boy's inaccuracy in arithmetic or his slovenly English to 'poor preparation in the grammar grades.' This may or may not be just, but I wish some one could find how much of the poor health in high school and college and during later life is due to the way in which our girls go to the grammar school.

'The way in which they go.' There is no especial fault in the content of their education, primary, secondary, collegiate or university. There is no need of making their curriculum feminine, lest womanly instincts be dulled. It is the way of taking the schooling, the physical demands of it, that have been

---

[2]Edward L. Thorndike, "The Decrease in Size of American Families," (May 1903) 63: 65. (Ed.)

responsible for most of the invalids that I have happened to know. Alumna's fate was sealed when she was in the grammar school.

When the bee larvae are about a week old, you remember, it is determined whether they shall become queens or workers. It is simply a question of nourishment; the queen has an abundance of the best food; the worker has a limited supply of inferior quality. The result is a stunting of the reproductive system of the worker bee.

May it not be possible that a similar effect comes in some degree to our women from our school system? The grammar school girl is a larva, if you please, at the age when she should develop a new system of her being, vital both to herself and to her race. To perfect these organs she needs all her rich red blood, all her nervous force. If the brain claims her whole vitality, how can there be any proper development? Just as very young children should give all their strength for some years solely to physical growth before the brain is allowed to make any considerable demands, so at this critical period in the life of the woman nothing should obstruct the right of way of this important system. A year at the least should be made especially easy for her, with neither mental nor nervous strain; and throughout the rest of her school days she should have her periodical day of rest, free from any study or overexertion. Most school girls have many unhygienic habits, all of which tend toward checking her development. Exactly these points were suggested in an editorial note in this magazine some months ago, I remember. The physical conditions and irregularities general among high school girls are appalling, in reference both to their own enjoyment and to the larger interests of the race.

But this is not an argument against our system of education in itself; the matter is not one for school boards to regulate. The intelligent fathers and mothers of our little girls of to-day are the only ones who can remedy these conditions. They can make the girl take one easy year, even though it means 'losing a grade,' that bug-a-boo of school girls; and they can keep for her her needed days of rest throughout her course. Even a year's delay in graduation is not so bad as a dwarfing of development. To hear a school girl speak to the question of her waiting a year, one would judge that existence out of her own particular class would necessarily separate her from all the desirable pupils in school. But those arguments—have you ever noticed?—are never employed when a girl is given a double promotion and advanced a class.

Losing a grade would not often be necessary, however. Ideal conditions would permit a mother to take her daughter to herself for that one year—to teach her the school work and all the other things in which she needs wise and loving instruction especially at this time, welding a companionship which will be the greatest possible barrier against future mistake and sorrow for the young woman in shaping her life.

That it is not always possible for a mother to follow this course I recognize, though it might be arranged much more often than it is if once the mother realized what it would mean. When the mother can not do it, perhaps she can arrange for a little time of private lessons, when her daughter, working at just the rate right for her, can accomplish a term's work with a minimum of study and with none of the nervous strain which comes from competition. I can think of nothing better worth a mother's time than to establish her daughter's health for the rest of her life and make possible for her all the blessed things that womanhood may mean.

Finally, there is no doubt that some husbands and wives limit their families to one or two that they may thus do more for those few children, or have none because they can thus do more for themselves — 'social ambition,' in other words. There may be to some extent a decrease in race fertility in certain racial groups without other signs of physical deterioration; yet there seems to me an amount of evidence too large to disregard which goes to show that the small families among schooled women are due to the physical weakness of the wives. Ask yourself how many really strong women you know. And while there are undoubtedly differing conditions operating in different classes and in different countries, and the contrasts between England and Germany (the birth rate is even lower for the English alumna than for the American), France and Italy, the United States and Canada, can by no means all be explained by this theory, yet I wish some investigation could be instituted to determine how much of the decrease of birth rate among native born American women comes from arrested development in our young girls — due in some classes to lack of proper food, to lack of sleep, to physical overwork, but in very many cases to their unwise manner of work and to untimely nervous strain in our grammar and high schools.

ALUMNA'S CHILDREN AGAIN

ANOTHER ALUMNA

July 1904

AN article in the May number of the POPULAR SCIENCE MONTHLY, entitled 'Alumna's Children,' was recently called to my attention by a woman who, though not a college graduate, is 'decidedly a schooled woman' and the mother of five girls. "I had planned," she said, "to let all my girls go to college, but I *do* want to be a grandmother some time."

For answer I heaped the anxious lady's lap with photographs, photographs of babies, babies large and small, babies masculine and feminine, asleep and awake, clean and dirty, elegantly dressed and not dressed at all, but every one a baby to exult over and all the children of my college friends.

Far be it from me to dispute the Massachusetts vital statistics, and farther yet to dissent from 'Alumna's' conclusion that girls in the 'larva stage' need an intelligent care which too few of them receive. But it is not that admirably sane and practical conclusion, nor yet the irrefutable official statistics forming the author's premise that strikes dismay to the mother of five and produces even in those less directly interested an uncomfortable impression of things being dreadfully wrong somewhere. No, it is that dismal array of tragic incidents drawn from the author's personal knowledge, and her consequent theory of causes for the officially vouched for 1.8. It seems only fair, then, to admit to consideration the personal experience of another alumna, an alumna of slightly later date, who may, from that very fact, be able to bring to the matter a slightly different point of view.

I, too, have known of just such brave struggles against physical odds as 'Alumna' reports. I, too, have known of heartrending defeat and dearly bought victory. But the women who have suffered them are not college women. Possibly my experience in this line with my college friends has been an exceptionally happy one, but in granting that possibility we must grant likewise that 'Alumna's' may have been exceptional as well. It is fair, therefore, to balance one against the other. I do not wonder that the mother of five was troubled by Alumna's article. It is obviously deeply sincere and genuinely thoughtful. But when I had read it I glanced up at the photograph of one of the sweetest, sanest mothers who ever presided wisely over the destinies of children, which shows her sitting on one end of a sea-saw with her baby in her lap, smiling up at four little redheads ranged in ascending scale at the other end of the board. Certainly neither college nor preparation for it has robbed of their dues her ten years of married life.

Naturally at this date I can tell of few such families, for most of the college women I know are younger than this one. It is only a few years since my graduation, and three quarters of the class are still unmarried. But all except a few predestined spinsters are still well on the youthful side of thirty, and the percentage of married members is likely to be considerably raised in the next ten years. And of those who are married not, only in my own class, but, with a single exception, among my other college friends as well, not one has failed to bear a healthy child within two years of marriage.

Naturally it is of my own class that I think first, the class whose average scholarship is the highest on the records of our *alma mater*, the class who did not use each other's christian names in the freshman year and never walked the halls in embracing couples, incurring thereby a reputation for utter lack of sentiment. But the bond that held us was all the stronger for not being flaunted abroad, and the children of the married members are all 'our babies.'

I put down the magazine and thought of our 'class baby,' our first born, with her splendid, sturdy little body and equally sturdy and independent mind. Not only her mother but her grandmother as well is a product — and a notable one — of the higher education, yet our class baby already possesses a baby brother, two years younger than herself and equally a model of physical and mental health.

Then came a picture of another of 'our babies,' 'the adorable,' with his sunny locks, his starry eyes and his gleesome laugh, always on tap. I thought of his mother as I used to see her crossing the campus, with her fine, high-bred face, her superb carriage and the movement that not even modern draperies could disguise, the very lines of buoyant grace which the 'Winged Victory' has made so familiar.

Then my thoughts strayed to our 'new baby,' the little daughter born in the west with two philosophers for parents and fair 'Mistress Wisdom' herself for godmother. The mother writes: 'My nurse says that health like mine is a thing to be conceited over,' and I know she will meet all the problems of wifehood and motherhood with the same serene clearsightedness that earned her college nickname of 'the Philosopher.'

Two others among my pictures I must mention, amateur photographs both, of sleeping babies. T— was about a year old when that was taken— tousled curls crushed on the pillow, lashes sweeping the rounded cheeks, soft pouting lips, bare dimpled arm and sleep-curled fingers—sweet tranquility and health breathing from every line of the relaxed little figure.

The other is the picture of a wee girl, just a week old. Her mother was the best biologist among the undergraduates of her college. This tiny maid and her sister two years older have spent all their long summers, both before and after birth, in a secluded camp on an Adirondack river shore and something of the woodland influence has entered into their being. They are as shy as young partridges—and as near to nature.

Now the question naturally arises whether the difference between my experience and that of 'alumna' is an accident, or whether it can be explained by a consistent theory. One suggests itself to me which may or may not be correct, that the difference is due to the different *kind* of girl who is going to college now. Only a short decade ago women's colleges drew practically all their students from what might be called the abnormally intellectual class. The girl who went to college, whether rich or poor, whether struggling to escape the demands of society life, or to scrape together money enough for her tuition, was the girl who made matters intellectual of paramount importance and was ready to sacrifice for them, from the grammar school up. College was either an outlet for insatiable mental activity or a technical preparation for the teacher. To girls of that sort domestic life was not imperatively attractive, a fact which may have some bearing on the low marriage rate. And when a woman of this type did marry she was too apt to furnish just such a woeful example as those cited in 'Alumna's' article.

It was the author of 'Harvard Stories,' I believe, who aptly classified all students, as 'grinds, sports—and just boys.' The 'just girl' was for a long time in the minority at women's colleges, but happily she is no longer so. On the contrary she is rapidly securing an overwhelming majority. 'Just girl' she is, 'just woman' she will be, and the four years of college life is beginning to assume its proper place in public estimation. To produce, not phenomenal scholars nor well-equipped teachers, but fine, strong, human women — that is the function of these precious four years.

Again I turn to my own class — and as I run down the familiar roll from B to W and glance back over the nine years that have made those names part of my life, I see that somewhere, somehow, among the jumble of 'prescribed' and 'elective' courses, we learned therewith the better things, to see largely, to judge temperately, to choose true values. They look but chilly infinitives, written so, but the class knows how they have wrought into the very fiber of our lives and made us the women that we are. And more and more, as the true function of college life becomes recognized, as popular expectation ceases to demand in justification of a B.A. anything but 'just woman,' the type which couples intellectual attainment with underdeveloped body will disappear. For some years the importance of proper attention to the physical well-being of school children of both sexes has been impressing itself upon the public, and no one will apply scientific principles to the nurture of her children more intelligently and with less danger of capricious 'fads' than the college bred mother. We do 'want more' of alumnae's children, and we are going to get them — an efficient and cumulative force toward those wide and beneficent ends which all true culture stands for.

A. Lapthorn Smith, a physician writing from Montreal, continued a tradition made infamous 30 years earlier by Dr. Edward H. Clarke, by arguing that higher education was harmful for women. While Clarke ostensibly limited his discussion to the biological risks of advanced education, Smith explicitly addressed the social problems resulting from a preponderance of highly educated women. Smith stated bluntly that it was impossible for men to please these women, and conversely, for these women to please men, since women's recognition of their own abilities brought "an aggressive, self-assertive, independent character which renders it impossible to love, honor, and obey the men of their social circles." The second selection is a response by Olivia R. Fernow. Fernow argued that the evils Smith deplored were not due to education, but to luxury and indolence. She supported college education as a constructive way to occupy young women until they were mature enough to marry.

HIGHER EDUCATION OF WOMEN AND RACE SUICIDE

A. LAPTHORN SMITH, M.D.

March 1905

THE author will limit himself principally to a discussion of the harm resulting from too high an education of women, because on that part of the subject he has had exceptional opportunities for observation and for drawing accurate conclusions; but, incidentally, he will take the liberty of questioning the advisability of affording *higher* education freely to the people at large, of the male, as well as of the female sex. . . . He will endeavor to show, as he believes to be the case, that the higher education of women is surely extinguishing her race, both directly by its effects on her organization, and, indirectly, by rendering early marriage impossible for the average man.

First of all, is education being carried on at present to such a degree as to at all affect the bodily or physical health of women? This is a very important question, because the duties of wifehood, and still more of motherhood, do not require an extraordinary development of the brain, but they must absolutely have a strong development of the body. Not only does wifehood and motherhood not require an extraordinary development of the brain, but the latter is a decided barrier against the proper performance of these duties. Any family physician could give innumerable cases out of his experience of failures of marriage, directly due to too great a cultivation of the female intellect, which results in the scorning to perform those duties which are cheerfully performed, and even desired, by the uneducated wife. The duties of motherhood are direct rivals of brain work, for they both require for their performance an exclusive and plentiful supply of phosphates. These are obtained from the food in greater or less quantity, but rarely, if ever, in sufficient quantity to supply an active and highly educated intellect, and, at the same time, the wants of the growing child. The latter before birth must extract from its mother's blood all the chemical salts necessary for the formation of its bony skeleton and for other tissues; and in this rivalry between the offspring and the intellect how often has not the family physician seen the brain lose in the struggle. The mother's reason totters and falls, in some cases to such an extent as to require her removal to an insane asylum; while in others, she only regains her reason after the prolonged administration of phosphates, to make up for the loss entailed by the growth of the child. Sometimes, however, it is the child which suffers, and it is born defectively nourished or rickety, and, owing to the poor quality of the mother's milk, it obtains a precarious existence from artificial foods, which at the best are a poor substitute for nature's nourishment. The highly educated woman seems to know that she will make a poor mother, for she marries rarely and late and, when she does, the number of children is very small. The argument is sometimes used that it is better to have only one child and bring it up with extraordinary care than to have six or eight children brought up with ordinary care because in the latter case the mother's attention is divided. But this is a fallacy. Everybody knows that the one child of the wealthy and highly educated couple is generally a spoiled child and has as a rule, poor health; while the six or eight children of the poor and moderately educated woman are exceedingly strong and lusty. But even supposing that the highly educated woman were able and willing to bear and rear her children like any other woman, she has one drawback from having a fairly large family, and that is the lateness at which she marries, the average being between twenty-six and twenty-seven years. Now, as a woman of that age should marry a man between ten and fifteen years older than herself, for a woman of twenty-seven is as old as a man of forty for the purpose of marriage, both she and her husband are too old to begin the raising of an ordinary sized family. Men and women of that age are old maids and old bachelors. They have been living their own lives during their best years; they have become set in their ways, they must have their own pleasures; in a word, they have become selfish. And, after having had one or at the most two children, the woman objects to having any more, and this is the beginning of the end of marital happiness. The records of our divorce courts show in hundreds of instances, that there was no trouble in the home while the woman was performing her functions of motherhood, but that trouble began as soon as she began to shirk them. Hundreds of thousands of men at the present day are married, but have no wives; and while this sad state of affairs occurs occasionally among the moderately educated, it exists very frequently among the highly educated.

Is the health of the women at the present day worse than it was in the time of our grandmothers? Are the duties of wifehood and motherhood really harder to perform now than they were one hundred years ago? Without hesitation the answer to both questions is 'Yes.' Not only are the sexual and maternal instincts of the average woman becoming less and less from year to year; the best proof of which is later and later marriages and fewer and fewer children; but, in the writer's opinion, the majority of women of the middle and upper classes are sick and suffering before marriage and are physically disabled from performing physiological functions in a natural manner. At a recent meeting of a well-known society of specialists for obstetrics and diseases of women, one of the fellows with the largest practise in the largest city on this continent stated that it was physically impossible for the majority of his patients to have a natural labor, because their power to feel pain was so great, while their muscular power was so little. On these two questions the whole profession is agreed, but I am bound to say that there is a difference of opinion as to the reason. Several of the most distinguished fellows of the above society claim that the generally prevalent breakdown of women is due to their inordinate pursuit of pleasure during the ten years which elapse between their leaving school and their marriage. This includes late hours, turning night into day, insufficient sleep, improper diet, improper clothing and want of exercise. The writer claims that most of the generally admitted poor health of women is due to over education, which first deprives them of sunlight and fresh air for the greater part of their time; second, takes every drop of blood away to the brain from the growing organs of generation; third, develops their nervous system at the expense of all their other systems, muscular, digestive, generative, etc.; fourth, leads them to live an abnormal single life until the age of twenty-six or twenty-seven instead of being married at eighteen, which is the latest that nature meant them to remain single; fifth, raises their requirements so high that they can not marry a young man in good health.

There is another aspect of the question, which is not often discussed, but which has an important bearing upon it. The very essence of cultivation of intellect to its highest point consists in raising the standard of one's requirements. A contented mind makes a man happy. Does a high education make one's mind contented, or does it make it discontented with the present, and ever struggle towards a higher ideal in the future? Is the woman who is versed in art and literature contented with a simple home, or must she be surrounded with objects of art and more or less costly books; and, if so, is she satisfied with her lot when she marries an average man, who is able to provide for her all the necessaries of life, but is not possessed of sufficient wealth to provide those things which would be useless luxuries to a woman of ordinary education, but which are necessities for her? . . .

We all want to be happy, and to that end we all want to be good and, I have already said, we want our children, especially our boys to be good and happy. But those who know anything about virtue in the male know that the marriage of our young men under twenty-five, to a woman with a sound body about eighteen years of age, is almost, if not the only, means of preserving the virtue of the rising generation of men. People, and even mothers, speak lightly of their daughters at twenty-six or twenty-seven marrying men who have sown their wild oats; but one must reap what he sows and do they realize what an awful misfortune such a harvest has brought to the character of the man, and will almost surely bring to the health of the innocent woman? . . .

Just as there are occasionally cases where a divorce becomes necessary, but very much fewer than those actually granted, so, occasionally, the life of an unborn child must be sacrificed to save the life of the mother. But will anybody pretend for a moment that there is any excuse for the two million of child-murders which is a fair estimate of the number occurring annually on the North American continent? The crime has become so general that public opinion has ceased to condemn it, and among the few who do condemn it we certainly do not find those women who claim that wifehood and motherhood are degrading and should be reserved to the lowest class of the population. It is well known that were it not for the enormous immigration pouring into America day by day and week by week, the population of this continent would have died out ere now. And it is generally admitted that the original American people have almost died out. Even the foreigners who are so quickly assimilated soon learn the practise of race-suicide, although never to the appalling extent of the native-born Americans. As far as my experience goes, the crime is most prevalent among the highly educated classes, while it is almost unknown among those with an ordinary education.

Another way in which the higher education is making people unhappy is in the cultivation of the powers of analysis and criticism. When the power of analysis is applied to one's own self it is especially unfortunate, for then it becomes introspection, a faculty which is carried so far with some women that their whole life is spent in looking into themselves, caring nothing for the trials or troubles of those about them. This produces an intense form of egotism and selfishness. These people are exceedingly unhappy, very often suffering from what is wrongly called 'nervous prostration,' but which should rather be called 'nervous prosperity.' When the wonderful power of criticizing is applied to others it takes the form of fault-finding. Such a woman must have many victims; will she make them happy?

One of the greatest objections to the higher education of women, namely, the interference with outdoor exercise, no longer can be raised, because the universities and boarding-schools have within the last ten years foreseen this danger and met it by special courses of instruction in athletics and the encouraging of girls to spend a good deal of time in outdoor sports. But even these universities and schools cannot avoid the charge of fostering a condition of intellectual pride, which is in exact proportion to the success of the school or college. There is no doubt that women can do everything that men can do, and a great deal more; but the knowledge of their ability brings with it an aggressive, self-assertive, independent character, which renders it impossible to love, honor and obey the men of their social circles who are the brothers of their schoolmates, and who in the effort to become rich enough to afford the luxury of a highly educated wife have to begin young at business or in the factory, and for whom it is impossible to ever place themselves on an intellectual equality with the women whom they should marry. These men are, as a rule, refused by the brilliant college graduate, and are either shipwrecked for life and for eternity by remaining single, or are only saved by marrying a woman who is their social inferior, but who, by reason of her contented mind, in the end makes them a much better helpmate than the fault-finding intellectual woman who is looking for an impossible ideal. . . .

Occasionally a college graduate goes through the ordeal of a high education, which has developed her intellect without ruining either her body or her natural instincts; but, as far as the writer can see, she is decidedly an exception. To the average highly intellectual woman the ordinary cares of wifehood and motherhood are exceedingly irksome and distasteful, and the majority of such women unhesitatingly say that they will not marry, unless they can get a man who can afford to keep them in luxury and supply them with their intellectual requirements. The gradual disappearance of the home, which any thoughtful observer must deplore, is, to a large extent, the result of the discontentment of the educated woman with the duties and surroundings of wifehood and motherhood, and the thirst for concerts, theaters, pictures and parties, which keep her in the public gaze, to the loss of her health and the ruin, very often, of her husband's happiness.

Fortunately, nature kills off the woman who shirks motherhood, but, unfortunately, it takes her a generation to do it; and in that short lifetime she is able to make one or many people unhappy.

What about the supply of female school teachers? Is not the very highest education possible necessary for them? From the writer's point of view most of the women who are now teaching school should have been married at eighteen and in a house of their own which might have been the schoolmaster's home. The profession of teaching was once exclusively in the hands of the men, and it can not be denied that they have achieved some great results. But as education rendered an ever-increasing number of women unsuited for marriage, that is, unwilling to marry the available men, they invaded the schoolmaster's rank to such an extent that his salary has been cut down one half, and now he is unable to marry at all. Two well-known consequences have followed this state of affairs; first it is impossible to get men in sufficient numbers to become teachers for the boys' schools; and secondly, even big boys being taught by women, the effeminization of our men is gradually taking place. Although there are some instances of a mother alone having formed her son into a manly man, yet as a rule the boys require the example of a man's character to make them manly men. This subject has recently been dealt with in several elaborate papers by well-known educationalists, to whom it appears to be a real danger to the coming generation.

What about the men? If the higher education prevents the women from being good wives and mothers, will it not prevent the men from being good husbands and fathers? To some extent it does, and in so far it is a misfortune, but to a much less extent than among women, for the simple reason that the man contributes so little towards the new being; while, on the other hand, high intellectual training enables him to win in the struggle for existence much better than if he were possessed of mere brute force. But nature punishes the man who has all the natural instinct cultivated out of him, just as it does the woman, namely, by the extinction of his race. For the struggle for existence among the highly educated men has become so keen, because there are so many of them, that great numbers of them are unable to earn a living even for themselves; while the supporting of a highly educated woman, with her thousand and one requirements, is simply out of the question. A president of a great company recently informed the writer that he had, in one month, applications from eighty-seven university graduates for a position equivalent to that of an office boy at fifteen dollars a month while out of one hundred millionaires, at least ninety-five of them are known not to have been highly educated; but, on the contrary, to have left school between fourteen and sixteen years of age. So there is such a thing as learning too much, without knowing how to do anything. Just as athletes may be overtrained, so men may be overeducated. . . .

DOES HIGHER EDUCATION UNFIT WOMEN FOR MOTHERHOOD?

OLIVIA R. FERNOW

April 1905

IN a very interesting article in the March number of the POPULAR SCIENCE MONTHLY, Dr. Smith discusses a subject which is vital to the well-being of our nation. The evils which he deplores are real and serious, but they are caused, I would submit, not by excess of *education*, as Dr. Smith has it, but by excess of *luxury* and *indolence*. I am inclined to agree with him that too many of our young people have the higher education bestowed on them, but my reason for thinking so is, not that their intellectuality and spirituality are too highly cultivated and that their ideals are consequently too high, but that they are unworthy of the great boon offered them and, not knowing how to use it aright, are injured instead of benefited: they leave college without having become cultured or intellectual.

There can be no greater mistake than to believe that intellectuality promotes extravagance. Are our college professors and men of science the extravagant members of society? Literature, botany, entomology — even music and art — are inexpensive and healthful pursuits as compared with balls, receptions, dinners and the whole round of social functions.

The most serious enemy to American family life is society — in the narrower meaning of the word. Not only among the wealthier classes do its functions absorb enormous amounts of money and time, but among the middle classes also. Let any of our middle class women who dare, sit down with paper and pencil and write on one side the amount of time given to calls, teas, clubs, golf, dinners and the dress necessary for these functions, and in a column opposite the time given to the intellectual and spiritual welfare of her children. Most of them would not like to show, or look at, the balance sheet. Many of them would say, if they were frank, 'Oh, school takes care of my children's intellects and Sunday-school of their religion.' Too true. And here lies one of the great evils of our modern primary education, to which I will just allude in passing, as it is too vast a subject to introduce here. The school is trying to do mothers' work and necessarily failing, but, owing to the time devoted to this failure, the school is prevented from doing the work which it should and could do and used to do. I will just give one example and pass to a more direct consideration of the subject in hand. For twenty years or more our schools have been trying to instill a love of English literature into their pupils. Can any one who knows the results doubt the failure? They are just launching out on what many of us believe will be a similar futile effort in regard to nature study.

A walk in the woods with a mother or father who has an enthusiasm for botany, entomology or mineralogy is worth ten lessons in the class room or even in the woods with a teacher and forty other children. The book or the poem that mother and child read together because they love it and each other, even if they do not know much about the unities or the functions of the various parts, is more likely to stimulate a love of reading than the most exhaustive and exhausting study in the class room.

Now, how are we to produce mothers who will love this work and hug it to themselves as their greatest blessing? Certainly, *only by education and culture*! If the college does not give these, reform but do not eliminate it.

Let us glance for a moment at the alternative opened to girls if they do not go to college. Our college girls come mainly, I think, largely at least, from our upper middle classes. Suppose our girl of seventeen or eighteen has just finished her school life, what is she to do? Dr. Smith says: 'Marry.' But really, it seems to me, she will have to wait until she is asked — and meanwhile she will have to occupy herself in some way. The fact of her entering college does not prevent any young man from asking her hand and many a girl leaves college to be married. The question to be solved is simply of how the time intervening between school and marriage is to be passed. If Dr. Smith will study carefully the girls who do not go to college, but who devote themselves to social functions, I think I may venture to predict that his choice of his future daughter-in-law would be more likely to fall upon an earnest college girl than upon one of the social denizens, and that the chances of her having extravagant tastes, indolent habits or poor constitution are less than they would be in the case of the less cultured girl. I believe, but am not sure, that statistics have shown that divorces are much less common among college women than among those who have not been to college.

That a girl's education should not be merely intellectual I readily admit. Sewing, cooking and housework are as much a part of their preparation for their life work as is shop work for our young engineers. But here again we find that mothers are too busy or too indulgent to undertake the task of teaching their daughters and so in the early years the school tries — again with questionable success — to do it for them, and in the later years as a rule it is not done at all — neither for the collegian nor non-collegian.

It is perfectly natural that a girl who has grown up without having done any manual work should not like it when the necessity for it arises, but I doubt very much whether it can be shown that the intellectual girl dislikes it more than the non-intellectual, and our college girl, if she has benefited by her course as she should, has learned the possibility of applying her powers to an uncongenial task and has many more powers to apply than her less educated sister. Were I a hungry husband, I should have more hope of a palatable dinner prepared by a college-bred wife ignorant of cookery, than from one equally ignorant who lacked the college training.

A word as regards the best age for a woman to marry. It seems to me that the chances for marital happiness are best where a woman marries at about 22, an age at which she can have easily finished her college course and at which she is certainly better qualified to judge of the qualities of the man who asks her to marry him than she was four years before. Suppose she marries a man but a few years older than herself, there is plenty of time for them to have a family of five or six children, which is as many as even an energetic mother can well attend to. Probably her demands in the choice of a husband will be more exacting, not as regards wealth, but as regards mind and soul. Could a better stimulus be found for the improvement of our young men?

As to the physical health of our college girls, I feel sure it will compare favorably with that of non-collegians. Neither class is as well as it should be, the reason being, in my judgment, not excessive intellectual exertion but undue excitement and anxiety. The college girl who is chairman of the dramatic association, or the non-college girl who is chairman of the entertainment committee in church, alike suffer from a mental strain which is much more likely to prove injurious than the steady and peaceful study of Greek, mathematics or anything else. If the college girl insists, as she sometimes does, upon filling the time which should be devoted to study with social amusements and activities and then crams for her examinations at a late hour at night and with a feeling of intense anxiety, she is unquestionably subjecting her health to an undue strain. But this is an *abuse* of the higher education not its *legitimate* pursuit. The girl who steadily pursues her studies as the business of her life, allowing herself reasonable time for exercise and recreation, is most unlikely to be found in the sanitarium among the victims of 'nervous prosperity.' The 'prosperity' shows where these come from — from the ranks of the luxurious and indolent. The best method of preventing egotism, selfishness, and consequently introspection, is to give our boys and girls such intense interests outside of themselves that they have neither time nor inclination for morbid self-study.

I can not go into the question of men's higher education, but should like to say in regard to Dr. Smith's statement about millionaires that I am quite ready to agree that the higher education is not likely to produce them. If a man wants to be a millionaire I think he does wisely to leave school early. But one of the things to be hoped from our higher education is that it will produce in our men nobler ambitions than those of mere money-getting, and in our women the desire for husbands whose aspirations are of a widely different kind.

Our whole notion needs to have the beauty of simplicity impressed upon it. As Emerson says: '*Things* are in the saddle and ride mankind.' What is the remedy? Less culture? No, more! So much that we shall see and taste the higher pleasure that comes from the intellectual and spiritual, and which is always more simple, more wholesome and less expensive than the material enjoyment which it replaces.

Chapter 4
THE PROFESSIONAL HOMEMAKER (1880–1910)

It is the aim . . . to elevate both the honor and the remuneration of all the employments that sustain the many difficult and sacred duties of the family state, and thus to render each department of woman's true profession as much desired and respected as are the most honored professions of men.

Catharine E. Beecher, *Principles of Domestic Science* (New York: J. B. Ford, 1870), p. 13.

In the United States, women have always had primary responsibility for maintaining the home and for the daily care of children, but their activities as homemakers and childrearers have been perceived and evaluated differently at various times. Some historians, among them Viola Klein and Gerda Lerner, argue that before the advent of industrialization, women enjoyed their greatest prestige and recognition as housewives and mothers in the small rural communities where their contribution to the economic welfare of their families was commonly acknowledged. According to this view, women lost social status and economic power in the nineteenth century, when industry took over their productive activities. Other historians, including Daniel Scott Smith, Julie A. Matthaei, and Carl N. Degler, emphasize that women augmented their autonomy and authority within the home in the nineteenth century and ultimately expanded the domestic sphere to include public activities.[1]

Historians who argue that women's status was greatest in the colonial era base their judgment on the range of activities performed by women, the relative frequency with which women did what was then considered "men's work," and the economic role they had within the family. To these historians, the division between the male and female sphere seemed less distinct in the eighteenth century than it would become in the nineteenth century. Thus, the argument goes, because colonial women performed more of the same kinds of work as their husbands and because their work was of equal economic importance, colonial women enjoyed greater social status than their nineteenth-century cohorts. Matthaei dissents, arguing that historians should not assess women's equality within a society on the basis of the content or economic value of their work, but on the work's "social meaning" as determined by the "constellation of social relationships within which the work takes place."[2] On these grounds, Matthaei concludes that colonial women occupied a subordinate position in their society because they lived under the formal authority of men and because their work was determined and controlled by their husbands. Thus, historians disagree not only about the period during which women enjoyed the greatest status, but also about the appropriate criteria by which to determine this status.[3]

Another approach concerns the *perception* of women's homemaking role within a given society, though not the accuracy of that perception. As is true today, Americans in the late nineteenth century disagreed about whether women were better or worse off than women of earlier generations. However, in the 1800s, most people agreed that men and women ought to occupy separate spheres: For men, there was the productive, competitive world of business, government, and law. For women there was the protected, nurturant environment of the family. By the end of the nineteenth century many people perceived a major change in the domestic sphere, for it seemed to them that factories had taken over many of women's "productive" domestic functions, leaving them with consumption-oriented activities. Some people welcomed women's new "consumer" role as an advance, interpreting it as a sign that women had "thrown off the yoke of economic production."[4] Others believed that industrialization had usurped many of women's useful functions, leaving them with little socially valued work to do at home. To the latter it seemed that domestic labor, directed to maintaining a home, was less important economically than men's work. This was the perception of many women who became active in the domestic science movement in the late nineteenth century, a movement that can be viewed as an attempt to restore the social importance of homemaking. In their efforts to elevate women's homemaking activities, domestic scientists stressed the importance to society of women's private role of homemaker and advocated that women devote themselves to managing the home and caring for the family. At the same time, however, domestic scientists began to see the public sphere as an extension of the private sphere—the world as an enlarged home—and began to conceive for women a significant public role. Women were to concern themselves with social reform, alleviating the ills of poverty, and working to reduce crime and intemperance. They envisioned the home as an efficiently run business enterprise. Motherhood took on new meaning, as women had to assume the grave responsibility of overseeing the social, intellectual, and moral development of children in a complex industrial environment. Domestic scientists believed women now required formal education to fulfill the responsibilities of "professional" homemakers and mothers.

The beginnings of the domestic science movement and the professionalization of homemaking can be traced to the late 1830s, when Catharine Beecher (1800–1878), a forerunner of the domestic scientists of the late nineteenth century, urged putting "domestic economy" on a par with other science courses at girls' schools. Although she never married, Beecher was nonetheless convinced that "the deplorable sufferings of the multitudes of young wives and mothers [resulted from] the combined influence of *poor health, poor domestics, and a defective domestic education.*"[5] Realizing that a shortage of servants would continue to plague middle- and upper-class Americans, Beecher argued that these women would have to learn to do their own homemaking and advocated that they be taught to apply "scientific" methods to domestic tasks. Thus, Beecher was among the first to insist that women be given a formal training in domesticity and to demand professional status for homemaking.

Beecher differed from the domestic scientists succeeding her in several important ways. First, she had no formal training in domesticity herself, nor did she have an affiliation to a university. While she firmly believed women ought to learn the principles of domestic science, Beecher also saw the tactical advantages of presenting domestic education as a means of persuading others to support newly emerging girls' seminaries.[6] The domestic scientists of the late 1870s, 1880s, and 1890s no longer needed to employ domestic science as a means of attracting support for women's education, but they did seek to establish themselves in colleges and universities. In an environment where it was difficult for women to obtain positions as general scientists, a group of women academics, including Ellen Swallow Richards (1842–1911), Marion Talbot (1848–1948), Helen Campbell (1839–1918), and Alice Norton (1860–1928) sought to apply science to "woman's domain" and dedicated themselves to making "household management," "scientific cookery," and "sanitary science" legitimate areas of scientific inquiry. By the end of the century, all except Campbell had obtained positions in universities teaching domestic science courses. From 1899 to 1907, Richards organized and presided over annual conferences for domestic scientists at Lake Placid. In 1908, the Lake Placid group officially became the American Home Economics Association and devoted itself to furthering the institutionalization of domestic science, or home economics, as it was by then called, as an academic discipline. The Association tried to persuade universities to offer advanced degrees in home economics and to make more teaching positions available for domestic scientists.[7]

The domestic science movement viewed itself as a corrective force to the social conditions threatening the family. On the one hand, domestic scientists interpreted the rising divorce rate and falling birth rate as evidence of a general deterioration in familial values among the "better classes" of society. They attributed this deterioration in part to misdirected women's education and to the unrealistic expectations deriving from that education. On the other hand, domestic scientists attributed the deplorable living conditions, the abuse of alcohol, and the high infant mortality rates among the poor to the ignorance of lower middle-class women concerning proper housekeeping. For all classes of women then, a scientific training in homemaking was believed to be necessary in order to promote familial values and happiness. Domestic scientists encouraged elite women, who had no housekeeping of their own to do, to study the principles of homemaking and to teach them to others. They advised rural women to attend state colleges to obtain an education that would have direct bearing on their daily lives. They convinced frustrated and bored middle-class women that a scientifically managed home would bring intellectual and personal fulfillment, and they attempted to reform the poor by giving them assistance and instruction in homemaking through settlement houses and charities.

For homemakers to be taken seriously as professionals, domestic scientists argued, women would have to treat the home as a business enterprise, managing it efficiently and economically, as men managed stores, offices, and factories. The precepts of scientific homemaking came from current theories in industrial scientific management.[8] As businessmen attempted to analyze and rationalize factory work, so homemakers were urged to analyze and rationalize theirs. Efficiency and increased productivity, the goals of capitalist production, became the goals of scientific homemaking.

In the interest of efficiency, domestic scientists urged middle-class women to use labor-saving appliances and to replace homemade items with store-bought goods. Domestic scientists went to great lengths to present consumption as an active, "productive" vocation of social value. In 1913, Mrs. Julian Heath, the President of the National Housewives League of New York, declared:

> If the American Woman has failed at any one point it has been to recognize her economic position as spender of the family income. We have demanded that the man be trained to produce, but we have not demanded that the woman be trained to spend, yet it devolves upon her to so spend what the man produces that the family shall be properly fed, clothed, housed, and educated to take their place in the World.[9]

In order to achieve professional status for homemakers, the domestic scientists adapted the goals of business to the home, but they did not foresee that efficiency and productivity would detract from the home's attractiveness as a distinctive social institution where members could take refuge from the monotonous, sterile environment of industrial production. Domestic scientists, criticized for reducing homemaking to impersonal, routinized activity, flatly denied the accusations. In October 1912, Christine Frederick defended the "new housekeeping":

> Some women may say that because I do sewing and other tasks in this apparently formal manner I am reducing them to bare mechanical processes, and robbing them of their beauty and that "home touch" which has been praised for ages. That

is just what I do not do. I put into these things all the inspiration and love and joy I feel as a home-maker and a mother. Because I do them more quickly and scientifically does not rob them of any ideal, esthetic touch which any task must receive in order to be other than factory work.

I don't want to run my home like an office or a factory, and I don't do it. I want it to be a home, and it is.[10]

Ironically, although the domestic science movement aimed to alleviate the boredom and dissatisfaction of middle-class housewives, its endorsement of time-saving devices contributed to the view of housework as boring and uncreative. Moreover if technological improvements made it theoretically possible for wealthier women to spend less time on housework, it appeared that many women did not reduce the number of hours so spent, nor did they fully embrace the consumer role. Instead they continued to perform many tasks using traditional methods, even when other alternatives were available. Time freed by the use of appliances was redirected to housework, not to leisure activities. As Hazel Kyrk explained in *Economic Problems of the Family* (1933), "The invention of the washing-machine has meant more washing, of the vacuum cleaner more cleaning, of new fuels and cooking equipment, more courses and more elaborately prepared food."[11] While increasing industrialization and economic prosperity changed the nature of homemaking, neither decreased the amount of time allocated to housework:

> One can easily explain why the home-maker with children and low income, especially if she lives on a farm, must work long hours. She can hire little or no assistance, she must make rather than buy so long as money-saving is effected thereby, she can buy little of the more expensive equipment. Her house is likely to be poorly planned. If she lives in the city it is likely to be in more grimy districts. What happens if the family income increases materially? They move to a cleaner district, into a house more conveniently and efficiently planned; they buy labor-saving utensils and equipment; they employ hired workers; they patronize dry-cleaner and laundry; they buy goods previously supplied at home. But their house is larger, the furnishings more abundant and elaborate; in time if not immediately their standards rise; time released at one point as in cooking and cleaning goes into care of children, buying and management, and the home-maker's week may not be far below that under the living conditions of earlier years.[12]

This trend has continued into the present century. Studies by Kathryn Walker and Ann Oakley confirm Kyrk's findings. Walker has found that the average amount of time spent weekly on homemaking in urban areas remained high throughout the 1900s, varying from 52 hours in 1926–1927 to 56 hours in 1967–1968.[13] Oakley gives much higher figures for the time spent on housework by urban housewives in the United States, claiming that average weekly hours rose from 51 in 1929 to 78–81 hours in 1945.[14] Obviously, much depends on how researchers define homemaking. Nonetheless, it is only recently that these figures have declined, and the decline has been attributed to the fact that now more than half of the married women in the United States hold full-time paying jobs.

What were the reasons for women's unwillingness to relinquish homemaking tasks throughout much of the twentieth century? The question is too complex to be treated fully here, but several factors should be mentioned. First, homemaking provided women with greater autonomy and authority and was more varied if not more interesting than most forms of paid work available to women at the time. Although the domestic sphere was narrow, women's freedom within that sphere was great, much greater than the freedom granted them outside the domestic domain. (See Chapter 6.) Second, although women's work in the home may have been devalued by society at large, it was of significant, recognizable value to family members. Third, homemaking enabled women to contribute to their families and through them to society. Finally, social pressures were against women leaving the home to work. Wives and others were expected to devote all their time to keeping house for husbands and caring for children.

During the nineteenth century, scientific homemaking was accompanied by a new attitude toward childrearing. Catharine Beecher again was among the first to insist that childcare required "practical science" and that women lacked "training" for motherhood. Previously women's maternal instincts were considered sufficient to make them adequate mothers; by the late nineteenth century formal education was thought necessary. Young women needed to be taught how to feed, clothe, and bathe infants and how to guide the moral and intellectual development of young children. As I mention in Chapter 2, many educational reformers based their pleas for women's education on the grounds that women would be better able to raise intelligent moral citizens. Childrearing eventually ceased to be simply one of many homemaking activities and became the central focus of domestic life. As children were made sacred, motherhood evolved into a holy office. Historian Bernard Wishy elucidates the significance of this new view of childrearing that occurred in the nineteenth century:

> In the new tide of "domestic reform," the American mother was placed under the sternest pressure. She was to give up wealth, frivolity, and fashion to conquer weaknesses and ailments, sloth and insensitivity, and acquire a discipline and knowledge preparing her for a great calling. The reason was simple enough: the mother was the obvious source of everything that would save or damn the child; the historical and spiritual destiny of America lay in her hands. Her own states of mind, body, and soul were of the utmost importance. The "new mother's" place was in the home as the most powerful figure in affecting American society.[15]

The widespread belief in the superior morality of women made it seem logical that women should act as the moral teachers of children. But domestic scientists, while never retracting their claim that women's foremost duty was to remain at home and raise children, also pointed out that women's moral strengths were needed in society. As historian Margaret Gibbons Wilson has observed:

> However, the idea of woman as the protector of society's morals was a two-sided argument used both by those who believed that woman should confine her activities to the home, and also by those who believed that she should enlarge her sphere of activity. The former insisted that to raise one's own children was the most important task, and the one for which women were especially suited. The latter argued that woman's higher morality made it essential that she not restrict herself to the home, but rather that she bring her moral superiority to bear upon finding solutions for society's woes.[16]

Domestic scientists used both sides of the argument. While they advocated that women, particularly working-class women, remain at home and devote themselves to their children, they encouraged single college-educated women to help mitigate the social evils resulting from industrialization and urbanization. Thus, middle- and upper-class women, many of them college-educated, formed settlement houses in urban slums to assist immigrants, ran homes for indigent women, orphans, and the insane, organized "female reform" societies and the "social purity" movement to reduce prostitution, participated in the temperance movement to protect women from the abuses of alcoholic husbands, and created the Women's Trade Union League to unionize female workers. These activities evolved out of small discussion groups and ultimately grew into large national movements. Some historians consider these various movements part of a larger phenomenon called "social feminism" or "social homemaking."[17]

The domestic science movement and the various reform movements mentioned above were predominately middle-class attempts to propagate middle-class values. Historians Barbara Ehrenreich and Deirdre English comment extensively on domestic science as a means of disciplining and "Americanizing" immigrants and the urban poor:

> The useful information—on cooking, shopping, etc.—which attracted neighborhood women necessarily came packaged with the entire ideology of "right living." And right living meant living like the American middle class lived, or aspired to live. It meant thrift, orderliness, and privacy instead of spontaneity and neighborliness. It meant a life centered on the nuclear family, in a home cleanly separated from productive labor (chickens and lodgers would have to go!) ordered with industrial precision—and presided over by a full-time housekeeper.
> Thrift, an obvious virtue, came with a host of assumptions about what represented worthwhile expenditures: soap, yes, but wine, the customary dinner beverage of many European immigrants, was outrageous intemperance. Cleanliness, a necessary virtue in the epidemic-ridden slums, was equated with Americanism itself. . . . Orderliness meant adherence to a family schedule. . . . Even cooking

lessons had a patriotic, middle-class flavor: emphasis was on introducing the poor to "American" foods like baked beans and Indian pudding and weaning them from "foreign" foods like spaghetti.[18]

Many domestic scientists and social reformers believed the poor suffered because they were unworthy or ignorant: thriftlessness, intemperance, laziness, and disorderliness were thought to derive from flaws in moral character or lack of education. They saw crime, disease, and poverty as manifestations of women's failure to meet the domestic needs of their families. Crime testified to the failure to inculcate high standards of morality in husbands and children. Disease signified the failure to maintain clean homes. Poverty implied an inefficiently run household.

Domestic scientists and social reformers rarely attributed the ills accompanying poverty to insufficient salaries or social prejudice, nor imagined that a lack of money, not immorality or ignorance, could be a primary cause of discomfort and disease in lower-class homes. As Laura Clarke Rockwood reasoned, housework was done badly, either because women did not care or because they were unprepared for domestic work. (See "Food Preparation and its Relation to the Development of Efficient Personality in the Home," pp. 181–191.) And because domestic scientists, believing in the superior moral nature of women, could not conceive of so many women as uncaring or malevolent, the only reasonable explanation seemed to them to be that lower-class women were ignorant—a failing easily remedied by education. Thus, domestic scientists and moral reformers who knew of the great hardships facing the working and immigrant classes overemphasized the importance of education in improving conditions in the home. As they were well educated and well off, they did not realize the extent to which it was their wealth, and not their education, that afforded them their domestic comforts.

If education was of limited use to working-class homemakers, it was of much greater use to middle-class women. The demand for educated homemakers became a way of justifying higher education for women; the professional homemaker needed to study chemistry, anatomy, physiology, hygiene, literature, and moral philosophy. It was difficult to ask to study these subjects for their own sake, impossible to study them for the sake of a "male career," but it was possible to demand them for the sake of the home. To a very large extent, the domestic science movement was a movement of college women, who not having access to other careers, made their allocated occupation into a profession.

After the turn of the twentieth century, the new science rapidly gained public recognition. By 1916, 20% of the public high schools offered courses in domestic science, or home economics as it was more commonly called by this time. In 1905, there were only approximately 200 students studying in home economics departments in coeducational universities; by 1910, there were almost 2,000 students, and by 1916 there were 18,000, most of them preparing to become teachers.[19] The Lake Placid conferences had expanded from a small private group into a professional organization.

The domestic scientists took a conservative view of women's role in the family and society. They believed women could contribute most to their society through their traditional functions as homemakers and mothers. They attempted to elevate the status of homemaking by introducing elements from business, while reasserting the importance of those qualities and values—piety, purity, and domesticity—befitting the female sphere. Sharing their century's faith in science and education as agents of social progress, domestic scientists continued in the Beecher tradition, advocating a "scientific" basis for homemaking and insisting upon formal training for housewives and mothers. Living in a society that looked to the trained physician to cure sickness, to the trained lawyer to uphold justice, and to the trained minister to safeguard morality, domestic scientists emphasized the opportunity for trained homemakers to act as physicians, lawyers, and ministers within the home. Although women had traditionally nursed the sick, arbitrated familial disputes, and set standards for piety and goodness, domestic scientists sought a new "social meaning" for women's domestic work.

Domesticity—motherhood and homemaking—were to be women's profession, a lifelong vocation that required the same type of preparation and commitment as "male" professions. Hence, domestic scientists saw no reason to struggle for access to medicine, law, business, government, or the clergy. There was enough for women to do within their own sphere, particularly as this sphere was expanding to include public activities.

In the process of reinforcing private homemaking, however, domestic scientists stepped outside the role of private homemaker to assume a public role, becoming teachers, professors, social workers, and reform organizers, and they encouraged other single women to do the same. Thus, the domestic science movement, which upheld homemaking as the only true profession for women, ultimately created other acceptable vocations for women, and in this fashion helped to expand society's conception of women's sphere.

NOTES

1. See Viola Klein, *The Feminine Character: History of an Ideology* (Urbana: University of Illinois Press, 1975; 1st ed., 1946), p. 10; Gerda Lerner, *The Majority Finds Its Past: Placing Women in History* (New York: Oxford University Press, 1979), p. 131; Gerda Lerner, ed., *The Female Experience: An American Documentary* (Indianapolis: Bobbs-Merrill, 1977), pp. xxix–xxii; Daniel Scott Smith, "Family Limitation, Sexual Control and Domestic Feminism in Victorian America," *A Heritage of Her Own: Toward a New Social History of American Women*, eds. Nancy F. Cott and Elizabeth H. Pleck (New York: Simon and Schuster, 1979), pp. 222–245; Julie A. Matthaei, *An Economic History of Women in America: Women's Work, The Sexual Division of Labor, and the Development of Capitalism* (New York: Schocken Books, 1982), pp. 28–29; Carl N. Degler, *At Odds: Women and the Family in America from the Revolution to the Present* (New York: Oxford University Press, 1980), pp. 26–29; 189–192.

2. Matthaei, *An Economic History of Women in America*, p. 29.
3. The status of the middle-class woman in American society is an immensely problematic concept. Most scholars of women's history attempt to determine this status through an analysis of women's position in the family and of the relationship between the family and society. However, once a distinction is made between women's role in the family (or private sphere) and their role in activities outside the family (the public sphere), the question of women's overall status is complicated still further. Klein and Lerner tend to view the doctrine of separate spheres as a successful, invidious attempt to prevent women from attaining power in the higher-status public sphere. Degler, Smith, and Matthaei argue that nineteenth-century women augmented their authority within the domestic sphere, gained greater autonomy within the family, and eventually enlarged their private homemaking activities to include public activities.
4. Bertha June Richardson, *The Woman Who Spends: A Study of Her Economic Function* (Boston: Whitcomb & Barrows, 1910; 1st ed., 1904), p. 37.
5. Catharine E. Beecher, *A Treatise on Domestic Economy for Young Ladies* (New York: Harper & Brothers, 1845; 1st ed., 1841), p. 5.
6. Beecher made the following comment on her reasons for pushing so hard for domestic education: "Inasmuch, then, as popular education was the topic which was every day rising in interest and importance, it seemed to me that, to fall into this current, and organize our sex, as women, to secure the proper education of the destitute children of our land was the better form of presenting the object, rather than to start it as an effort for the elevation of woman. By this method, many embarrassments would be escaped, and many advantages secured." Catharine E. Beecher, *True Remedy for the Wrongs of Women* (Boston, 1851). Cited in Thomas Woody, *A History of Women's Education in the United States* (New York: Science Press, 1929), 1: 465.
7. See *Lake Placid Conferences On Home Economics: Proceedings of the Annual Conference* (Lake Placid, 1901).
8. See Frederick Winslow Taylor, *The Principles of Scientific Management* (New York: Harper & Row, 1911).
9. Mrs. Julian Heath, "The Work of the Housewives League," *Annals of the American Academy of Political and Social Sciences* 48 (July 1913): 121. Cited in Matthaei, *An Economic History of Women*, p. 166.
10. Christine Frederick, "The New House-Keeping," *The Ladies Home Journal* (October 1912): 100. See Matthaei, *An Economic History of Women*, p. 162.
11. Hazel Kyrk, *Economic Problems of the Family* (New York: Harper & Row, 1933), p. 99. Cited in Matthaei, *An Economic History of Women*, p. 167.
12. Kyrk, *Economic Problems of the Family*, p. 100. Cited in Matthaei, *An Economic History of Women*, pp. 167–168.
13. Kathryn Walker, "Home-Making Still Takes Time," *Journal of Home Economics* 61 (October 1969): 8. Cited in Matthaei, *An Economic History of Women*, p. 168.
14. Ann Oakley, *Woman's Work: The Housewife, Past and Present* (New York: Vintage Books, 1976), p. 7.
15. Bernard Wishy, *The Child and the Republic: The Dawn of Modern American Child Nurture* (Philadelphia: University of Pennsylvania Press, 1968), p. 28. See also Sheila Rothman's discussion of the same phenomenon, which she calls the ideology of educated motherhood in *Woman's Proper Place: A History of Changing Ideals and Practices, 1870 to the Present* (New York: Basic Books, 1978), pp. 97–132.
16. Margaret Gibbons Wilson, *The American Woman in Transition: The Urban Influence 1870–1920* (Westport: Greenwood Press, 1979), p. 19.
17. Matthaei uses the term "social homemaking" in *An Economic History of Women*, pp. 173–186. William L. O'Neill uses the term "social feminism" in *Everyone Was Brave: The Rise and Fall of Feminism in America* (Chicago: Quadrangle Books, 1969), pp. 77–106.
18. Barbara Ehrenreich and Deirdre English, *For Her Own Good: 150 Years of the Experts' Ad-*

vice to Women (New York: Doubleday, 1978), p. 156.
19. T. Woody, A History of Women's Education in the United States, 2: 61.

ARTICLES ON HOMEMAKING APPEARING IN *THE POPULAR SCIENCE MONTHLY*

(Arranged by Date of Publication)

Editor. "Cookery and Education," (September 1878) 13: 620–621.
Editor. "Cookery and Education," (October 1878) 13: 748–749.
Editor. "Improved Domestic Economy," (January 1879) 14: 395–396.
"Woman as a Sanitary Reformer," (July 1881) 19: 427–428.
Oliver E. Lyman. "The Legal Status of Servant-Girls," (April 1883) 22: 803–815.
"On Science Teaching in the Public Schools," (June 1883) 23: 207–214.
W. Matthieu Williams. "The Chemistry of Cooking," (June 1883) 23: 214–223. This was the first of many articles of the same title written by Williams and published in The Popular Science Monthly from June 1883 to July 1885.
Editor. "Progress and the Home," (July 1883): 23: 412–416.
H. Brooke Davies. "A Kitchen College," (November 1887) 32: 96–100.
Edward Atkinson. "The Art of Cooking," (November 1889) 36: 1–19.
Editor. "Science in Domestic Economy," (November 1889) 36: 123–124.
Editor. "Everyday Science," (December 1889) 36: 264–265.
Alice B. Tweedy. "Homely Gymnastics," (February 1892) 40: 524–527.
M. A. Boland. "Scientific Cooking. A Plea for Education in Household Affairs," (September 1893) 43: 653–662.
Frederick A. Fernald. "Household Arts at the World's Fair," (October 1893) 43: 803–812.
Louise E. Hogan. "Milk for Babes," (August 1894) 45: 491–495.
Vivian Lewes. "The Chemistry of Cleaning," (November 1894) 46: 101–111.
Clare de Graffenried. "The 'New Woman' and Her Debts," (September 1896) 49: 664–672.
Kate Kingsley Ide. "The Primary Social Settlement," (February 1898) 52: 534–547.
Mary Roberts Smith. "Education for Domestic Life," (August 1898) 53: 521–525.
Charlotte Smith Angstman. "College Women and the New Science," (September 1898) 53: 674–690.
Laura Clarke Rockwood. "Food Preparation and its Relation to the Development of Efficient Personality in the Home," (September 1911) 79: 277–298.

SUGGESTIONS FOR FURTHER READING

Primary Sources

Beecher, Catharine E. *A Treatise on Domestic Economy for the Use of Young Ladies At Home, and At School*. New York: Harper & Brothers, 1845; 1st ed., 1841.
Bevier, Isabel, and Usher, Susannah. *The Home Economics Movement*. Boston: Whitcomb & Barrows, 1906.
Calhoun, Arthur W. *The Social History of the American Family from Colonial Times to the Present*. 3 vols. Cleveland: Arthur H. Clark, 1919.
Dealey, J. *The Family in its Sociological Aspects*. Boston: Houghton Mifflin, 1912.
Gilman, Charlotte Perkins. *Concerning Children*. Boston: Small, Maynard, 1901.
Gilman, Charlotte Perkins. *The Home: Its Work and Influence*. Urbana: University of Illinois, 1972; 1st ed., 1903.
Gilman, Charlotte Perkins. *Women and Economics*. New York: Harper & Row, 1966; 1st ed., 1898.

Goodsell, Willystine. *A History of the Family as a Social and Educational Institution.* New York: Macmillan, 1915.
Lake Placid Conference on Home Economics. Proceedings of Annual Conferences. Lake Placid, 1901.
Richards, Ellen. *Euthenics: The Science of Controllable Environment.* Boston: Whitcomb & Barrows, 1910.
Salmon, Lucy. *Domestic Service.* New York: Macmillan, 1897.
Salmon, Lucy. *Progress in the Household.* Boston and New York: Houghton Mifflin, 1906.
Sanitary Science Club of the Association of Collegiate Alumnae. *Home Sanitation: A Manual for Housekeepers.* Boston: Ticknor, 1877.
Taylor, Frederick Winslow. *The Principles of Scientific Management.* New York: Harper & Brothers, 1911.
Veblen, Thorstein. *The Theory of the Leisure Class.* New York: Viking, 1899.

Secondary Sources

Ideals of Womanhood and Motherhood

Rothman, Sheila M. *Woman's Proper Place: A History of Changing Ideals and Practices, 1870 to the Present.* New York: Basic Books, 1978, pp. 97–132.
Welter, Barbara. "The Cult of True Womanhood: 1820–1860," *American Quarterly* 18 (Summer 1966): 151–174. Reprinted in Gordon, Michael, ed. *The American Family in Social-Historical Perspective.* New York: St. Martin's Press, 1973, pp. 224–250.
Wishy, Bernard. *The Child and the Republic: The Dawn of Modern American Child Nurture.* Philadelphia: University of Pennsylvania Press, 1968.
Also see readings in Chapter 7.

Social History of the Family

Aries, Philippe. *Centuries of Childhood: A Social History of Family Life.* Robert Baldick, translator. New York: Vintage Books, 1962.
Gordon, Michael, ed. *The American Family in Social-Historical Perspective.* New York: St. Martin's Press, 1973.
Lasch, Christopher. *Haven in a Heartless World: The Family Besieged.* New York: Basic Books, 1977.

The Domestic Science and Home Economics Movement

Branegan, Gladys Alee. *Home Economics Teacher Training Under the Smith Hughes Act 1917 to 1927.* New York: Teacher's College Press, 1929, pp. 7–31.
Hunt, Caroline. *The Life of Ellen H. Richards, 1842–1911.* Washington, D.C.: American Home Economics Association, 1980; 1st ed., 1912.
Lopate, Carol. "Ironies of the Home Economics Movement," *Edcentric: A Journal of Educational Change* 31–32 (November 1974): 40–57.
Lubove, Roy. *The Professional Altruist: The Emergence of Social Work as a Career 1880–1930.* Cambridge: Harvard University Press, 1965.
Matthaei, Julie A. *An Economic History of Women in America: Women's Work, The Sexual Division of Labor, and the Development of Capitalism.* New York: Schocken Books, 1982, pp. 101–186.
Sklar, Kathryn Kish. *Catharine Beecher: A Study in American Domesticity.* New York: W. W. Norton, 1976.
Weigley, Emma. "It Might Have Been Euthenics: The Lake Placid Conferences and the Home Economics Movement," *American Quarterly* 26 (March 1974): 79–96.

From 1877 to 1887 when the two Youmans brothers, Edward Livingston (1821–1887) and William Jay (1838–1901), coedited The Popular Science Monthly, *six editorials appeared arguing for improved domestic education for women. The Youmans strongly believed that science applied to the home could significantly improve the domestic lives of all Americans and considered it women's responsibility to promote the new home sciences.*

IMPROVED DOMESTIC ECONOMY

EDITOR'S TABLE

January 1879

THE Princess Louisa received an address from a deputation of the Ladies' Educational Association of Montreal, and they got her from a very sensible reply. But what will the operators of our fashionable girls' schools and of our female colleges and normal schools say to the closing observation of this royal and court-bred lady, which was as follows: "May I venture to suggest the importance of giving special attention to the subject of domestic economy, which properly lies at the root of the highest life of every true woman?" This is a momentous truth; none the more true, of course, because uttered by a princess, but perhaps some will be induced to reflect upon it on account of the distinguished source from which it comes. For if what the princess here says is correct, our schools for the education of women are very far from what they should be. Domestic economy, in its full significance as a foundation of the highest life of woman, happens to be just the one particular thing which our female boarding-schools, colleges, and normal schools systematically avoid. They learn languages, and history, and algebra, and music, and many other fashionable things, but the science of domestic life and the art of home-making find no place in the feminine scheme of studies. Here and there a little attention is paid to it, but it nowhere has the rank and importance which is rightfully its due, and which this most sensible princess claims for it.

The term "domestic economy" has been hitherto used in so narrow and misleading a sense that there is considerable prejudice in regard to it. Its common implication is a mere improved mechanical housekeeping, or domestic drudgery made methodical, with a chief view to economy in home expenditures. The term has participated in the vulgarity that attaches to the menial and servile associations of the kitchen, so that little books upon domestic economy are thought the proper things to put into the hands of cooks and hireling housekeepers. But domestic economy, as something "which properly lies at the root of the highest life of every true woman," is a very different thing, implying culture and intelligence in the whole circle of home duties and responsibilities, and the consequent renovation and elevation of the domestic sphere. This view happily begins to be more clearly and widely appreciated. We have just read with great interest a lecture on scientific housekeeping delivered by Mrs. Arthur Bate before the Popular Science Society at Milwaukee College, which explains in an admirable way how many new subjects and questions there are upon which women require to be trained in order to make them competent and skillful administrators of home affairs. Mrs. Bate shows conclusively that science has exactly the same office to perform in guiding domestic art that it has had to perform in giving efficiency to all the other arts, and that it will confer the same interest and dignity upon household affairs that it has already conferred upon other departments of activity. She well observes that in gaining the knowledge necessary to make the home a sanitarium—the house of health—educated housekeepers would do more to emancipate the world from fleshly ills than doctors have ever done or ever can do.

Mrs. Bate makes an important point in showing that the ignorance of women is a fatal hindrance to the introduction of many improvements by which domestic operations could be greatly facilitated, if only housekeepers knew enough to make them available. An illustration of this is just now at hand. The use of gas-stoves for cooking is one of the most important ameliorations that have been conferred upon the kitchen in a long time; but, as that realm is given over to tradition and blind habit, but little advantage has been taken of the improvement. Gas-stock holders are losing their sleep for fear Edison is going to destroy their business, but, if the benefits to be gained by the consumption of gas in cooking were generally understood, there would be but little occasion to fear from a diminished consumption of the article. A lady trained in the South Kensington Cooking-School, and who has taught in the Culinary College of Edinburgh, has recently come to this country and given a course of demonstrative lessons in cookery in New York. Her mode of working has been a sort of new revelation to the large class of ladies which has attended her instructions. Cooking has hitherto been associated with dingy kitchens and fiery ranges that evoked the free perspiration of the attendant; but Miss Dods uses a gas-stove, and does her work so neatly that it might be carried on in a parlor. In the dozen lessons she gave, scores of dishes of all kinds were prepared rapidly by the use of gas, and that they were well made was sufficiently evinced by the eagerness of the ladies to purchase them at the close of each lecture. To a curious inquirer she said, that in her practical demonstrations she had cooked by gas alone for years in preparing hundreds of dishes of a great variety in teaching. This is but one example, of which many might be cited, showing how people suffer in their domestic life because women are not properly instructed in the principles of practical household art, and in the resources that might be commanded for its improvement.

Mary Roberts Smith Coolidge (1860–1945), daughter of I. P. Roberts, received a bachelor's (1880) and master's degree (1882) from Cornell University. From 1886 to 1890 she was an instructor of economics at Wellesley, leaving to study sociology at Stanford University, whose president at the time was David Starr Jordan, also a Cornell graduate and active proponent of coeducation. (See "The Higher Education of Women," pp. 96–104.) While she was working toward a doctorate, Mary Roberts Smith (she had married by this time) became assistant professor of sociology (1894). She received her doctorate in 1896 and continued to teach at Stanford until 1903. In 1906 she married Dane Coolidge and then worked as an assistant for the Carnegie Institute in Washington, D.C. (1906–1909) and as a professor of sociology at Mills College, California (1918 until retirement). She published several influential studies on women, including Almshouse Women (1895); "Statistics of College and Non-College Women" (see Chapter 3) and Why Women Are So (1912). In the following selection, Smith argued for the need for college courses in domestic science.

EDUCATION FOR DOMESTIC LIFE

MARY ROBERTS SMITH

August 1898

THE need of some kind of education as a basis for every activity is constantly emphasized to-day; but this emphasis is rarely applied to the need of training for domestic life, for which it is usually supposed that any kind of preparation will do. One million six hundred thousand women in the United States are engaged in domestic service, and eleven million one hundred thousand more are married and presumably have some kind of domestic duties. Several writers have called attention recently to the fact that a woman does not necessarily have an instinct for home-making; that while her instinct for the care of children may be strong, yet she may lack the skill to make a fire properly or to mix the ingredients of wholesome food. Or she may be skilled in handling modern kitchen appliances, but may lack the knowledge of the effect of exercise, regular hours, wholesome food and clothing adapted to climate, upon the future health and mental development of her children. It seems to be only just now dawning on women that domesticity—that is, the care of the household and children—is in itself a profession for which the best training and the fullest development attainable are not too much.

The education of women has tended to develop along the same lines as that of men. The classical education for the gentleman has changed to the general education for the average man, and to the specialized education for the industries as well as for the professions. A similar change is taking place in the education of women, but has reached only the second stage. Those who first insisted upon the value of a higher education for women thought it sufficient that they should have the same opportunities as men. This experiment has been tried now for a generation, and it is found that all women do not need the same kind of training as men any more than all men need a purely classical or a purely scientific education. In other words, individualism is breaking up all the accepted lines of education for women as it has for men. As a result, differentiation of courses within the higher training is demanded to meet the practical needs of a life in which no two individuals can possibly do precisely the same things. The fact that one third of all women in the United States are married sets them aside as needing a peculiar training for their profession.

In domestic life women need at least two things: first, the greatest general culture attainable to enrich the home life and to retain the sympathies of children, as well as to store up for themselves resources in hours of difficulty, loneliness, or sorrow; second, they need an education adapted to the everyday business, especially to the emergencies, of domestic life. No education is complete nor, indeed, of great permanent value that does not teach how to live contentedly and to economize nerve energy. To be contented, one must feel sure that one is in the right place, and must have spiritual and intellectual resources to tide over life's emergencies whose end one can not see. To be economical of nerve energy, one must learn a finely balanced self-control and a large-minded discrimination between the values of competing duties and attractions.

It is a significant fact that of one hundred and eighty-four living children of two hundred and twenty-eight almshouse women, less than one third are self-supporting. One fourth are lost — that is, they have been separated from the mother in one way or another, and she no longer knows where they are. The women themselves give all sorts of plausible reasons why their children do not support them; but the fact is, as the stories show, that nineteen women were cast off by their relatives or children because of their drunken, vicious, or filthy habits, and nearly as many because their children were ashamed of them; five have quarreled with daughters or grandchildren. These facts show that the home life was defective in those characteristics which tend to bind the family together. In the Elmira Reformatory seven per cent had had a good home, thirty-nine per cent fair, fifty-four per cent poor, showing the preponderance of bad home conditions. In conducting a student employment bureau it was found that there was an undue supply of those who wished to do bookkeeping, typewriting, and clerical work, while there was great difficulty in securing any one who could do mending, plain sewing, or ordinary housework satisfactorily. There was a large amount of this domestic work to be performed in the community, but the young women who were obliged to earn a part of their living in college were quite incapable of doing what was needed. Charity and settlement workers continually testify that the women of the laboring classes lack proper training and skill in making home comfortable and wholesome. Without additional illustration, it appears that women are being prepared for everything else than domestic life—the life which, as statistics show, nearly one half of them are living.

What, then, does the average woman need? In the first place, a thorough manual training. She needs to know how to cook a wholesome meal properly, to put it on the table appetizingly, and to do this with the minimum expenditure of energy. It is one of the most hopeful signs in elementary education that kitchen gardening and household training are being introduced into those schools which the children of the general population attend. The need of this practical domestic training for girls has probably been sufficiently emphasized, but in the general readjustment of occupations and duties going on between men and women, it is more and more apparent that boys as well as girls need a certain amount of elementary domestic training. It is a mere fetich, for instance, that women should do all the mending or even have all the care of children. There are many families in which family happiness, comfort, and prosperity would be greatly promoted if the husband and father could, at least in an emergency, take a competent share in the routine work of the household. There are many generous and kindly husbands who would be glad to help, but who are incapable through lack of elementary training. Since the bearing and rearing of children is the most important function of women, the mother must be relieved, at least at times, from many of her ordinary household cares. If there be not money enough to hire extra service, it is inevitable that the father should take, at least temporarily, some of these duties, if the family is to be maintained in comfort.

Again, the average mother needs a thorough grounding in elementary physiology and hygiene. Five per cent of all children born in the United States die under five years of age. When this occurs, the waste of human energy both before and after birth is something appalling. Prof. A. G. Warner estimates that it costs about a hundred dollars in loss of labor on the part of the mother, in doctor's bills, medicine, and nursing to bring a child into the world, in a laboring-class family; while in families where a higher standard of living prevails this may amount to hundreds or even thousands of dollars. From a purely economic standpoint it is of the utmost importance to society that a child which costs so much, not merely in money but in vital energy, should be reared to maturity. The appalling mortality of children that are born fairly normal and vital is chiefly to be accounted for by the ignorance of mothers. The average woman may not need to know how many bones there are in the body, but she does need to know the connection between rich gravies, indigestion, and bad colds. She may not need to know how to bandage a broken arm, but she does need to realize the effect of sudden changes of temperature upon the delicate infant organism. The value of applied physiology in preserving infant life and diminishing hereditary and individual disease can not be overestimated; and no woman is fit to be married who has not had a training which gives her the elements of this essential knowledge.

Finally, women need a training in ethical standards. One of the curious anomalies disclosed by the entrance of women into industrial life is that while they have higher standards of purity than men, they frequently have much lower standards of honor and honesty. They do not hesitate to outwit, deceive, and "manage" difficult husbands; they train children in dishonesty by continually violating the most common standards of sincerity and directness. Children learn far more by example than by precept: the mother who continually promises, but always finds excuses for not performing; who threatens, but does not punish; who suppresses the child's frank comments on evil actions in others, while herself gossiping about her neighbors; who pretends to dress and to live above the scale of the family income, gives an education in dishonesty and sham which can not be overcome by any amount of so-called moral training.

If to all these practical and utilitarian attainments the mother can add the graces of culture in music or art or literature, she may give the child a background for education and a resource in life beyond the power of statistics to estimate. The elevation, enrichment, and sweetening of the family life by these contributions from the mother's own storehouse of culture are a safeguard against temptation from without not to be matched by legislation or training, nor even by church influence. To make the household sweet, wholesome, dignified, a place of growth, is certainly a profession requiring not merely the best training, but a specific training adapted to these ends.

The following selection is taken from a detailed and sympathetic review of college women's activities in settlement houses and kitchens during the 1880s. The author, Charlotte Smith Angstman, remains an unknown figure, but her treatment of her subject suggests that she was part of the movement she described: one of the many, now anonymous, college alumnae who worked to assist the urban poor.

COLLEGE WOMEN AND THE NEW SCIENCE

CHARLOTTE SMITH ANGSTMAN

September 1898

IT is only after many years of earnest work on the part of comparatively few that it is beginning to be understood that domestic science is something definite, reducible to forms, capable of being studied comprehensively, and worthy of a place beside the other sciences in the curriculum of important universities and colleges.

Women have gone to college and heard lectures on physiology in an atmosphere of eighty-five degrees, heavy with carbonic-acid gas, and then passed to others where the thermometer read sixty-five degrees and the chill air from without blew upon their heads, wondering that such things could be side by side with perfect theoretical instruction.

They have gone from new knowledge of bacteria to a certainty of the existence of unwholesome germs in the improperly-cared-for furnishings of their student apartments.

They have learned the composition of the blood, bone, and muscle of human beings, and what substances contain their chemical elements, and then have asked what better use could be made of this knowledge than in securing diets which should perfectly nourish.

They have studied political economy and sociology, and have returned to reflect and observe that their principles are applicable to the social and domestic problems which are now before their eyes.

In the study of mathematics they have learned that nothing wrong can be righted without going to its root, and so have naturally turned their minds to the causes of the complications in domestic machinery which are apparent on every hand.

The study of history has made them realize that any plan for improvement in any condition of things, in order to rest upon a sure foundation, must be based upon a knowledge of the past. . . .

With the thirty thousand girls who have already graduated from colleges, according to Alice Freeman Palmer, carrying these reasonings into innumerable towns and hamlets, the outcome must be something definite, and it is no source of surprise to find that some of them have gradually collected the present knowledge on all topics relating to the welfare of the home, under the comprehensive title of household economics or domestic science, and that great numbers of them are working hard in various lines of this subject.

Let us examine this work of some of our college graduates who have done most in this direction.

The active interest of college women in the subject of household economics was shown as long ago as 1883, when the Boston branch of the Collegiate Alumnae organized its Sanitary Science Club, the first organization of distinctively college women for the study of any branch of household economics. The report at the end of its first year's work says: "The members of the Sanitary Science Club can not too strongly urge upon the Association of Collegiate Alumnae the importance of giving thought and attention to the hygiene of the home. This duty falls more or less upon all women, but with none should it be more exacting than with college graduates. . . . "

After five years' study, a manual for housekeepers, called *Home Sanitation*, was prepared by this club and edited by two of its members, Mrs. Ellen H. Richards and Miss Marion Talbot. This manual has been for some time one of the standard works upon the subject, and used as a basis for study in home science clubs.

One of these editors, Miss Marion Talbot, who has the degrees of A.B. and A.M. from Boston University, and of S.B. from the Massachusetts Institute of Technology, after having first realized the importance of the subject in this club, began lecturing regularly upon domestic science in 1886 at La Salle Seminary, and continued till 1890, when she took charge of this department at Wellesley College. In 1892 she was called to the University of Chicago as dean of the woman's department, where she is now carrying on courses in sanitation and the study of foods.

The first interest of the other editor of *Home Sanitation*, Mrs. Ellen H. Richards, in domestic science dated from a much earlier period. Having graduated at Vassar in 1870, she went to the Massachusetts Institute of Technology to work in the chemical laboratory preparatory to taking the degree of S.B., which she received from that institution, as well as the degree of M.A. from Vassar, in 1873. While working in the chemical laboratory in 1871, a prominent educator, now deceased, made the sneering remark to her, "What good do you expect all this will do you in the kitchen?" "As if," as she says, "I was necessarily to spend my life in the kitchen, or as if there was no chemistry to be used in the kitchen!"

Even sneers have their value, since, as we shall see in this case, they are often the spurs to great achievements.

Shortly after the culmination of the work of the Sanitary Science Club in *Home Sanitation*, in the fall of 1889, Mrs. Mary Hinman Abel, a graduate of Elmira College, returned from a six-years' residence in different European cities with the idea that something might be done toward the better nourishment of the working people, such as she had seen in Germany and Austria in the *Volksküche*, and in the *Fourneau Économique* in France.

During her husband's prolonged absence in Europe, she went for six months to stay with Mrs. Ellen H. Richards, now professor of sanitary chemistry in the Massachusetts Institute of Technology, who had become especially interested in her through being one of the judges in the matter of the prize of five hundred dollars offered by Mr. Lomb, of Rochester, N.Y., for the best essay on practical sanitary and economic cooking. Mrs. Abel won this prize, and her little volume bearing this title is considered the simplest and still the most scientific presentation of the subject yet made. The fruit of this six-months' companionship was the now famous New England Kitchen, started under Mrs. Abel's direct charge. . . .

By January 24, 1890, the kitchen was opened for the sale of food, it offering at that time beef broth, vegetable soup, pea soup, corn mush, boiled hominy, oatmeal mush, cracked wheat, and spiced beef. Since then other preparations have been added.

Every dish offered by the kitchen is what would be known as a standard dish as compared with one suited only to occasional times. In order to be able to offer standard dishes, the requirements which they must meet were several. "First, the cost of material must not go beyond a certain limit. Second, the labor of making it must not be too great. Third, it must be really nutritious and healthful. Fourth, it must be in a form that it could be easily served, and kept hot without loss of flavor. Fifth, it must suit the popular taste enough to be salable."

One of the first things accomplished by the New England Kitchen was the making of beef broth from the cheaper cuts of meat which was unvarying in nutriment and flavor. To this end the broth was frequently analyzed under the supervision of Mrs. Richards at the Massachusetts Institute of Technology. . . .

Other kitchens have already been started upon the plan of the New England Kitchen—Chicago, Philadelphia, and New York each having similar ones.

It is not too much to expect to see in a few years such kitchens in every large city in our Union, all the outcome of the practical application of the scientific training of two college women to the betterment of, primarily, the physical condition of their fellow-creatures, and, secondarily, their mental and moral condition.

In October, 1890, while yet busy with the development of scientific principles in connection with the New England Kitchen, Mrs. Richards wrote a forceful paper urging upon college women the study of domestic science. In this paper, which was published for the Association of Collegiate Alumnæ, she says: "What is our education worth to us if we can not order our houses in peace and comfort? You say, 'Modern life makes so many demands upon us.' True; but no demand can supersede that of home.... Let each young college graduate begin her housekeeping in a simple way, feeling keenly that all her future happiness and the welfare of her family depend on the thoroughness with which she masters at the very beginning the essentials of a home.

"But not only in her own home is there a call for this knowledge of the fundamental principles of healthful living and domestic economy. In all work for the amelioration of the condition of mankind, philanthropic and practical, there must be a basis of knowledge of the laws and forces which science has discovered and harnessed for our use."

In alluding to the New England Kitchen she says: "In this experiment the training of the college woman showed. No mere enthusiasm would have patiently waited, understanding that success is reached only through failure and after a most careful study of every detail, and is maintained only by constant vigilance."

Urging the study of domestic science in colleges, she says: "First, the subject should be put in the college curriculum on a par with the other sciences, and as a summing up of all the science teaching of the course, for chemistry, physics, physiology, biology, and especially bacteriology, are all only the stepping-stones of sanitary science.

"Therefore, in the junior or senior year, after the student has a good groundwork of these sciences, there should be given a course of at least two lectures a week, and four hours of practical work.

"The lectures should treat of—

"1. The house and its foundations and surroundings from a sanitary as well as an architectural standpoint.
"2. The mechanical apparatus of the house, heating, lighting, ventilation, drainage, etc., including methods of testing their efficiency.
"3. Furnishing and general care of a house, including what might be called applied physiology, chemistry of food and nutrition, and the chemistry of cleaning.
"4. Food and clothing of a family.
"5. Relation of domestic service to the general question of labor, with a discussion of present conditions and proposed reforms.

"The practical work should include:—

"1. Visits of inspection, accompanied by the instructor, to houses in process of construction, of good and bad types, both old and new.
"2. Visits to homes where the housekeeper has put in practice some or all of the theories of modern sanitary and economic living.
"3. Conferences with successful and progressive housekeepers.

"4. Practical work and original investigation in the laboratory of sanitary chemistry."

This was the outline originally prepared by Miss Talbot, and describes the course as she gave it at Wellesley in 1891 and 1892, entering upon her work there at about this time.

To show the respect which Mrs. Richards's attainments as a scholar and scientist have won with the world, as well as giving added significance to the fact of her doing so much in the field of domestic science, *The Outlook* for September, 1897, is quoted: "Her contributions to science have placed her at the head of the domestic science department of one of the leading educational institutions of the country, and have established her as an authority in her own field, a woman whose advice, investigations, and decisions are accepted by the leading scientists and authorities."

Mrs. Richards and Miss Talbot have made themselves felt in connection with this work in still another direction. In the University of Chicago, domestic science is not only taught but practiced. When Mrs. Alice Freeman Palmer and Miss Marion Talbot consented to go there as deans of the woman's department, it was with the understanding that they should have an opportunity to carry out their convictions that college trustees and professors have done their whole duty by their students only when they see that they are properly fed as well as properly taught. Accordingly, when the three halls for the accommodation of the women students were completed, they undertook, with Mrs. Ellen H. Richards as expert adviser, to furnish a dietary which should be kept within a certain cost, be of the best quality, prepared in the best manner, and at the same time furnish the known scientific requirements of proper nutrition. . . . Every day's *menu* was planned with direct reference to supplying the chemical requirements in their proper proportions, at the same time meeting the other stated requirements. . . . The most of this work has been and is now continuing under the direction of Dean Marion Talbot.

The other dean of the woman's department of Chicago University, Alice Freeman Palmer, a graduate of Michigan University, and later the honored president of Wellesley College, afterward, as one of its trustees, was chiefly instrumental in having the new science introduced there. Later she has been lending her strength to this subject as a member of the Massachusetts State Board of Education; as president of the Woman's Education Association of Boston, which last year had an important exhibit of domestic art and science, and which now has a strong committee on domestic science; as a member of the committee on domestic service investigation of the Boston branch of the Association of Collegiate Alumnae; in introducing something of this work into the vacation schools in Cambridge. She is also identified with the movement in Boston to introduce domestic science into public and private schools and colleges, and which last fall established a school of domestic service to attack the problem in another way. That band of twelve college women who organized the Sanitary Science Club in November, 1883, builded [sic] better than they knew. . . .

That there is a strong demand for courses in which the study of chemistry shall be applied to food, economics of the household, and its kindred subjects, is evinced by the number of colleges where these subjects are now taught. This age is awakening to the fact that women need special opportunities as *women*; and after the first blind rush for equal opportunities with men for higher education, it is demanding courses of instruction which shall include full credit-earning courses in that combination of sciences which is woman's own.

Important coeducational institutions besides Chicago University give instruction now in domestic science, while others are considering the matter. Wisconsin State University has already been mentioned [as having benefitted from the services of Helen Campbell, a brilliant teacher in household economics]. The Leland Stanford, Jr., University has lately done admirable work under the able direction of Mrs. Mary Roberts Smith, a graduate of Cornell, and for some years professor of history at Wellesley college.* These, with the Boston Institute of Technology and Ohio State University, are a few which have already been teaching the subject, while inquiries are continually coming from many more, as well as from large seminaries.

To-day a study of cooking is required of every girl in the Boston common schools, as well as in other schools in Massachusetts. In the Providence Manual Training High School, Miss Abbie L. Marlatt, a graduate of the Kansas Agricultural College, conducts a most admirable course in domestic science, covering a period of four years. In the Brookline (Massachusetts) public schools, Mrs. Alice P. Norton, a graduate of Smith College, has arranged and conducts a comprehensive course of study in this department, beginning with the sixth grade and continuing through the high school. This course, as arranged and conducted by her, is a good example of what might be done in the public schools of every city, without crowding out anything of importance or overburdening the pupils. In the sixth grade, where it begins, it only occupies one hour per week; in the seventh and eighth grades, two hours per week; and in the ninth, one hour per week. . . . In Brookline, when the pupil reaches the high school, she has already been instructed in many more things concerning the house and the preparation of food, with the reason *why*, than the majority of young ladies know when they enter upon the life occupation of mistress of a home. . . .

The *American Kitchen Magazine* shows how college women are giving of their best to put before the public scientific and practical knowledge upon all matters pertaining to the home. Home is the magnet to which their thoughts and efforts are continually drawn. . . .

The idea of duty and obligation to give to others less fortunate something

---

*See "Education for Domestic Life," pp. 170-173. (Ed.)

from the riches of opportunity and training enjoyed by college women so impressed itself upon the mind of a graduate of Smith College, Miss Vida D. Scudder, that she succeeded in imbuing the minds of six other graduates of that institution with her own conviction. Her plan was to establish a home in the midst of a densely populated, ignorant, and wicked district, from which they could reach the homes of their neighbors and add something of pleasure and knowledge of their dull lives full of ignorance and vice. . . .

Upon maturing their plans, they moved into quarters at 95 Rivington Street in September, 1889, a locality, according to Frances J. Dyer, "said to be more densely populated than any part of London. One half of all arrests for gambling and one tenth of all arrests for crime in New York come within the limits of the election precinct in which they (the residents) live. Five churches vainly try to meet the spiritual needs of fifty thousand people, and there is one saloon for every hundred inhabitants. These facts sufficiently indicate the character of the neighborhood in which these young collegiates, representing the highest type of American womanhood, elect to spend a portion of their time. . . . "

The undergraduates of Smith, Swarthmore, the Woman's College of Baltimore, and Bryn Mawr have assisted the residents of the Philadelphia Settlement in many ways; while Barnard, Elmira, and Packer students consider the New York Settlement as their care, and Wellesley and Radcliffe girls are very helpful at the Boston Settlement. . . .

Thus, by addressing popular audiences, by writing magazine articles and books, by demonstration lectures upon the science and art of cookery, by teaching the subject in high schools, grammar schools, and colleges, by the establishment of depots for the sale of scientifically prepared as well as savory food, by practically demonstrating her knowledge in different ways in the homes of her poor neighbors in connection with college settlements, by working upon practical problems connected with domestic science in strong committees connected with education associations and branches of the Association of Collegiate Alumnae, and in many other ways do we find the college woman working in the field of domestic science, reaching thousands of homes and home makers. All her intellectual training, which it has been feared might divert her energies from home duties which by nature and opportunity she is especially fitted to discharge, has but made her the more eager to discharge them, but with a new and different interest, along better lines thought out as a natural consequence of her new opportunities.

. . . The college woman is giving us methods as a permanent basis for results, and with that largeness of vision and special understanding born of her special opportunities, yet true to her woman's instinct, which nothing can eradicate, has seen what might be bettered, and is bettering it in that place which is most potent for all that is good or evil in life—the home.

Laura Clarke Rockwood (1870–1922) received a bachelor's (1892) and master's (1896) degree from the University of Iowa. In 1894, she married Elbert W. Rockwood (1860–1935), a chemistry professor at the university, and lived in Iowa City throughout her life. Upon her death, the Iowa City Press-Citizen called Mrs. Rockwood a "splendid representative of the best in woman citizenry; and an ideal wife and mother in her greatest domain, the home." Indeed, Laura Clarke epitomized the ideal middle class social homemaker of the late nineteenth century: she was active in community affairs, the Congregational Church and numerous women's clubs. Among the many unpaid service positions she held were the vice presidency of the Iowa City Library Board; the chairmanship of the Home Service Committee of the Red Cross; directorship of the Iowa City Child Welfare station; county chairmanship of the Woman's Bureau of the United War Work Campaign; managing directorship of the Iowa Memorial Union and the chairmanship of the economics committee of the Iowa State Federation of Women's Clubs.

FOOD PREPARATION AND ITS RELATION TO THE DEVELOPMENT OF EFFICIENT PERSONALITY IN THE HOME

LAURA CLARKE ROCKWOOD

September 1911

. . . .

THE home is a human institution, and as such is capable of change and improvement. Other human institutions have come and gone, but the home of to-day in many particulars remains essentially the same as it was when our cave-dwelling ancestors left the woman at home to guard the fire, care for the children and, incidentally, to prepare the food, while they roamed abroad in search of game and new experiences. Those who insist that a woman's place is at home by divine decree need only to study the life of primitive man to find out how very human are some of our domestic customs, for they will then see this distinction, that while nature has specialized woman for child-bearing, it is society which has specialized her for housework. To be sure, a custom which has continued as long as this one of woman remaining by the fireside must have as its foundation something which is fundamentally right for social development. Yet so many myths have grown up about this custom that we are in danger at times of mistaking these additions for the good of the original idea. Because of these erroneous ideas current, popular opinion is inclined immediately to conclude, if for economic reasons a woman strays from the fireside, that she and her husband are incompatible, or that he is unwilling to support her and that in some way she is neglecting what is popularly supposed to be her divinely appointed mission. Of course such a conclusion is often unjust and not warranted by the facts.

In discussing the subject of the preparation of food, we accept conditions as they are; that woman is specialized by society as the housekeeper and food purveyor to the various groups called families.

In spite of the fact that many of the home industries have developed into world-wide industries there still remains for the housemother much of the food preparation for her family. The necessary meals for one's family are the essential fact which confront each housewife anew each day. No matter what other home activity can be neglected, that one duty of preparing food can not be omitted. How often we hear women make some such remark as this, "I was too sick to do anything else but get barely enough for the family to eat." This shows that every other home duty can in extremity be left undone except that one. Food preparation then is the fundamental duty of the housewife to-day. "A good stomach kept in a healthy condition is the foundation of all true greatness," says Dr. Tyler, professor of biology at Amherst College.

The housemother's first duty then after bearing her family is to provide them with food for their growing needs which shall give them the best endurance for life's conflict. But her responsibility is much greater than this, for closely connected with the necessity for food are the other hygienic necessities for survival — air, water, sunshine, shelter, rest, all in due season; and depending upon these vital necessities are the opportunities for the development of the mental, moral and social personality or the completely social man. Because woman is specialized for home work, these wider responsibilities for individual growth have, in a large measure, become hers also. . . .

Although woman is specialized for this important business of home work upon the outcome of which depends first the vital, then the mental, moral and social, welfare of mankind, she receives in but few cases any preparation for her important task. She takes the methods of housework which are traditional in her environment and makes use of them according to her individual training and aptitude. The exceptional woman may use all the data of the scientists in the administration of her home. What she is able to do ought not to be made the measure of what the average woman may be expected to do. The average woman goes blindly on in her specialized work, laboring hard at a task for which she has no aid but home traditions and the columns of the home magazines.

To be sure the home magazines are contributing much toward better conditions of home living, for many women would not know that scientists are at work continually on problems of home betterment were not some of the results of their investigations made available through the medium of the home magazines.

The bulletins of the Department of Agriculture, the farmer's institutes throughout the country, the home economics movement, and many writers — Mrs. Richards, Miss Barrows, Miss Bevier, Mrs. Rorer, Miss Kinne, Mrs. Hill, Miss Farmer, Miss Hunt, Mrs. Lincoln and others — are doing much by a crusade of enlightenment to improve conditions of foods, their preparation and uses in our homes. That is, they are trying to make available for women the knowledge of the scientists, which either is not available for the average woman or else is beyond her mental reach.

For instance, two noteworthy books on the subject of nutrition by Professor Chittenden, of Yale University, have been published in recent years.[1] They are noteworthy because by scientific tests upon different classes of people they show the exact needs of the body for protein, the material for building up and replacing tissue. Consequently some very definite conclusions can be drawn concerning food customs which should be known to every woman and prevail in every household. The essential parts of the books are so plainly and entertainingly written that a woman with at least a high school education could understand them and profit by them. Yet inquiry reveals very few housekeepers who have read them. The reason for this ignorance can be attributed to the fact that as yet woman in her evolution has not reached the state of mind when she expects to enjoy or understand anything scientific, so she does not seek out that class of reading. Meanwhile, nevertheless, she is in charge of the vital development of the race. A strange inconsistency! Thus we see that in most cases she is not fitted for her work and hence can have no sense of satisfaction in it.

As previously stated the purpose of this paper is to show the connection between woman's work in the home in preparing and serving food for her family and the harmonious development of personality; and also to indicate that as conditions now prevail in society she is unable to make use of the accumulated knowledge of scientists as to what is best for human welfare.

. . . [W]e need to keep in mind the image of the home as an intermediate agent between the scientific knowledge of what is best for human development, on the one hand, and on the other the finished product of personality as we find it in society. Since the home stands as the connecting link between these two — the knowledge and the result, it must of necessity help or hinder the harmonious development of the social personality or the individual in his relations to his fellow men. Hence anything which will improve home conditions is of great importance to society in its attempts at individual and general progress. . . .

Statistics tell us that under the present home system one fourth of all deaths for the United States during the year 1908 were of children under five years of age.[2] The infant mortality of England was higher for the three years 1896-1900 than for 1861-65.[3] Of the total deaths in Iowa in August, 1910, about

---

[1]Chittenden, "Physiological Economy of Nutrition," Chittenden, "The Nutrition of Man." The F. A. Stokes Company.
[2]"Mortality Statistics," 1908, p. 8, Special Census Report, Department of Commerce and Labor.
[3]*Ibid.*, p. 9.

one fifth were under one year of age and of these over 80 per cent. were from cholera infantum, a disease largely preventable through hygienic measures.[4]

It is stated of all diseases of infancy between the ages of 2 and 6 sixty-seven per cent. may be prevented on the basis of our present knowledge of sanitary measures, were they widely used.[5]

Statistics tell us that under the present home system, the prevalence of disease greatly impairs efficiency.

In the United States, 500,000 people are constantly ill from tuberculosis alone, which is in a large measure a preventable disease in the homes where it is understood and the necessary diet and food are provided.[6]

The minor ailments, such as colds and sore throats, cause even well men to lose at least five days a year from their work.[7] Yet investigation and research are showing that these minor ailments can be controlled in a large measure by diet.[8] Professor Fisher says that he knows scores of cases in which the tendency to take cold has been almost completely overcome by diet. Such being the case it will be necessary for those who prepare and serve the food to be cognizant of what kind of food the body requires for its highest efficiency, and since as society is now organized, the preparation of food is woman's work, it must be possible for women to know in some way the latest results of scientific research in foods and general hygiene in order to prevent disease. . . .

This "unpreparedness" of women in our homes is not confined alone to a knowledge of foods. Because woman is specialized by society for housework, and the care of children, the general responsibility for the health of her family is hers also. But so often through ignorance she is not equal to this responsibility. For instance, statistics and the reports of the various state charitable institutions show that a large per cent. of blindness among children is due to diseased conditions which might be remedied by intelligent care at birth,[9] or might have been prevented years before by a proper knowledge of sex hygiene.

Intermarrying is also given by experts as a cause of physical degeneracy such as deaf mutism. Both of these social errors might be diminished by greater knowledge on the part of the guardians of the homes of the vital necessities of the race in the matter of sex and reproduction. The economic necessity which presses so hard to-day upon the man as bread-winner of the family lays all the more responsibility upon the woman in the home not only as in the

---

[4]Rockwood, POPULAR SCIENCE MONTHLY, March 1911.
[5]Irving Fisher, "Bulletin of the Committee of One Hundred on National Health."
[6]*Ibid.*, p. 3.
[7]*Ibid.*, p. 39.
[8]*Ibid.*, p. 40.
[9]See report for 1908 Illinois Institution for the Blind.

first essential of food, but in all hygienic matters which pertain to vital efficiency.

Statistics tell us that with the present home system divorce is increasing much faster than the population. Divorce was about three times greater in 1905 than in 1870.[10]

The special census report of the Department of Commerce and Labor shows that the most important ground for divorce is desertion and of the divorces granted to the husband nearly one half had desertions as their cause. That is, one half of the husbands who sued for divorce had for a cause the desertion of the wives. It would look from this fact as though women are growing weary of home conditions.

We have then the fact that with our present system of homes, one fourth of all deaths are of children under five—those who are entirely dependent on the home whose diseases in 67 percent. of the cases could be prevented by proper diet and care. We have the fact that 500,000 people in the United States are ill of tuberculosis and that such prevalent diseases can, in many cases, be cured by diet and fresh air. We have the fact that as estimated even well people lose at least five days a year from colds and minor ailments which might have been largely prevented by a proper diet. We have the fact that one half of the women who are divorced by their husbands desert home voluntarily. We have the fact that many charity workers give as their testimony that much social misery is caused by the "unpreparedness" of the homemakers. We have statistics to show the great waste of infant life in mansion as well as in humbler home. We have the statement of Professor Devine, the well-known charity expert, that in many homes we find the beginnings of tendencies which often lead to crime and disabling disease. We have the statements that innutritious food is a prolific cause of intemperance, which of itself leads to crime. We have the facts that much blindness and physical degeneracy might be prevented by a proper knowledge made available to the masses through the housemother in the home.

In the face of all these facts it would certainly appear that woman, who is the guardian of the home, is either ignorant of the proper consumption of wealth in the home in serving human welfare or else she is remiss in her duty. It is safe to say that not all women would be consciously negligent or remiss in their life's work, so the natural conclusion is that woman for the most part is ignorant of many of the essentials of the great mission assigned her. This ignorance is not strange, since in our educational system she receives slight preparation for this her real life's work. As has been said before, all the aid she has are home traditions and the home magazine, and in the ma-

[10]"Marriage and Divorce," p. 11, Special Census Report, Department of Commerce and Labor.

jority of instances she is ignorant of the noteworthy investigations along her own line of work. . . .

1. THE NEED OF EDUCATED HOMEMAKERS

Let us bear in mind the fact stated before that the home and its caretakers stand as a connecting link between the knowledge of what is best for the individual and the finished product of personality as we find it in society. Or, in other words, the function of the home and of those who minister there is the adjustment of the individual to society through the utilities of the home which are both economic and cultural. Certainly such important work should require some special training. While primarily woman's work is that of dietitian in the home, she must not specialize in this capacity at the sacrifice of her effectiveness as a teacher or the cultural si[d]e of home life will suffer. Neither must her training as a dietician and a teacher exclude the training necessary as a financier and as an employer of labor in a broad sense, for the home is in direct touch with the labor problem on all sides. The liveliest imagination and inventive genius she will need also to develop to enable her to meet with discrimination and equanimity the daily complex problems of home life.

It is to be hoped that as yet woman's education is in a transitional stage. The last half century has been given up to proving that women can learn the same lessons as men if they wish to do so. It is very desirable that the next half century may mark a much greater triumph in woman's education by making plain and popular the fact that although she can learn the same lessons as a man she, as a woman, has more important ones to learn of an entirely different nature, bearing on her profession of home making.

No man without special training is likely to be employed as consulting engineer on so important an enterprise as the Panama Canal, yet women the country over are intrusted with the vital, the mental, the moral and social welfare of the individuals who make up the state, without any preparation whatever beyond an inheritance of tradition and such additional information as can be gained from home magazines or such other literature as their minds are able to grasp.

There is a sentiment current in respect to woman's higher education that if she is given culture studies, ability will in some way come to her for her life's work. Culture studies are good, but they are only part of the preparation needed by her. She needs for her life's work much more preparation than is ordinarily included under the subject of domestic science in order that she may be competent to develop the personalities of her family in every way. A suitable course for her in order to be of wide value should be a bringing together of all the work of her college days in its bearing upon individual adjustment to society through the medium of the home. The studies of literature, art, music, psychology, pedagogy, child study, emergency nursing, chemistry, physiology, biology, bacteriology, botany, sociology, *in their bearing upon home life*, will all be necessary for this ideally equipped homemaker. She needs, in other words, to be taught how to control her environment by making use of the knowledge which is available for race progress.

It is not going too far for the state through its educational institutions to require that each woman graduate who goes out from their walls shall be thus equipped for the work which is of greatest value to the state through the home. For instance, the state of Iowa now requires a course in pedagogy extending through much of two years for those who wish to have a state certificate to teach. Iowa might well go much farther and require that each woman who graduates from her university shall be prepared by suitable studies for the position of homemaker which sooner or later she may assume. . . .

Most state agricultural colleges and state normal colleges have schools of domestic science to train teachers. The state universities as the crown of the educational system need training schools for wives and mothers, with all the advantages which higher education can give. This training must be of the very highest grade and no expense should be spared to make and keep it thus, so that every woman who goes forth from their halls shall be a center of light in the broadest way on the subjects of the hygiene of environment, of nutrition and of activity. If these conditions can be made to prevail in all our educational institutions, it is safe to say that the coming century will mark great vital, mental, moral and social advancement of the human race. Statistics tell us that at the present time 74,908[11] women are enrolled in the higher institutions of learning in this country. If each of these 75,000 college women and all who succeed them could go forth from their college life thoroughly prepared for their duties as women a great increase in individual and national efficiency might be expected.

The fact that many women say they "hate house work" does not lessen their responsibility for doing it well since they undertake to do it. A proper education in the fundamentals, the purposes, the methods and the results of home work would no doubt go far to lessen the dislike for that form of labor. . . .

Until by a process of evolution some plan of farther combination or socialization of home industries is worked out, in order to raise the standard of home industries it is necessary to take immediate steps to improve conditions in our homes. A wholesale campaign of education of our girls, beginning with the grade schools and extending through our colleges, is the most hopeful means available for improving our homes in their work of developing and maintaining individual and social welfare through the proper adjustment of the individual to society.

[11] Wm. G. Curtis, "Ages of Universities," *Record Herald*, April 15, 1910.

2. HOME METHODS OF DEVELOPING PERSONALITY

One of the difficulties of our educational system to-day is that educated women seem to feel that when they assume the responsibilities of home work they have been made to surrender to disuse whatever mentality they possess. Indeed, one very capable woman said to the writer soon after her marriage, when she was wrestling with the difficulties of domestic management, "Do you not feel your mind becoming atrophied with all this petty round of duties?" In order to be loyal to our little home nest, I replied, "No indeed, I find that my domestic science takes as much mentality as my political science did." I have been trying ever since to live up to that remark, and I have found that the homemaker by following out her path of duty can have an opportunity for mental, moral and social development in proportion to her desire for growth.

The whole world of science centers around the daily work of preparing food. The housewife who wishes mental expansion in this line can begin by perusing the numerous food bulletins of the Department of Agriculture which are provided free. She can hang her kitchen walls with the food charts furnished by the Department of Agriculture for a consideration and learn while at her work the relative values of foods. This knowledge will prepare her to be interested in the many food experiments of the large experiment stations, and while following them through the publications she can add to them by carefully kept records of similar experiments in her own home, and if she is of a literary turn of mind she will find a ready sale for such articles as she chooses to write on the subject. Personal experience has verified this in the study of the food requirements of growing children.[12]

Her interest in her own growing family can lead her to a study of the development of the family, primitive culture and the development of the home and the different phases of what Baldwin calls "the dialectic of personal growth" which occurs in the socialization of each of her children. While she is increasing her own stock of information in this way she will find that she is becoming a much more interesting companion for her children and is able to show them their real place and share in the world's work. In this way she can exert a much wider influence over them than by merely providing for their physical wants. . . .

Not long ago several hundred club women in one of the eastern states were asked to reply to this question, "Who is the greatest woman in history?" Numerous replies were received and a great many women known to history were named. The prize was given to this answer: "The wife of a man of moderate means who does her own cooking, washing and ironing, brings up a large

[12] E. W. and L. C. Rockwood, *Science*, XXXII., p. 351.

family of boys and girls to be useful members of society, and finds time for her own intellectual and moral improvement — she is the greatest woman in history."

Alarmists tell us sometimes that the home is disintegrating, that with the invasion of some women into industry and the indifference of others in regard to their home responsibilities, the death knell of the home is sounded. However, as long as the home can contribute anything to the development of personality and race progress, it will remain. With advancing education and civilization, no doubt, several of the functions which we now consider necessary will pass from the home, but our homes as the sanctuary of family life are not in danger of disintegration. While there are father-love and mother-love and dependent childhood there will be homes where the physical, the mental, the moral and the social personality can be developed. . . .

3. A PREVENTION FOR DIVORCE

The subject of divorce is one which much concerns the sociologists and theologians to-day because of its demoralizing influence on the development of efficient personality in the family.

Statistics tell us that during the period from 1900–1905, while the population increased 8.7 per cent., divorces increased 22.1 per cent.[13]

It is not the purpose of this paper to enter upon a discussion of divorce except in so far as it is affected by woman's specialized industry in the home of food preparation and resulting necessities. . . . For one thing a woman who must work sixteen hours a day at unspecialized industry with the attendant fatigue, is unable to compete in charm oftentimes with the leisure parasitic class whose lives are devoted to pleasing men. Such overworked women are too tired to be interested in men's affairs or themselves interesting. Oftentimes because of this lack of leisure the discordant note is struck which later grows into utter lack of harmony.

Sometimes too the duties of married life are so taxing in the early years when the children are small that women, because of their excess of physical work, begin to feel a mental deterioration, and this consciousness of a lack of growth or of cumulative happiness often is the pathway leading to the divorce court. On the other hand, if a woman by means of any previous training is enabled to keep in touch with the mental life of her family as well as the physical life, she has in her work of motherhood found the one thing in life worth while, and in her work then she can feel a sense of satisfaction in her own growth and activity or "cumulative happiness," for she has found her share of the world's work. . . .

[13] "Marriage and Divorce," Special Census Report, Department of Commerce and Labor, 1909.

We know, however, that divorce frequently comes in families where women are really idlers in the economic field, who have no responsibility beyond a good time and to be supported by their husbands. In these cases idleness, discontent, desertion and divorce are the result. A socialization of industry would be a good thing for this type of women by compelling them to have some definite share in the social service, for we all know that there is no greater cure for the blues and discontent than rational activity. By the socialization of domestic industry, if it could be brought about, the monotony of isolated home labor would be removed. . . .

A good broadening course in our schools and colleges with a proper presentation of the duties of adult life would do much to lessen divorce, because after all the home is just what men and women make of it as a public utility in the development of efficient personality. Such a course would serve also to establish a tradition in favor of the homemaker and prevent in some degree the rush of women into outside industries which to many now appear attractive.

One of the difficulties with our educational system to-day from the kindergarten upward is that it seeks to make hard things easy, from the learning of the multiplication table to easy helps for Latin and kindred subjects. The only really satisfactory way of mastering the multiplication table is by definitely learning it. All through our educational system this spirit of helping children to avoid the hard things which require persistence and application is shown in our home life when neither men nor women are able to endure the hardships and unpleasant factors which do come up in every home at some time.

Instead of facing these difficult problems and bringing to bear upon them a rational mentality which will restore order from chaos and strengthen the bond of helpfulness between husband and wife, the husband and wife brought up by our educational system to look for easy things, drift from not knowing how to assume responsibility in the home to avoiding it altogether by means of the divorce court.

There are no statistics to prove the statement, yet careful observation in a good many cases has shown that a cheerful common sense and ability to turn defeat into victory through perseverance would have kept many homes intact to-day.

And so, our educational system seems wrong when it permits our boys and girls to grow up looking only for easy places. Too many girls look forward to matrimony as a life of surcease from the disagreeable surroundings before marriage.

In consequence when the water pipes burst and the furnace grate falls out and the refrigerator springs a leak and the baby is teething and fretful and the meals must be prepared and the husband, owing to a belated breakfast, has not had time to be as affectionate as usual in his farewells as he ran for his car — when such a combination as this happens, as we housekeepers all know it can do — unless we are trained to listen for the eternal harmonies behind some of the discords of life, we are apt to grow discontented in home life because of our own inability to make a success of it and to bring to ourselves "cumulative happiness."

However, if we do have a sufficiently high ideal of our mission as homemakers and the spirit and necessary training to inform our task, we can set to work on our domestic problem with a cheerful courage, for we know that ice can be thawed and leaks mended, furnace grates repaired, cross, fretful babies can become the joy and light of a whole household, and belated husbands if only given a chance can more than atone for their seeming indifference. . . .

The natural conclusion from this fact is that our educational system should provide some way of showing every girl that she must expect serious conditions in dealing with the serious problems of life and that she must have some training for her fundamental task of developing vital personality with its resultant mental, moral and social responsibilities. Otherwise our whole industrial system must change so that domestic industries can become socialized and women do their share, specializing for home work according to inclination. But in either case for human evolution we must have trained guardians of the personality, whether they be natural mothers or selected ones. . . .

Chapter 5
THE SUFFRAGE DEBATE (1890–1900)

> *"Government is in fact the government* of men by men. *It is men who do things, and, among other things, they are the most frequent lawbreakers. It takes men to govern men, and what governs the greater force will control the lesser. It is not necessary to cut two holes in the gate, the one for the large, the other for the small cat. The small cat can go through the large hole."*
>
> E. D. Cope, The Popular Science Monthly 34 (February 1889): 558–559; emphasis in original.

In the United States, the struggle for woman suffrage formed part of a broader social movement demanding equal rights for women. This struggle began formally in 1848 at the Seneca Falls Convention, in New York, organized by Elizabeth Cady Stanton and Lucretia Mott.[1] For the convention, Stanton wrote a statement of purpose, which she modeled after the Declaration of Independence. Her "Declaration of Sentiment" asserted women's inalienable right to "life, liberty, and the pursuit of happiness," reiterated the definition of a just government as one that governed with the "consent of the governed," and listed the injustices then facing women:

> The history of mankind is a history of repeated injuries and usurpations on the part of man toward woman, having in direct object the establishment of an absolute tyranny over her. To prove this, let facts be submitted to a candid world.
>
> He has never permitted her to exercise her inalienable right to the elective franchise. . . .
>
> He has made her, if married, in the eye of the law civilly dead.
>
> He has taken from her [the married woman] all right in property even to the wages she earns. In the convent of marriage, she is compelled to promise obedience to her husband, he becoming, to all intents and purposes, her master—the law giving him power to deprive her of her liberty, and to administer chastisement.
>
> He has so framed the laws of divorce, as to what shall be the proper causes,

and in case of separation, to whom the guardianship of the children shall be given, as to be wholly regardless of the happiness of women. . . .

He has monopolized nearly all the profitable employments, and from those she is permitted to follow, she receives but a scanty remuneration. . . .

He has denied her the facilities for obtaining a thorough education, all colleges being closed against her. . . .

He has created a false public sentiment by giving to the world a different code of morals for men and women, by which moral delinquencies which exclude women from society, are not only tolerated, but deemed of little account in man. . . .

Now, in view of this entire disfranchisement of one-half the people of this country, [and of] their social and religious degradation . . . we insist that they [women] have immediate admission to the rights and privileges which belong to them as citizens of the United States.[2]

From its inception, the woman's rights movement included female suffrage as one of its goals, but the denial of the franchise was only one of many injustices with which women had to contend. They could not attend most colleges. They could not hold public office. Social injunctions prevented them from practicing most professions. Upon marrying, women legally relinqished to husbands the right to control their property and earnings. Grounds for divorce were strict and generally limited to desertion and lack of support.

Many devoted suffragists* argued that suffrage was the key issue; once women were enfranchised, they would be able to change other unfair laws, thus modifying their inferior legal status and correcting existing political, social, and economic ills. Consequently, the right to vote was not just an "individual right," although it was that too. It was the right of collective political action and thus represented a direct challenge to the social practice of separate spheres. In demanding the right to vote, suffragists were insisting that women be given a voice in public affairs, which was until then an exclusively male domain. Moreover, the franchise had immense symbolic importance. By withholding suffrage, the society that claimed to revere womanhood, which it equated with motherhood, demonstrated its actual opinion of women: They were men's inferior and hence denied equal citizenship, classified with lunatics and criminals, who also were deprived of the franchise.

In the first section of this chapter, I review the major political events in the history of woman suffrage. In the second section, I examine the arguments

---

*I use the terms *suffragist* and *antisuffragist* in their most general sense to indicate a person who does (or does not) support female suffrage. One must be careful not to equate "suffragist" with "female suffragist," nor "antisuffragist" with "male antisuffragist." The suffragist movement had dedicated male leaders and supporters, while some of the most virulent opponents of suffrage were women. In this respect, the suffrage debates resembled debates on the proposed Equal Rights Amendment today.

made by suffragists and antisuffragists in view of the assumptions they contained concerning women's nature and equality.

OVERVIEW OF THE SUFFRAGE MOVEMENT, 1848-1920

During the early years of the woman's rights movement, most people viewed suffrage as the most radical element in women's protest against oppression. Female suffrage leaders, Elizabeth Cady Stanton (1815-1903), Susan B. Anthony (1820-1906), Lucretia Mott (1793-1880), Lucy Stone (1818-1893), Sarah Grimke (1792-1873), and Angelina Grimke (1805-1879), experienced severe social censure and occasional physical harm as they took the unprecedented and disreputable step of addressing mixed audiences on women's and slaves' rights. During the Civil War, some women suspended their annual suffrage conventions to devote themselves to the abolitionist cause. They were prepared to put aside their own claims to fight for the abolitionist cause on the understanding that once abolition was won, the Republican party would support woman suffrage. After the Civil War ended, the party leaders informed them that "'this is the Negro's hour,' and that women must wait for their rights."[3]

In 1865, Stanton and Anthony, unwilling to wait, petitioned Congress directly for the first time. Up until this point, they had treated suffrage as a state issue, attempting to secure the franchise for women through appeals to state legislatures. The decision to go to the federal government demonstrated a growing awareness among suffragists of its power to institute reforms. One amendment had abolished slavery; another could enfranchise women.

The first major setback to the suffrage movement occurred with the ratification of the Fourteenth (1868) and Fifteenth Amendments (1870). Section two of the Fourteenth Amendment (See Document 5.1 in Appendix) contained the terms "male inhabitant" and "male citizens" in clauses meant to protect the freed slaves' right to vote, the first time the Constitution had used the word "male" to differentiate citizens. By protecting the franchise of male citizens only—not the franchise of all citizens—the Constitution legitimized the principle of discrimination by sex and implied that women were not equal citizens. Until the passage of the Fourteenth Amendment, suffragists had not needed an explicit sanction from the federal government guaranteeing women's right to vote because nothing in the Constitution had cast doubt on that right. But the Fourteenth Amendment, Stanton believed, meant that suffragists would require another amendment before women would be allowed to vote in federal elections.

In 1869, the Republicans introduced the Fifteenth Amendment, which stated that the franchise could not be denied or abridged on account of race, color, or previous condition of servitude. Stanton and Anthony, wanting to add "sex" to the list of conditions, fought against its ratification. But Lucy Stone, her husband, Henry Blackwell (1825–1909), and the male contingent of the American Equal Rights Association (AERA), Wendell Phillips (1811–1884), Horace Greeley (1811–1872), and Gerrit Smith (1797–1874), believed that a partial extension of the franchise was better than no extension at all. They thus reconciled themselves to the wording of both amendments and favored their passage.

In May 1869, Stanton and Anthony, incensed over the abolitionists' and Republicans' betrayal of women, established their own organization, the National Woman Suffrage Association (NWSA), open to women only, and began publishing a radical feminist newspaper, entitled the *Revolution*. The Boston contingent within the AERA, led by Lucy Stone and Henry Blackwell, formed the American Woman Suffrage Association (AWSA), which published its own journal, the *Woman's Journal*. The two organizations adopted different tactics and were not reconciled until 1890, when they were reunited as the National American Woman Suffrage Association (NAWSA).

The early 1870s witnessed what historian Ellen Carol DuBois has described as a spread in "militant suffragism."[4] In January 1871, Victoria Woodhull (1838–1927) argued in front of a Congressional committee that the Fourteenth and Fifteenth Amendments unwittingly had enfranchised women. The Fourteenth Amendment specified that "all persons born or naturalized in the United States" were citizens. The Fifteenth Amendment guaranteed that *citizens'* right to vote would not be denied on account of race. Thus, Woodhull claimed, women as citizens were entitled to vote. Woodhull's argument was taken up by the NWSA, which encouraged women to participate in the 1871 and 1872 elections. In 1872, Anthony, leading a group of Rochester women to the polls, was arrested for "illegal voting." Anthony, along with Matilda Joslyn Gage (1826–1898), thoroughly canvassed the county to inform prospective jurors of the legal issues involved in her trial. The prosecuting attorney succeeded in having the venue moved on the grounds that their tour had made it impossible to select an unbiased jury. Again, Anthony delivered speeches across the county. Ultimately, the judge, Justice Ward Hunt, would not permit the jury to decide the case and directed them to return a verdict of guilty. Unable to appeal the decision because of a technicality that the judge upheld, Anthony was forced to admit defeat. In 1875, the Supreme Court definitively repudiated suffragists' claims that the Constitution enfranchised women. In *Minor v. Happersett*, it ruled that suffrage was not a right of national citizenship, but a privilege that each state could grant to those whom it defined as deserving.[5] Suffragists then had little choice but to return to their earlier tactics: struggling for the franchise state by state and petitioning Congress for a separate amendment to grant women suffrage.

In 1872, Woodhull's publication of an illicit affair between Henry Ward Beecher (1813–1887), an eminent liberal minister, and Elizabeth Tilton, wife of Theodore Tilton (1835–1907), editor of the reform paper the *Independent*, caused a scandal whose repercussions were felt throughout the woman's rights movement.[6] The NWSA's earlier public support of Woodhull was well known, and Theodore Tilton, reputedly Woodhull's lover, was a friend and political advisor to Stanton and Anthony. Beecher was president of their rival organization, the AWSA. Theodore Tilton sued Beecher, who denied everything. The trial resulted in a hung jury, but public opinion sided with Beecher.

At the time, many Americans saw the scandal as confirmation of suffragists' immoral views on marriage and used the incident to discredit the NWSA, which had taken up Mrs. Tilton's defense. Despite the fact that Beecher was its president, the AWSA lost less prestige than the NWSA because it had always disdained Woodhull and had given the incident little coverage in the *Woman's Journal*. Historian William L. O'Neill argues that the scandal contributed to the stifling of public debate on marital conventions. While Stanton never gave up her fight for more liberal divorce laws, many suffragists grew more cautious about involving themselves in issues concerning women's marital and sexual rights.[7]

In the 1870s, more conservative women's reform organizations emerged, including the Young Woman's Christian Association, 1871; the Women's Christian Temperance Union, 1873; the Association for the Advancement of Women, 1873 (precursor to the General Federation of Women's Clubs, 1879); and the Women's Educational and Industrial Union, 1877. Some former suffrage activists, among them Julia Ward Howe and Mary Livermore, began to concentrate their efforts on other more popular reforms. Anthony tried to unify the women's organizations around suffrage, but most of them, with the notable exception of the WCTU under Frances Willard, did not support suffrage until much later.

Between 1869 and 1916, suffragists organized 41 state campaigns with 9 victories and 32 defeats. By 1890 only one state had fully enfranchised women: Wyoming, which had just received statehood, had granted women suffrage in 1869. The territory of Utah enfranchised women in 1870, and Kansas gave women municipal suffrage in 1887. Colorado enfranchised women in 1893. From 1896 until 1910 there were no more victories until the state of Washington granted women suffrage in 1910. Other successes soon followed: California, 1911; Oregon, 1912; Kansas, 1912; Arizona, 1912; and Illinois, 1913. (See Document 5.2 in Appendix.) In 1913, when the General Federation of Women's Clubs endorsed female suffrage, suffragists no longer had to educate middle-class women on the importance of the franchise. They now only needed to translate mass female support into political pressure.

By the early twentieth century, the movement had changed substantially. Suffrage had become a respectable issue, supported by conservative white middle- and upper-class women. The leadership of the movement reflected this change: The first generation of suffragist leaders, the radicals who had defined the issues in the 1840s, were no longer active. Stanton retired from the presidency of the NAWSA in 1892 at the age of 77. Anthony, who served as president after Stanton, retired in 1900 at age 80, choosing Carrie Chapman Catt (1859–1947) to succeed her. Catt served two years, resigning because of her husband's ill health, but returned in 1915 to mobilize the NAWSA for its ultimate victory in 1920.

THE ARGUMENTS

The foremost argument of the early suffragists was based on justice or rights. Drawing on the eighteenth century's conception of enlightened government, suffragists argued that the franchise was a basic or "natural right" of citizenship. The federal and state governments had given this right to male citizens; suffragists demanded that it be extended to women.[8]

Although suffragists argued for a brief period in the early 1870s that the Constitution implicitly guaranteed women, as citizens, the right to vote, after the Supreme Court ruled in 1875 that this interpretation was invalid, the debate in the late 1870s, 1880s, and 1890s centered around whether women were "equal" to men. Equal meant different things, depending on who used the term and in what context. Antisuffragists tended to equate equality with identity. Since men and women were not identical, they were not equal; and if they were not equal, then one sex must be superior to the other. Suffragists, however, while acknowledging that sexual differences existed, felt that un-identity could still mean equality. They argued that sexual differences balanced out: Women were obviously weaker in certain areas, but they were stronger in others and thus equally deserving of the vote.

The antisuffragists believed women were unfit to vote because of their physiological and psychological constitutions, their place in the socio-evolutionary hierarchy, and their fundamental duties as Christian wives and mothers. In a word, *nature* explained why women could not and should not vote. But nature contained three strands of thought that were often inextricably woven together: One referred to women's biology—their inherent nature; another referred to women's subordinate place in nature, as dictated by social-Darwinist principles of evolution; and a third involved a theological analysis of women's natural duties.

The antisuffragists drew theological evidence from Genesis' emphasis on the division of labor between Adam and Eve. Adam tended the garden, while Eve, created from his rib (a symbol of her subordination), kept him company. Furthermore, there was a division of responsibilities contained in their expulsion from Eden. Adam was condemned to toil by the sweat of his brow to provide for himself and his dependents, while Eve was punished with the rigors of childbirth. Antisuffragists found additional testimony pertaining to women's natural place in St. Paul's Epistles to the Corinthians and Galatians. St. Paul insisted that women should remain silent in the Church (hence, by extension in all public activities) and show total obedience to their husbands. For some antisuffragists, the Bible demonstrated men's superiority; for others it indicated only that men should represent their wives in public affairs. But one thing was clear: The Bible dictated what was the natural—God-determined and thus unchallengeable—division of labor by sex: Men should support, govern, and protect their women, while women should bear and raise children.

Suffragists countered the antisuffragists' theological claims in several ways. The most radical women, including Stanton, agreed with antisuffragists that the Bible opposed any change in the status of women and attacked it directly for subjecting women to male authority. In the 1890s, when Stanton was in her seventies, she formed a committee of women to write a work interpreting the Bible. The first volume of the *Woman's Bible* (1895) contained commentaries on Genesis, Exodus, Leviticus, Numbers, and Deuteronomy and stirred a great controversy among suffragists. The National American Woman Suffrage Association (NAWSA) voted to disassociate itself publicly from the book despite the entreaties of Anthony, who felt such a move would be a terrible insult to Stanton. This incident revealed the internal division within the suffrage movement. Stanton, challenging fundamental religious beliefs, met with active resistance from more conservative and status-conscious women, who considered Stanton's religious views blasphemous and dangerous to the suffrage cause.[9]

A more conservative and common response of suffragists was to assert that antisuffragists misinterpreted the spirit of the Bible. As Christian women, these suffragists were unable or unwilling to repudiate Scripture. Instead they reinterpreted it. In 1895, Alice Stone Blackwell explained how Scripture could be made compatible with suffrage:

> The one unanswerable and all-sufficient text of Scripture in behalf of woman suffrage, as in behalf of all other reforms is the Golden Rule. Every man knows at the bottom of his heart that he would not like to be taxed without representation, and governed by law makers whom he had no voice in choosing. If he would not like others to do it to him, he ought not to do it to them.[10]

Evolutionary evidence lent support to the theological case that women were unfit for politics. Women, for centuries restrained by their "reproductive functions from taking the same active part in the world's life as does man," had inherited an evolutionary handicap: Their physical and mental capacities could never equal those of men. They suffered "deficiencies of endurance" of the "rational faculty" and were more emotional than men. (See Cope, "The Relations of the Sexes to Government," pp. 210–216.) This view of women's nature was not incompatible with the general belief that women had a higher faculty, "woman's intuition," which permitted them to surpass men in some forms of moral understanding. But woman's intuition, antisuffragists believed, operated only in the domestic sphere.

Women's emotionality and irrationality made it impossible to trust them with the government of themselves or others. Their physical weakness made it impossible for them to bear the immense strain of political activity. Professor Edward Cope even argued that women should be denied the franchise because their physical inferiority made them unable to execute or enforce legislation, the assumption being that American government did not operate with the consent of the governed but relied upon physical coercion. Men would not comply with women's laws since women were not strong enough to make sure they did.

In sum, the antisuffragists claimed that the biological nature of women was the root cause of their inferiority and made them unfit for political activities on three accounts: the physical, intellectual and affectual. Women were too weak, stupid, and emotional to be enfranchised. Here, too, suffragists adopted various responses to these antisuffragist claims. Some denied outright that women were inferior to men, pointing out examples of gifted or accomplished women. Others emphasized that women's minds were different, but not inferior. Women had special qualities—sympathy and gentleness—that warranted and justified women's enfranchisement. Carrie Chapman Catt argued that it was "because of the differences between men and women that the nineteenth century more than any other demands the enfranchisement of women." Women ought to be allowed to vote because they had "in greater perfection the gentler traits of tenderness and mercy, the mother heart, which goes out to the wronged and afflicted everywhere, with the longing to bring them comfort and sympathy and help."[11] However, some suffragists acknowledged that women's intellects were less "advanced" but attributed this weakness to society's subjugation of women, not to any inherent defect in female nature. As women were educated and participated in a greater range of activities, these differences would diminish by virtue of the progressive evolutionary process.

Most suffragists did not deny the existence of sexual differences, nor did they dispute the genesis of these differences, only the assumption that they were immutable. For how could differences that evolved over time suddenly have become fixed? If natural selection meant change and progress, how could women's development be arrested or deteriorating? For most antisuffragists, the critical sexual differences derived from the one immutable, all-defining feature of femininity: the female reproductive system. The inescapable, inexorable demands of the ovaries and womb diverted the necessary energy for development from other structures, particularly the brain. Thus, women were caught in an irresolvable dilemma: Short of not having children, they could not develop intellects equal in power to those of men. By the turn of the twentieth century, however, suffragists were less willing to concede mental or intellectual inferiority. They admitted that a majority of their class might still be uneducated, but that was no reason to bar women as a group from the franchise. By this standard, they pointed out, a majority of men would be "undeserving" of the vote, and yet the franchise extended to all men. (See "The Extension of the Suffrage to Women," pp. 216–217.)

With the advent of public schools and the increase in numbers of women attending high school and college, suffragists no longer felt the need to prove that women were the intellectual equals of men. Society had grown more accepting of the principle of equality between the sexes, although not of identity or androgyny. The debate now turned on whether equality could exist between the separate spheres.

For many suffragists, separate spheres precluded equality in practice, but what they understood by equality was not always clear. Sometimes, they seemed to mean complete development; at other times, the same rights and privileges as those accorded men. But while these suffragists sensed that so long as separate, unidentical spheres of activity divided the sexes women would never achieve full equality, they rarely if ever advocated that the differences between the male and female spheres be eradicated entirely. They agreed that women alone had a special moral obligation to rear children.

Another antisuffragist argument was sociological in form, social-Darwinist in substance. Society would "progress" only if women remained in their natural place, the home, caring for children and husbands. Extension of the franchise would result in removing women from the home, disrupt the "harmonious relations" between the sexes, threaten the family as a social institution, and lead to moral degeneration. While suffragists concurred that childrearing constituted women's most important social duty, they did not concede that voting would interfere with familial responsibilities. Instead, they argued that the franchise and political activity would help women improve themselves, make them better wives and mothers, and enable legislatures to pass laws that would strengthen the family and improve society's morals.

The debate on suffrage was notable more for its emotional intensity than for its logical consistency. Suffragists disagreed with each other as to whether men and women were equal and argued that the franchise was a natural right to be granted in accordance with justice, a privilege to be bestowed because women were equally deserving as men, and a responsibility women shared in view of their maternal responsibilities and special role as moral reformers. When suffragists claimed that justice alone mandated women be given the vote, they could insist upon or concede the point of equality; it did not matter because natural rights were not predicated on equality. If female fitness was required to prove women were responsible or deserving, they could still concede that the sexes were *unequal* if what they meant was *unidentical*, without admitting that women were inferior. Suffragists could even acknowledge feminine inferiority and demand suffrage on the grounds that the franchise would improve women's nature, eradicating the weaknesses. Even here, suffragists did not believe that the sexes would become identical, only that they would become equal, since the franchise would further develop feminine strengths, compensating for feminine failings. Hence, the franchise would eradicate female inferiority without endangering sexual differences. The more common argument was to assert overall equality—unidentity—along with moral superiority: women merited suffrage and should be enfranchised because they had superior moral characters, they would legislate more humanely, and their votes would help reform society.

Sometimes suffragists complained of the injustice of maintaining separate spheres; at other times, they upheld separate spheres as justification for women's enfranchisement: Women's own sphere gave them special interests, which they had a right to protect since men could not or would not legislate justly for women. Antisuffragists replied that while it was true that women had their own sphere, they had no special interests. Women's interests were family interests, which men shared. Men's votes simply represented these common interests.

Antisuffragists also adopted whatever argument was most expedient. They argued that certain feminine traits (emotionality, inability to reason) were immutable and so debilitating that woman *could not* participate in government, and they pointed to feminine characteristics (altruism, piety) that were so necessary and fragile that woman *should not* participate, for fear of destroying what was best in womanhood. Antisuffragists felt government was too great a responsibility for women to assume, but as compensation and comfort they offered the assurance that motherhood was the greatest responsibility of all. Some antisuffragists expressed incredulity that any woman could want suffrage—it was so insignificant compared to the power of women to influence public opinion through husbands and children. Few antisuffragists seemed to wonder why women were fit enough to influence, but not fit enough to vote.

As historians Ellen Carol DuBois and Carl N. Degler have argued, suffrage was perceived by most Americans in the nineteenth century as one of the most radical demands of the woman's rights movement.[12] It was considered radical because it threatened to undermine the social ideal of separate spheres for men and women by offering women power in the male sphere of politics. (See "Women and Politics" and "Let Us Therewith be Content," pp. 239–242 and pp. 316–322.) It is easy to understand why many men would object to women voting: fear of encroachment into an arena in which they had a monopoly; reluctance to relinquish authority and share power; anxiety that women would then use the vote to challenge men's power and privileges in other domains. (Male antisuffragists rarely gave these as reasons, however.) Men also feared that suffrage would interfere with domestic life, that women would neglect domestic duties, meddle in men's affairs, make a mess out of politics, and ruin the country. (They did give these as reasons.) But suffrage was not a popular reform among *women* and did not attract the large female following that the temperance movement did, for example. Why *women* feared enfranchisement—and many did—is an intriguing question, and one that, with the exceptions of Jane Camhi and Carl N. Degler, has been largely ignored by contemporary historians.[13]

Degler suggests that because suffrage was perceived by nineteenth-century women as a threat to the family, many of them did not support the movement. Indeed, women were the only group in American history to actively oppose their own enfranchisement. Contrary to the claims of nineteenth-century feminists, many women did not see their position in the family as oppressive and did not desire changes that would alter their position. Indeed, it is difficult for us in the twentieth century to understand how suffrage *could* have represented a real threat to the family, particularly since once suffrage was attained, it had little impact on familial roles. Moreover, the kinds of changes that suffragists argued might come about with the help of women's votes—prohibition, illegalization of gambling and prostitution, and improvement in the material conditions of the urban poor—could be seen as strengthening, not undermining, domestic relations. So how did suffrage conflict with familial values?

First, many suffragists insisted upon the recognition of women as a group having interests distinct from those of men. Frequently they were forced into taking this position by antisuffragists who argued that there was no need to give women the vote, since doing so would only double the number of voters without giving representation to new interests. Some suffragists even insisted that whether women shared men's interests was beside the point: Women should be allowed to vote because they had the right as individuals to direct representation. Thus, for many Americans, woman suffrage entailed an acknowledgment of female autonomy and of tension and conflict between husband and wife—a reality that called into question the ideal of harmonious domestic relations.

Secondly, suffrage was seen as "self-serving," as indeed it was—women wanted the vote for themselves. It was because suffrage conflicted with the ideal of the selfless, altruistic woman that many women did not support it. Suffragists frequently were held up to public scorn for being aggressive, unattractive, and selfish. They were accused of neglecting their children, or perhaps worse yet, not having any, of promoting their own interests instead of serving others. As suffragists changed their tactics, no longer representing the franchise as a challenge to conventional sex roles, instead emphasizing how voting would make women better mothers and strengthen the family, the suffrage movement attracted more supporters among men *and* women and eventually achieved its goal.[14]

Nonetheless, many women clearly felt some dissatisfaction with the strict restrictions placed upon their actions and with the vulnerable, dependent position they held within the domestic sphere. Evidence for these points can be ascertained from women's involvement in a variety of reform movements, including the temperance, social purity, and voluntary motherhood movements, in which they tried to protect themselves from the abuses of men. But women's protests (or lack of them) must be considered in light of their economic dependency upon fathers, husbands, and brothers. Without men, women had few adequate means of earning livings. Moreover, the image of a financially or emotionally independent woman was presented in unappealing terms: This "new woman," as she was sometimes called, was commonly depicted as an overworked and lonely person, who had tried to prove that she could do something more "noble" than marry and raise children and had failed. Of course, women who wanted to have children had to marry, and once married it was extremely difficult to overcome the social pressures and barriers that made it unusual for white middle-class, married women to work outside the home. These are the issues I examine in the remaining chapters.

APPENDIX

Document 5.1.
Amendment XIV (Section 1 and 2)
(Proposed June 1866; Ratified July 1868)

Section 1.   All persons born or naturalized in the United States, and subject to the jurisdiction thereof, are citizens of the United States and of the State wherein they reside. No State shall make or enforce any law which shall abridge the privileges or immunities of citizens of the United States; nor shall any State deprive any person of life, liberty, or property, without due process of law; nor deny to any person within its jurisdiction the equal protection of the laws.

Section 2.   Representatives shall be apportioned among the several States according to their respective numbers, counting the whole number of persons in each State, excluding Indians not taxed. But when the right to vote at any election for the choice of electors for President and Vice President of the United States, Representatives in Congress, the Executive and Judicial officers of a State, or the members of the Legislature thereof, is denied to any of the *male inhabitants* of such State, being twenty-one years of age, and citizens of the United States, or in any way abridged, except for participation in rebellion, or other crime, the basis of representation therein shall be reduced in the proportion which the number of such *male citizens* shall bear to the whole number of *male citizens* twenty-one years of age in such State. (Emphasis added.)

Amendment XV (Section 1 only)
(Proposed February 1869; Ratified March 1870)

Section 1.   The right of citizens of the United States to vote shall not be denied or abridged by the United States or by any State on account of race, color, or previous condition of servitude.

Amendment XXIX
(Proposed June 1919; Ratified August 1920)

The right of citizens of the United States to vote shall not be denied or abridged by the United States or by any State on account of sex. Congress shall have power to enforce this article by appropriate legislation.

Document 5.2.
Calendar of Suffrage Campaigns

Date	State	Electoral Vote
1890	WYOMING was admitted to statehood with woman suffrage, having had it as a territory since 1869.	3
1893	COLORADO adopted a constitutional amendment after defeat in 1877.	6
1896	IDAHO adopted a constitutional amendment on its first submission.	4
1896	UTAH, after having women suffrage as a territory since 1870, was deprived of it by the Congress in 1887, but by referendum put it back in the constitution when admitted to statehood.	4
1910	WASHINGTON adopted a constitutional amendment after defeats in 1889 and 1898. It had twice had woman suffrage by enactment of the territorial legislature and lost it by court decisions.	7
1911	CALIFORNIA adopted a constitutional amendment after defeat in 1896.	13
1912	OREGON adopted a constitutional amendment after defeats in 1884, 1900, 1906, 1908, 1910.	5
1912	KANSAS adopted a constitutional amendment after defeats in 1867 and 1893.	10
1912	ARIZONA adopted a constitutional amendment submitted as a result of referendum petitions.	3
1913	ILLINOIS was the first state to get presidential suffrage by legislative enactment.	29
1913	Territory of ALASKA adopted woman suffrage. It was the first bill approved by the governor.	

Date	State	Electoral Vote
1914	MONTANA adopted a constitutional amendment on its first submission.	4
1914	NEVADA adopted a constitutional amendment on its first submission.	3
1917	NORTH DAKOTA secured presidential suffrage by legislative enactment, after defeat of a constitutional amendment in 1914.	5
1917	NEBRASKA secured presidential suffrage by legislative enactment after defeats of a constitutional amendment in 1882 and 1914.	8
1917	RHODE ISLAND secured presidential suffrage by legislative enactment after defeat of a constitutional amendment in 1887.	5
1917	NEW YORK adopted a constitutional amendment after defeat in 1915.	45
1917	ARKANSAS secured primary suffrage by legislative enactment.	9
1918	MICHIGAN adopted a constitutional amendment after defeats in 1874, 1912, and 1913. Secured presidential suffrage by legislative enactment in 1917.	15
1918	TEXAS secured primary suffrage by legislative enactment.	20
1918	SOUTH DAKOTA adopted a constitutional amendment after six prior campaigns for suffrage had been defeated, each time by a mobilization of the alien vote by American-born political manipulators. In that state, as in nine others in 1918, the foreign-born could vote on their "first papers," and citizenship was not a qualification for the vote. The last defeat, in 1916, had been so definitely proved to have been caused by the vote of German-Russians in nine counties that public sentiment, in addition to the war spirit, aroused a desire to make a change in the law, which resulted in victory.	5
1918	OKLAHOMA adopted a constitutional amendment after defeat in 1910.	10

Date	State	Electoral Vote
1919	INDIANA secured presidential suffrage by legislative enactment in 1917. Rendered doubtful by a court decision, the law was re-enacted with but six dissenting votes.	15
1919	MAINE secured presidential suffrage by legislative enactment after defeat of a constitutional amendment in 1917.	6
1919	MISSOURI secured presidential suffrage by legislative enactment after defeat of a constitutional amendment in 1914.	18
1919	IOWA secured presidential suffrage by legislative enactment after defeat of a constitutional amendment in 1916.	13
1919	MINNESOTA secured presidential suffrage by legislative enactment.	12
1919	OHIO secured presidential suffrage by legislative enactment after defeat of a referendum on the law in 1917 and of a constitutional amendment in 1912 and 1914.	24
1919	WISCONSIN secured presidential suffrage by legislative enactment after defeat of a constitutional amendment in 1912.	13
1919	TENNESSEE secured presidential suffrage by legislative enactment.	12
1920	KENTUCKY secured presidential suffrage by legislative enactment.	13

Total of presidential electors for whom women were entitled to vote before the nineteenth amendment was adopted: 339; full number 531.

Source: The National Woman Suffrage Association, *How Women Won It* (New York: H. W. Wilson, 1940), pp. 161–164.

NOTES

1. Stanton and Mott attended the London antislavery convention in 1840, at which the female delegates were relegated to the gallery and barred from taking part in the conference's debates. Incensed at this treatment, Stanton and Mott vowed that they would actively strive to secure the rights of women, in addition to those of slaves. Eight years later, they called a convention in Seneca Falls to discuss women's grievances. For a personal account of the suffrage move-

ment, see Elizabeth Cady Stanton's autobiography, *Eighty Years and More: Reminiscences 1815–1897* (New York: Schocken Books, 1975; 1st ed., T. Fisher Unwin, 1898).
2. Elizabeth Cady Stanton, Susan B. Anthony, and Matilda Joslyn Gage, *The History of Woman Suffrage* (Rochester, New York: Susan B. Anthony, 1876–1885), 1: 70–71.
3. Most early suffragist leaders were also active abolitionists. Men such as William Lloyd Garrison, Gerrit Smith, Henry Blackwell, Wendell Phillips, Horace Greeley, Thomas Wentworth Higginson, and Frederick Douglass supported both the abolitionist and women's rights movements. Susan B. Anthony and Lucy Stone were paid organizers for the American Anti-Slavery Society and Elizabeth Cady Stanton, whose husband Henry Stanton and cousin Gerrit Smith were prominent abolitionists, lectured against slavery in tours of western New York state. Quotation cited in Aileen S. Kraditor, *The Ideas of the Woman Suffrage Movement, 1890–1920* (New York: Columbia University Press, 1965), p. 3.
4. See Ellen Carol DuBois, ed., *Elizabeth Cady Stanton/Susan B. Anthony: Correspondence, Writings, Speeches* (New York: Schocken Books, 1981), pp. 101–109.
5. Ibid., p. 107.
6. See Ibid., pp. 105–106.; O'Neill, *Everyone Was Brave*, pp. 154–155.
7. O'Neill, *Everyone Was Brave*, p. 29.
8. Elizabeth Cady Stanton still argued this position as late as 1894 in *Suffrage a Natural Right* (Chicago: Open Court, 1894). Cited in Kraditor, *The Ideas of the Woman Suffrage Movement*, p. 45.
9. See Kraditor's excellent chapter "Woman Suffrage and Religion" in *The Ideas of the Woman Suffrage Movement*, pp. 75–95.
10. *Woman's Journal*, October 26, 1895. Cited in Kraditor, *The Ideas of the Woman Suffrage Movement*, p. 93.
11. Carrie Chapman Catt, "Evolution and Woman's Suffrage," manuscript of a speech delivered May 18, 1893 in Catt Collection, New York Public Library; cited in Rosalind Rosenberg, *Dissent From Darwin, 1890–1930: The New View of Woman Among American Social Scientists* (Ph.D. dissertation, Stanford University, 1975), p. 46.
12. See Carl N. Degler, *At Odds: Women and the Family in America From the Revolution to the Present* (New York: Oxford University Press, 1980), pp. 328–361; Ellen Carol DuBois, *Feminism and Suffrage: The Emergence of an Independent Women's Movement in America 1848–1869* (Ithaca: Cornell University Press, 1978), pp. 40–47; and Ellen Carol DuBois, "The Radicalism of the Woman Suffrage Movement: Notes Toward the Reconstruction of Nineteenth-Century Feminism," *Feminist Studies* 3 (Fall 1975): 63–71.
13. Jane Jerome Camhi, *Women Against Women: American Antisuffragism 1880–1920* (Ph.D. dissertation, Tufts University, 1973) and Carl N. Degler, *At Odds*, pp. 342–355.
14. Degler, *At Odds*, pp. 357–359.

ARTICLES ON WOMAN SUFFRAGE APPEARING IN *THE POPULAR SCIENCE MONTHLY*

(Arranged by Date of Publication)

Luke Owen Pike. "Woman and Political Power," (May 1872) 1: 82–94.
Goldwin Smith. "Female Suffrage," (August 1874) 5: 427–443.
J. E. Cairnes. "Woman Suffrage as Affecting the Family," (November 1874) 6: 87–90.
Editor. "Professor Cairnes on Woman Suffrage," (November 1874) 6: 112–114.
"Biology and 'Woman's Rights,'" (December 1878) 14: 201–213.

Alfred Fouillee. "The Problem of Universal Suffrage," (December 1884) 26: 194–204.
Mrs. L. D. Morgan. "Equality or Protection," (July 1888) 33: 410.
Edward D. Cope. "The Relations of the Sexes to Government," (October 1888) 33: 721–730.
Lucy S. V. King. "Women in Business," (October 1888) 33: 842–843.
Frank Cramer. "The Extension of the Suffrage to Women," (January 1889) 34: 415.
E. D. Cope. "Woman Suffrage," (February 1889) 34: 558–559.
Therese A. Jenkins. "The Mental Force of Woman," (April 1889) 34: 841–843.
George F. Talbot. "The Political Rights and Duties of Women," (May 1896) 49: 80–97.
Alice B. Tweedy. "Woman and the Ballot," (June 1896) 49: 241–253.
Editor. "Women and Politics," (August 1896) 49: 556–559.
Grace A. Luce. "Occupations, Privileges, and Duties of Woman," (September 1896) 49: 698–699.
Alice B. Tweedy. "Woman's Claims to the Ballot," (October 1896) 49: 842.
Helen Kendrick Johnson. "Woman Suffrage and Education," (June 1897) 51: 222–231.
Ellen Coit Elliott. "Let Us Therewith be Content," (July 1897) 51: 341–348.
Editor. "A Woman on Woman Suffrage," (September 1897) 51: 700–704.
"Women Opposed to Woman Suffrage," (July 1898) 53: 429–430.
"The New Zealand Experiment in Woman Suffrage," (June 1899) 55: 279–280.

SUGGESTIONS FOR FURTHER READING

Primary Sources and Documents

Buhle, Mari Jo, and Buhle, Paul, eds. *The Concise History of Woman Suffrage: Selections from the Classic Work of Stanton, Anthony, Gage, and Harper*. Urbana: University of Illinois Press, 1978.

DuBois, Ellen Carol, ed. *Elizabeth Cady Stanton/Susan B. Anthony: Correspondence, Writings, Speeches*. New York: Schocken Books, 1981.

Scott, Anne F., and Scott, Andrew M. *One Half the People: The Fight for Woman Suffrage*. Philadelphia: J. B. Lippincott, 1975.

Stanton, Elizabeth Cady. *Eighty Years and More: Reminiscences 1815–1897*. New York: Schocken Books, 1975; 1st ed., 1898.

Stanton, Elizabeth Cady; Anthony, Susan B.; and Gage, Matilda Joslyn, eds. *The History of Woman Suffrage*. 6 vols. Rochester: S. B. Anthony, 1891–1922.

Secondary Sources

Anthony, Katharine. *Susan B. Anthony: Her Personal History and Her Era*. New York: Doubleday, 1954.

Banner, Lois. *Elizabeth Cady Stanton: A Radical for Woman's Rights*. Boston: Little, Brown, 1980.

Blackwell, Alice Stone. *Lucy Stone: Pioneer of Woman's Rights*. Boston: Little, Brown, 1930.

Degler, Carl N. *At Odds: Women and the Family in America From the Revolution to the Present*. New York: Oxford University Press, 1980, pp. 328–361.

DuBois, Ellen Carol. *Feminism and Suffrage: The Emergence of an Independent Woman's Movement in America 1848–1869*. Ithaca: Cornell University Press, 1978.

DuBois, Ellen Carol. "The Radicalism of the Woman Suffrage Movement: Notes Toward the Reconstruction of Nineteenth-Century Feminism," *Feminist Studies* 3 (Fall 1975): 63–71.

Flexner, Eleanor. *A Century of Struggle: The Woman's Rights Movement in the United States*. Cambridge: Belknap Press, 1959.

Grimes, Alan P. *The Puritan Ethic and Woman Suffrage*. New York: Oxford University Press, 1967.

Kraditor, Aileen S. *The Ideas of the Woman Suffrage Movement, 1890–1920*. New York: Columbia University Press, 1965.

Lerner, Gerda. *The Grimke Sisters from South Carolina: Rebels Against Slavery*. Boston: Houghton Mifflin, 1967.

Morgan, David. *Suffragists and Democrats: The Politics of Woman Suffrage in America*. East Lansing: Michigan State University Press, 1972.

O'Neill, William L. *Everyone Was Brave: The Rise and Fall of Feminism in America*. Chicago: Quadrangle Books, 1969.

Edward Drinker Cope (1840–1897), born to wealthy members of the Society of Friends, was educated at the Friends' School of Westtown, then spent one year at the University of Pennsylvania, and received the remainder of his schooling at the hands of private tutors. Later he was granted honorary degrees from Haverford College (master's, 1870) and the University of Heidelberg (Ph.D., 1885). In 1864, Cope accepted a chair of comparative zoology and botany at Haverford, which he resigned in 1867, due to ill health. In 1889, financial considerations led him to accept a professorship of geology and mineralogy at the University of Pennsylvania, and in 1895 he also assumed the chair of zoology and comparative anatomy. Cope was known nationally as an eminent zoologist and paleontologist and was recognized (at the age of 22) as the leading authority in reptiles. He was a prolific writer, publishing as many as 50 papers in a year, and leaving a legacy of 600 separate titles, most of them monographs on specialized scientific topics.

In "The Relations of the Sexes to Government," Cope presents an essentially Darwinian analysis of the physical and mental inferiority of women, which he felt would make unqualified woman suffrage a serious mistake. Cope's article elicited two responses, one from Frank Cramer, a professor at Lawrence University in Appleton, Wisconsin, and another from Therese A. Jenkins, a resident of the territory of Wyoming, where women had gained the right to vote in 1869.

THE RELATIONS OF THE SEXES TO GOVERNMENT

EDWARD D. COPE

October 1888

AS is well known, the diversity of sex is of very ancient origin. It appeared in the history of life before the rise of any but the most rudimental mentality, and has at various points in the line of development of living things displayed itself in the most pronounced manner. Great peculiarities of sex structure are witnessed in the higher forms of life, as in birds and mammalia. The greatest peculiarity of mental sex character can only be seen where mind is most developed—that is, in man.

. . . Being free from the disabilities imposed by maternity, the male could acquire a greater mastery over his environment than the female. His time would be less occupied, and his opportunity for physical exertion greater, and he could and would take a more active part in the struggle for existence. Hence, of the two sexes the male became the fighter and the provider, and necessarily, from the increasing muscular strength acquired in this more active life, the master of the two. He, therefore, became more specialized in some respects, particularly in those necessary to success in his various undertakings. His part in reproduction became a specialization as compared with that of the female, which more nearly resembles the asexual method. So the male became the author of variation in species in two ways: first, by adding to the sources of inheritance; and second, by his own more numerous specializations.

In man the mental organization of the sexes expresses these facts in various ways. The sexual mental characteristics of men and women have been described by Lecky, Delaunay, Ladd, P. G. Hamerton, and others, and with a unanimity that would of itself be authoritative if they did not confirm the belief of thoughtful observers generally. Woman is not only restrained by her reproductive functions from taking the same active part in the world's life as does man; but, what is more important, she inherits a greater disability from thousands of ages of equal and in some cases greater disability in the countless generations of man's animal ancestors. This nature is thoroughly ingrained, and is as permanent as any other part of her organism. In considering these mental peculiarities, it must be borne in mind that she inherits from her father as well as from her mother, so that she has benefited by the general progress of the race, but her relation to the male remains the same in each family taken by itself. Thus it has resulted that the women of a higher race or family will display superior traits to men of a lower race or family, even in some of the endowments which are the especial field of the male. And it is comparisons of this sort which frequently cause the question to be raised, whether the supposed superior rationality with which men are credited is ascribed to them justly. In the great variety of history and origin possessed by the people who are thrown together by our modern civilization, it must often happen that the women of superior lineage provoke favorable comparison with men whose ancestors have emerged from semi-savagery within a comparatively recent period. Nevertheless, in these cases also, sex qualities of mind are well marked, though more or less limited on the part of the inferior type.

. . . We find in man a greater *capacity* for rational processes, a capacity which is not always exercised to its full. We find in men a greater capacity for endurance of the activity of the rational faculty. We find in men a greater capacity for work in those departments of intelligence which require mechanical skill of a high order. In the aesthetic department, we find incapacity more general than in women, certainly in the department of the aesthetics of the person. In woman we find that the deficiency of endurance of the rational faculty is associated with a general incapacity for mental strain, and, as her emotional nature is stronger, that strain is more severe than it is in man under similar circumstances. Hence the easy breakdown under stress, which is probably the most distinctive feature of the female mind. This peculiarity, when pronounced, becomes the hysterical temperament. But in all departments of mental action that depend on affection or emotion for their excellence, woman is the superior of man; in those departments where affection should not enter, she is his inferior. . . . Beginning with the maternal instinct, woman has become, by constant exercise, a being of affections. Her long protection by the male has reduced her capacity for defense; while the mastery by him has accustomed her to yielding, and to the use of methods of accomplishing her desires other than force. There are apparent exceptions to these definitions, but they are generally more apparent than real. For one of the characteristics of the female of man, acquired by long practice, is a capacity for keeping up the appearance of possessing qualities in which she is more or less deficient. A ready capacity for acquisition of knowledge, and skill in language, are important contributors to this result.

It would seem, then, that Nature has marked out very clearly the relative positions of the sexes of man. This relation is beneficial not only from a natural but also from a social standpoint. The sex affection or passion has the greatest influence in compelling evolution of unwilling lives, and of driving where nothing can lead. The best emotions are aroused in the man who finds a woman dependent on him for support, and the infant's breath will awake that woman to serious thought and exertion who never had a serious thought before. . . .

It must be here premised that the progress of civilization has thus far emphasized and not diminished the peculiarities of sex. The civilized woman is more refined, more tender, more intelligent, and more hysterical than her savage representative. Her form is more different from that of the male, and her face more expressive of her distinctive character. There is good reason to believe that this development has been due to the increased immunity from the severity of the "struggle for existence" which woman enjoys in civilized communities, and the greater opportunity thus given her to develop her own especial excellences.

The first thought that strikes us in considering the woman-suffrage movement is, that it is a proposition to engage women once more in that "struggle" from which civilization has enabled them in great measure to escape; and that its effect, if long continued and fairly tried, will be to check the development of woman as such, and to bring to bear on her influences of a kind different from those which have been hitherto active. And it becomes an impartial thinker to examine the question more closely, and see whether investigation bears out these impressions or not. We inquire, then, in the first place, is government a function adapted to the female character, or within the scope of her natural powers? We then endeavor to discover whether her occupation of this field of action is calculated to promote the mutual sex interest which has been referred to above, and thus to subserve the natural evolution of humanity.

In endeavoring to answer the first question we are at once met by the undoubted fact that woman is physically incapable of carrying into execution any law she may enact. She can not, therefore, be called on to serve in any executive capacity where law is to be executed on adults. Now, service in the support of laws enacted by those who "rule by the consent of the governed" is a *sine qua non* of the right to elect governors. It is a common necessity to which all of the male sex are, during most of their lives, liable to be called on to sustain. This consideration alone, it appears to me, puts the propriety of female suffrage out of the question. The situation is such that the sexes can not take an equal share of governmental responsibilities even if they should desire to do so. Woman suffrage becomes government by women alone on every occasion where a measure is carried by the aid of woman's votes. If such a measure should be obnoxious to a majority of men, they could successfully defy a party composed of a minority of their own sex and a majority of women. That this would be done there can be no question, for we have a parallel case in the attempt to carry into effect negro suffrage in some parts of the South. We know the history too well. Intimidation, deception, and the manipulation of the count, have nullified the negro vote. How many Governors, Legislatures, and even Presidents have attained their positions in violation of the rights of the ballot during the last twenty years, we may never know. In times of peace and general prosperity these things have excited indignant protest, but nothing more. But when serious issues distract the nation or any part of it, frauds on the ballot and intimidation of voters will be a more serious matter, and will lead to disastrous consequences. We do not want to increase possibilities of such evil portent. Unqualified negro suffrage is, in the writer's estimation, a serious blunder, and woman suffrage would be another. And it is now proposed that we have both combined. . . .

On account of their stronger sympathies girls always think themselves the moral superiors of boys, who are often singularly devoid of benevolence, especially toward the lower animals. Some women imagine, for this reason, that their entire sex is morally the superior of the male. But a good many women learn to correct this opinion. In departments of morals which depend on the emotional nature, women are the superior; for those which depend on the rational nature, man is the superior. When the balance is struck, I can see no inferiority on either side. But the quality of justice remains with the male. It is on this that men and women must alike depend, and hence it is that women so often prefer to be judged by men rather than by their own sex. They will not gain anything, I believe, by assuming the right of suffrage, that they can not gain without it, and they might meet with serious loss. In serving the principle of "the greatest good of the greatest number," man is constantly called on to disregard the feelings of particular persons, and even to outrage their dearest ties of home and family. Woman can not do this judicially. After the terrors of the law have done their work, woman steps in and binds up the wounds of the victims, and the world blesses both the avenger and the comforter.

In the practical working of woman suffrage, women would either vote in accordance with the views of their husbands and lovers or they would not. Should they do the former habitually, such suffrage becomes a farce, and the only result would be to increase the aggregate number of votes cast. Should women vote in opposition to the men to whom they are bound by ties sentimental or material, unpleasant consequences would sooner or later arise. No man would view with equanimity the spectacle of his wife or daughters nullifying his vote at the polls, or contributing their influence to sustain a policy of government which he should think injurious to his own well-being or that of the community. His purse would be more open to sustain the interests of his own political party, and if he lived in the country he would probably not furnish transportation to the polls for such members of his family as voted against him. He would not probably willingly entertain at his house persons who should be active in obtaining the votes of his wife and daughters against himself; and on the other hand the wife might refuse entertainment to the active agents of the party with which she might not be in sympathy. The unpleasantness in the social circle which comes into view with the advent of woman suffrage is formidable in the extreme, and nothing less than some necessity yet undreamed of should induce us to give entrance to such a disturber of the peace. We need no additional causes of marital infelicity. But we are told by the woman-suffrage advocate that such objections on the part of men are without good reason, and are prejudices which should be set aside. But they can not be set aside so long as human nature remains what it is. Men may grant women anything but the right to rule them, but there they draw the line. Is it not on questions of rule that the wars of men are mostly fought, and will men yield to the weak what they only surrender to irresistible force? In the settlement of all questions by force, women are only in the way.

The effect of sexual discord is bad on both sexes, but has its greatest influence for evil through woman. While it does not remove her frailties it suppresses her distinctively feminine virtues. This suppression, continued for a few generations, must end in their greater or less abolition. The lower instincts would remain, the flowers which blossom on that stem would wither. No matter what their intellectuality might be, such women would produce a race of moral barbarians, which would perish ultimately through intestine strife. The highest interests and pleasures of the male man are bound up in the effective preservation of the domestic affections of his partner. Where these traits are weak, he should use every effort to develop them by giving them healthy exercise. As in all evolution, disuse ultimately ends in atrophy, and the atrophy of the affections in woman is a disaster in direct proportion to its extent. It may be replied again that woman suffrage carries with it no such probable result. But I believe that it does, unless the relations of the sexes are to be reversed. But it will be difficult to reduce the male man to the condition of the drone-bee (although some men seem willing to fill that *rôle*);

or of the male spider, who is first a husband and then a meal for his spouse. We have gone too far in the opposite direction for that. It will be easier to produce a reversion to barbarism in both sexes by the loss of their mutual mental hyperaesthesia.

If women would gain anything with the suffrage that they can not gain without it, one argument would exist in its favor to the many against it; but the cause of women has made great progress without it, and will, I hope, continue to do so. Even in the matter of obtaining greater facilities for divorce from drunken or insane or brutal husbands than now exist in many States of the Union, they can compel progress by agitation. A woman's society, with this reform as its object, would obtain definite results. The supposition that woman would improve the price of her labor by legislation is not more reasonable than it is in the case of men, who have to yield to the inexorable law of supply and demand.

When we consider the losses that women would sustain with the suffrage carried into effect *bona fide*, the reasons in its favor dwindle out of sight. The first effect would be to render marriage more undesirable to women than it is now. A premium would be at once set on unmarried life for women, and the *hetoera* would become a more important person to herself and to the state, than the wife, because more independent. The number of men and women who would adopt some system of marriage without obligation, would greatly increase. Confidence and sympathy between married people would be in many instances impaired; in fact, the first and many other steps would be taken in the process of weakening home affection, and there would follow a corresponding loss of its civilizing influences and a turning backward of the current of moral progress. The intervention of women into public affairs is to be dreaded also by those who desire peace among men. Both women and their male friends resent treatment for them which men would quite disregard as applied to themselves; and woman suffrage would see the introduction of more or less numerous women into public life. The extreme and irresponsible language used by Mrs. Stanton and Mrs. Lathrop at the last woman's congress in Washington effectively illustrate this aspect of the question.

The devotional nature of women must not be left out of the account in considering this question. While this element is of immense value to that sex and to society when expended upon ethical themes, when it is allied to theological issues it becomes an obstruction to progress of the most serious nature. Were woman suffrage granted, theological questions would at once assume a new political importance, and religious liberty and toleration would have to pass through new perils and endure the test of new strains. What the effect would be we can not foresee, but it could not be good. The priest would acquire a new political importance, and the availability of candidates would be greatly influenced by the question of their church affiliations. . . .

What I have written does not include any reference to supposed inherent right to the suffrage or to any principles of representative government. This is because the view that suffrage is not a right but a privilege appears to the writer to be the most rational one, and because any system of government which tends to disturb the natural relations of the sexes I believe to be most injurious. In the absolute governments of Europe the home is safe whatever else may suffer; but a system which shall tend to the dissolution of the home is more dangerous than any form of absolutism which at the same time respects the social unit.

What America needs is not an extension, but a restriction of the suffrage.

THE EXTENSION OF THE SUFFRAGE TO WOMEN

FRANK CRAMER

January 1889

IN Prof. Cope's "Relation of the Sexes to Government," in the October number of the "Monthly," he makes intellectual inferiority, physical inability, and the social position of woman the practical objections to granting her the "privilege" of suffrage, and favors its restriction rather than an extension.

But even if men are on the whole superior to women, the difference is not so great but that, if the same restrictive process were applied to women and men, a considerable minority of the women would fulfill the conditions which a not very large majority of the men could fulfill. Although any system of suffrage can only be an approximation to what might be best, it is a poor approximation indeed that will shut out a large minority of one sex because the majority of that sex fail to fulfill the qualification for suffrage. That is majority-rule with a vengeance.

It is declared that "woman suffrage becomes government by women alone on every occasion where a measure is carried by the aid of woman's votes." Then government by a successful party, whose candidate is elected by a majority of one thousand in a "deciding State," becomes government by five hundred and one men; and government everywhere becomes government by the smallest possible majority of the majority by which a party elects its candidate. What becomes of popular government? It is further declared that, if women vote with their husbands, suffrage becomes a farce. It is a very plain social fact that men who associate much come to think alike, especially on subjects that are much thought upon. Like teacher, like student; like father, like son. Politics runs in families almost as much as features do. If all who acquire their political leanings from their constant associates shall not vote, a very large majority of the sons of the country must be disfranchised, and in a generation there will be no voters at all. And if the women of the land, by exercising suffrage, run the danger of becoming the mothers of a "generation of moral barbarians," are the fathers of the race so entirely different in quality from the mothers that the transmission of a very large amount of barbarism might not be prevented by a wholesale restriction of the suffrage?

Physical inability to execute the laws when they are made, and to defend them in a military capacity, is made a principal objection to the granting of suffrage to women. "This consideration alone, it appears to me, puts the propriety of female suffrage out of the question." But only a small proportion of men are willing to be executors of the law, as policemen and sheriffs; and, as for the judicial positions, an even smaller proportion is *fitted* to fill them. Restriction of the suffrage would be a good thing; let it be applied under the principle of immunity from military service, and who would be disfranchised? War demands able-bodied men; only men that are perfectly regular in form and sound in health can be soldiers. If immunity from service is to form the boundary-line of suffrage, all the rest, a vast number, would be shut out. This excluded list would include perhaps the best class of voters the nation has — the older men — because they are exempt from military duty. But I am sure the professor himself would be unwilling to begin restriction under the principle he has enunciated, and reduce the elders of the nation to the condition of Gulliver's Luggnaggian *struldbrugs*.

THE MENTAL FORCE OF WOMAN

THERESE A. JENKINS

April 1889

NO article, perhaps, that has lately appeared in print has called out a more decided difference of opinion than the one entitled "The Relation of the Sexes to Government," which appeared in "The Popular Science Monthly" for October. Especially has this been marked in Wyoming, for it is here, I believe, that we find the nearest approach to a relation of both sexes with the Government. In the outset of his article Prof. Cope stated that, "being free from the disabilities imposed by maternity, the male could acquire a greater mastery over his environment than the female." Now, in all observations of animal life lower than man, the contrary appears to be the case. We find the female taking the most active part in the struggle for the existence of the young, and certainly doing as much for her own existence as the male for his. The lioness, in providing for and protecting her young, which in animal life represents the home, exerts a much greater "mastery over the environment" than the male, which only for a brief period shows a care for the female, and neither affection for nor government over the young. The horns of the female kine in defense of the calf are to be dreaded as much as those of the male. We do not find the male cat feeding or protecting the kittens. The hen not only provides for and defends but also chastises and governs her brood. In the insect world we find that the female spider eats her husband, bees kill theirs, and female ants make slaves of theirs. Coming to man, we find that among the Indians the female does the drudgery, and also the providing, with the exception of the hunting. In the wild Kurdish mountains we find women doing labor that the beasts of burden fail in, bringing great bundles of fire-wood down those terrible mountain-sides. We find them protecting their fields from the ravages of bears, fighting and slaying them with as much fury as the men, hindered neither by lack of physical strength nor by maternity. Macaulay speaks of a scene in the Scottish Highlands where aged mothers, pregnant wives, and tender girls are harvesting oats, while the men bask in the sun or angle in the streams.

Prof. Cope claims that women would be irresponsible voters, as they can not assist in the execution of the laws that they help make. Does their physical nature prevent them from doing this? In the riots of Ireland, Canada, and the United States does woman stand back hindered by physical weakness from throwing stones, beating the magistrates, or barricading street-car lines? Can it be proved scientifically that man had rather meet infuriated woman in preference to a male antagonist? In the pioneer days did not woman's bullet speed as true to the mark as man's in the protection of her home? Where has woman failed? In the exhausting marches of exiles to Siberia do the facts show that man stands the journey better than the Russian woman?

Again, the professor says, "The mastery by him has accustomed her to yielding, and to the use of methods of accomplishing her desires other than force." This amounts to saying that, while man is superior in force, woman is superior in diplomacy. Now, if it can be proved that in government the latter is as important as the former, then will be shown the absolute necessity for co-operation of the two sexes in political affairs. In the garden of Eden we find, instead of Adam choking the apple down the throat of Eve, Eve persuading Adam to partake, and here diplomacy wins. It can not be denied by our most adverse opponents that during the last half-century woman has taken possession of educational government. The teachers of the United States today are women. Our sex governs the schools throughout this broad land, and we maintain this government, not through force, but through tact or diplomacy.

Here in Wyoming some experience with woman suffrage has been acquired, though in a Territory of course there can not be as wide scope for its exercise as under Statehood. Now, if it could be believed all over this land that women would allow themselves to be "loaded" into wagons by their man, and driven to the polls to vote his ticket, as the writer of the article in question rudely states it, this would give a mighty impetus to woman suffrage. But this is false. Suffrage is not denied woman because she will vote as man dictates, but because she *will not*; and man knows full well that force would very quickly succumb to diplomacy. It is true we go to the polls in carriages placed at our disposal by the candidates, but is this any proof of disloyalty to our convictions? Are the members of a choir who attend the services of the G.A.R. in carriages provided for them to be accused of having no patriotism nor respect for the honored dead? Is it to be supposed that, in spite of birth, education, or culture, we would become as ignorant vassals to the husbands and fathers whose love, respect, and protection we had possessed, or that our male associates are so debased that they would wish us to become such willingly, or compel us to become so unwillingly? There are women, no doubt, who vote as their husbands vote; but, having been a resident and a voter eleven years in Wyoming, I have yet to find one case where a woman has voted as the *force* of man dictated. There are women in Wyoming who do not vote, but it is not because their male associates compel them to remain at home, and they resent such an imputation. Neither is the woman-suffrage movement condemned by them. The majority of women in Wyoming vote, and vote according to their own preferences, and the men so desire and expect them to vote. It has been stated, rather coarsely, that woman, for the sake of remaining her own master politically, would be tempted to refrain from *legal* marriage. But were this to prove true, and were woman without a legal protector to step up to the polls to deposit her ballot in opposition to the males, we might look more confidently for exhibitions of force, and, instead of finding woman submitting privately to the maltreatment of her husband, we should see her obliged to suffer publicly the brutality of many men.

I wish that women everywhere would study the one argument that can be brought against woman suffrage. It is this: Woman may reform man. He has shown us clearly that he will not reform himself. Now, unless woman will interest herself in this reformation, she has no business with the ballot. So far woman has done as well as man in the use of the ballot; she has done no better; but she can, if she will. Man has no right to expect woman to take up issues that he ignores, nor has he any right to withhold the suffrage for fear she will do so. But woman in asking for the ballot ought to say to man, We will make better use of it than you have. This is the ground on which we must demand the suffrage. Not the use of the ballot simply to make our own importance greater, but the ballot as it could be used to raise politics out of its filthiness, corruption, and ignorance, and to bring in the reign of purity, patriotism, and intelligence.

George F. Talbot (1819-1907), a lawyer practicing in Maine, set forth in "The Political Rights and Duties of Women" what were at the time prevalent arguments against woman suffrage. Talbot's first point was that women were disqualified from the franchise because voting (to the antisuffragist) necessitated officeholding, which was clearly incompatible with homemaking, the occupation of 80% of all women. His second point was a common response to the claim that women required the vote in order to rectify social ills under which they suffered. Talbot argued simply that women were a privileged class and could alter existing ills without the franchise. His final argument was that women should not have the same "rights" as men because they did not share the same responsibilities: Talbot referred specifically to women's exemption from military service and from the obligation to earn their own livings.

Talbot's article should be read in conjunction with two replies that appeared in subsequent issues of The Popular Science Monthly: "Woman and the Ballot," by Alice B. Tweedy, a New York writer, and "Occupations, Privileges, and Duties of Women," by Grace A. Luce, a popular novelist.

THE POLITICAL RIGHTS AND DUTIES OF WOMEN

GEORGE F. TALBOT

May 1896

IT is avowed by all these persons, who speak frankly, that women want the ballot in order that they may become candidates and officeholders, and so be able in the interest of their own sex to affect local, State, and national legislation. We may, therefore, lay on the table the specific question of giving the ballot to women — leave it unsettled — conceding that, if it were only that, the matter might be arranged to meet the wishes of the petitioners, and confine ourselves in this discussion to the rights and qualifications of women to be the administrators of political power, and the effect which the exercise by women of those political functions now performed exclusively by men would have upon the welfare and character of women generally.

I. To the complete performance of such political functions there is this serious natural impediment: four fifths of the women all the world over, between the ages of twenty and sixty, are occupied with paramount domestic obligations quite incompatible with that integrity of devotion to public duties which all the great executive, judicial, and legislative offices demand of those who fill them. Under this disability of Nature, or closely related to it, all the objections to the exercise of political functions by women may be classed, so that no other objection need be considered. If the mother of a family of young children should give to the office of President, Governor, judge, or sheriff that entire devotion of energy, time and thought which her official oath exacted of her, she would be obliged to do it at the expense of that assiduous care, watchfulness, and service which her wifely and maternal relations demand. . . .

I know the answer to this objection generally made: Yes, there are many women, as there are also some men, whose health, whose business, whose domestic cares, render them averse to office and exempt them from its responsibility. The good sense of the voters may be trusted not to select such engaged persons as candidates; and if they should be selected the good sense of the candidate can be trusted to decline the office, and that will end it.

But is this answer quite satisfactory? It is a question of reconsidering and readjusting the occupations respectively of men and women, which all civilized and uncivilized peoples, without concert among themselves, have established and built into their social institutions. An arrangement of this permanence and universality may be considered an arrangement of Nature. Nature evidently regards as of supreme importance the perpetuation of the race, and imposes presumably, and at least potentially, upon all women a paramount duty in accomplishing this purpose. The political disability, whether extending actually to four fifths of womankind or potentially to all womankind, is one irrevocably connected with that very office and *raison d'être*, which called woman into existence. An objection to employment in public office good as against four fifths of the female sex ought to be good as to the whole sex, just as if it were a question of enlisting women as soldiers, or shipping them as seamen, or engaging them as miners or engineers—a disability affecting the greater number would be likely to disenable the whole.

This is the situation. The great body of men—the men in the prime of their physical and mental powers—have no employments or duties imposed upon them by Nature incompatible with the strict performance of the obligations of public office. A man may be a punctual and industrious executive officer, a studious judge, a commanding general successfully conducting a campaign, and be no whit less a faithful and helpful husband, a wise and provident father. This very excellence in these purely private and domestic virtues, while it would add to his popularity, would never be thought of as impairing his efficiency as a public servant.

Now it happens that women during the same period of their physical and mental prime are by their ruling instincts and their dominant sentiments assigned to duties which leave neither time nor faculty for any absorbing and responsible public station. It might be invidious to say that the best women are in this category of disability; it must be said, however, that the women whom men think the best—at least the best to be wives and the mothers of their children—are not eligible to public office.

In this actual condition of things what will be the probable result of sharing with all women, by a sweeping enfranchisement, the privileges of all political offices? Only those will be likely to be proposed as candidates, or at least will consent to be candidates, who have no incompatible domestic duties — unmarried women, who have no pleasant homes, or fathers, brothers, or sons with whom they can live harmoniously, and all the forlorn class, who have failed to come into agreeable relations with other persons, or who have made shipwreck of their domestic ventures. I question whether the great body of virtuous and intelligent women, the mothers, wives, and sisters of the citizens, would be so well satisfied to be represented by such persons as by those citizens themselves. . . .

Let us suppose that in this congressional district, under the *régime* of full woman suffrage, some brilliant, educated, and accomplished lady, whose eloquence on the stump in a political campaign had electrified thousands of listening voters, had been nominated and elected as representative to Congress. Among the auditors whom she had fascinated might not one every-way eligible man have been bold enough to make confession of a personal attachment before the eloquent pleadings of this young Jeanne d'Arc of politics should find herself compelled to forego her ambition for public distinction, and take upon herself the humbler but sweeter duty of consecration to a single man? The same accident would befall everywhere. Only the intelligent and agreeable women would be popular, and only the popular women would be candidates and elected. To put them in office would of itself expose them everywhere to appropriation by men brought by the occasions of public business into the circle of their acquaintance. I dare not pursue further this dangerous argument.

Will the female suffragists consent to a *self-denying ordinance* that shall exclude from office? I apprehend not. That is the very thing they will not listen to with patience. They have avowed that what they demand is that women shall have an opportunity to try their hands at law-making and law-administering, with a view of bettering both. They wish to vote in order that they may vote for each other, and no way has been proposed or seems practicable of making women electors that will not also make them potentially the elected.

As soon as the naturalized Irishmen in Portland became an appreciable element of the voting population, they began to be put upon the electoral tickets for municipal and State offices by both parties. In Boston and in New York, where they compose a majority of the voters, they get the majority of candidacies for places under the city government. When by a heroic effort we lifted a million of ignorant and degraded slaves to the rank of citizens and electors, the immediate result was negro justices of the peace, negro judges of the courts, negro members of Congress, and in more than one State negro Legislatures, which proceeded in a summary way to confiscate the property of the late masters by taxation, ostensibly expended in public works and largely wasted by private plunder. If the effect of raising to the grade of voters the whole mass of illiterate slaves was to give them the whole political control of several States, why will not this complete enfranchisement of women give them the political control in all the older States, where they will be in the numerical majority? . . .

II. What wrongs are there affecting society which the women's vote and the political power it gives will set right? What disability or oppression does woman suffer at the hands of man, which she must rise in her physical might to redress? Every other agitation for social or political reform now rife, or that has been rife in my day, has been able to justify itself by a flagrant abuse repugnant to the universal sentiments of mankind. Slavery, intemperance, the poverty and privations that have been caused by unjust distribution of the products of industry—all these are palpable evils that denunciation can not exaggerate nor eloquence winged by strong emotion overstate. But the woman's grievance against man in these modern times, in any civilized country, what is it? The moment you begin to sum it up, the moment you undertake to tabulate and itemize it, you provoke the indignation of all generous and intelligent women. The moment you attempt to inflate its emptiness with the breath of invective you have to deal with hysteric fancies rather than hard facts, or consciously to enact a *make-believe*.

I am careful to say that woman has no grievance *against man*, I do not say she has no grievance. In common with all sentient creatures she does complain of the hard conditions of universal existence—conditions which it has been the long, slow effort of what we call civilization to amend and improve. The special hardship of the lot appointed to her by Nature is, that the pains, burdens, and weary cares that parentage imposes upon each generation, in order to provide for the succession of the race, have been unequally and cruelly laden upon one of the sexes. . . . Civilization increasing the leisure, lightening the toil of man, and relieving him in a large degree from the wars in which he was mutilated and slain, has not been able, in any appreciable degree, to redeem woman from the primitive sufferings by which she consecrated her motherhood; and so the unequal fortunes of the two branches of the human race have become under the improved fortunes of the race more pronounced. . . .

Leaving out, then, whatever offense a cruel Nature has committed against woman, let us see if men have fairly acquitted themselves of their natural obligations to her. Take the present legal status of woman. Since men began to make laws they have made them for women, and in what situation has their deliberate sense of justice left women before the law? One after the other they have obliterated from the statute book all laws that discriminated against women in respect to their personal rights, and to the acquirement, possession, and disposal of their property—old laws that had their origin in the barbaric spirit that made woman the slave of man, and, it must be confessed, which found no little sanction in that dogma of our accredited Christianity, which taught, too plainly to be misunderstood, that woman was as much below man in the scale of being as man was below the angels, her paramount duty being to be subject to him. But all this barbarism, Christian and un-Christian, has been swept away, and that too not by woman's suffrage, actual or prospective, nor by woman's petition or any political agitation prompted by her, but by man's own sense of equity and right.

No, women are not an oppressed class, least so in the United States, in England, in any country whose people have inherited the Teutonic sentiment which in the ancient Germany described by Tacitus made women counselors and advisers in the affairs of war, government, and business, as well as in matters purely domestic. *Women are a privileged class.* . . .

III. Women are exempted from the perils, wounds, and deaths incident to war. . . .

To what is the exemption of women from military service due? The hasty answer may be, to their physical and mental unfitness. The physical strength of the average woman is perhaps twenty per cent less than that of the average man. This disparity could be readily adjusted by adapting the labor and discipline of the two classes of recruits to it. Make the regulation musket for the female regiments twenty per cent lighter than the standard, and so the personal baggage; and if twenty miles is a fair day's march for men soldiers, require of the women soldiers but fifteen miles. The nerves of women might more quickly than those of men succumb to the terror of shot and shell, or of a bayonet charge, but actual wounds and mutilation they would endure with more patience.

In the few instances of an exceptional custom preserved in history, natural disability had ceased to be a factor. There is a Greek legend of the Amazons, a race of women in Asia, so formidable as to terrorize all the early Grecian settlements, and to require such valorous heroes as Bellerophon and Hercules to subdue them. Travelers more trustworthy than Baron Munchausen tell us of an African king, whose standing army is of women—a fierce and terrible array.

It is even possible that in this judgment an effect has been mistaken for a cause. Women have not been exempted from military service on account of their congenital delicacy, but their characteristic delicacy is the slow result of their exemption from this and other hardships submitted to by men. . . .

To what, then, must the anthropologist attribute that custom, almost immemorial and universal, which savage and civilized men have concurred in establishing, of exempting women from the service of fighting? To that which will be found to be the spring and source of most of their customs and institutions—sentiment; a sentiment in man of mingled pity, respect, and affection—a sentiment, like most others of somewhat low origin, beginning, it may be, in selfishness and the promptings of instinct, but flowering out in its complete evolution into a noble and divine virtue. . . .

IV. The next great immunity which, in recognition of the offices they exercise in the social system, and having its origin in the same sentiment, women enjoy, is exemption from all kinds of labor dangerous to life or exposing to hardship and privation. . . .

All the hard, repulsive, life-wearing work under ground in coal, mineral, and metallic mines is generally assigned to men, and they alone are exposed to those perils which beset engineers, train-men, the handlers of explosives, and the tenders of machinery.

It is certainly apparent that man, as the stronger sex, has not made an ungenerous use of his strength in his assignments. Having, in the right of his strength, the opportunity to determine the customs of society, he has taken upon himself, and exempted his mate from, all those vocations that expose to premature death or to great physical suffering, as well as those which segregate men from the social enjoyments of home and doom them to long exile in cold, storm, and darkness.

V. The last privilege of the sex—for only the great, cardinal privileges of womanhood need be enumerated—is woman's virtual exemption from the care of earning her livelihood and that of her offspring. Here, . . . I have disregarded exceptional and abnormal instances, and traced industrial and social customs, as men following the promptings of their dominant sentiments have been able to establish them. . . . I do not forget to have seen women tugging baskets of manure to their miserable fields, or buried under burdens of hay, in Switzerland; nor the Sunday I drove by women mixing mortar and carrying hods of bricks up shaking ladders in the elegant city of Vienna; nor how all-prevalent poverty has equalized the lot of the two sexes in Russia.* . . .

But even in our times and among the most advanced races there are many exceptions to the general assignment to men of the primal care for daily bread. Very many men, through accident, sickness, and mental or moral incapacity, get disabled in the struggle for life, and the burden they were appointed to carry falls unnaturally upon a wife, a sister, a daughter. Widows carry on successfully the farms which a dead husband had cleared. Sisters take places in stores, in schools, engage as copyists, and contribute to periodicals, and so bring to the family fund the stipend which a deceased father, a dissipated husband, or an invalid brother ought to have earned. The social organism gets mutilated and wounded, and these are Nature's efforts at recuperation and supply. In the healthy normal society—such as man establishes wherever

---

*"The women are more diligent than the men; and the hardest work is often turned over to them, as is generally the case where peasant properties prevail. They are only 'females of the male,' and have few womanly qualities. They toil at the same task as men in the fields, ride astride like them, often without saddles; and the mortality is excessive among the neglected children, who are carried out into the fields, where the babies lie the whole day with a bough over them, and covered with flies, while the poor mother is at work. Eight out of ten children are said to die before ten years old in rural Russia."—(From a Review of F. P. Verney's *Rural Life in Russia*, in the *Nineteenth Century*, of January, 1887.)

he can—the true order seems to be that "man must work and woman must weep"—unless a cheerful temperament shall convert her weeping into a song, while waiting on the weariness of her yoke-fellow with affection and the ministry of a lighter service. . . .

VI. Now, putting any just valuation upon these great exemptions which the majority of women enjoy, and which all women enjoy so far as the appointment of men as a whole can predetermine their lot, can it be for a moment claimed that the position of women is one of *oppression*, and not of *privilege*?

I do not wish to say that men and women ever made a formal compact that the latter should surrender to the former all their natural political rights, as a condition for enjoying these privileges which have been awarded to them. None of the established customs of society, not even the forms of political government, were made in this artificial way. They were not made at all, but grew spontaneously. Rousseau's doctrine of a social contract is less in vogue than it was in the political philosophy of our fathers. But what I do wish emphatically to say is that, bargain or no bargain, the weaker sex has not been taken advantage of, and that its immunities and privileges are a full equivalent for all the political rights of which it may have been deprived. . . .

We may as well consider what changes in human society, and especially in the character and fortunes of woman, the new order of things sought to be inaugurated will be likely to bring about. . . .The inevitable ultimate result of subjecting the two human sexes to the same labors, the same employments, the same cares, will be just the same as when domestic animals have been subjected for long periods to the same conditions: sexual differences, physical and mental, will tend to disappear, and the two branches of the race will approximate a common type.

He has inadequately considered the nature of the demands made by that section of womanhood in insurrection against the present social order, and the implications which lie behind their specific demands, who does not see the radical changes that will come finally as a result of conceding these demands. However disastrously the experiment may issue, the dificulties of either turning back or arresting the movement will be nearly insurmountable.

Certain discontented women say they want the ballot, in order that they may with it open to themselves, on the same terms and for the same compensation, a free career in all the professions and occupations in which men are engaged. They want to place all women in the condition of service and hardship in which the casualties of life and the precarious fortunes of business now place a few women. They wish to make wounds which the present social structure now receives here and there parts of its normal status. For they want to be lawyers and physicians charging the same fees, ministers having the same salaries, artisans and workmen having the same wages as men. The greater competition among the many women as against the few men in the occupations now open to women they propose to counteract by a statutory equalization of wages for the same kind of work.

The great labor crises and the imperiled industrial equilibrium in the whole civilized world being confessedly due to the excessive number of competititors for such paying work as machinery has left to be done, it is proposed to aggravate the situation by turning into the competition the whole mass of able-bodied women, not hitherto generally reckoned among the working class.

In the woman-suffrage movement the "insurgent women" virtually serve notice upon us men, that they do not desire any of our courtesies, which are a badge of their servitude, and that our politeness in giving them the best places in the concert room and the horse-car is superserviceable and compromises their sense of independence. They do not longer care to be petted or exempted from perils and hardships or to be maintained by labor not their own. They only want an equal chance to *"paddle their own canoe"* in quest of their own fortunes.

Whatever the answer to this demand may be, it will not be likely to be this: Very well, please yourselves; rough it with us in the struggle for life, asking no favors if such a contest invites you. Enlist in the military companies and stand the drill, and when the next war comes, go to the front. Join the fire company in your ward, and run with the machine, when the next fire calls you out at midnight. There is a ship in port bound round Cape Horn, on a year's voyage; the owners have had such bad luck with drunken men, that they mean to try a crew of athletic girls. Go up the Penobscot and live next winter in a camp, and come back next spring balancing yourself with a pickpole on the floating, slippery logs you have cut. Go down into the mines, and with your pickaxe and shovel dig coal and iron. Offer your services at the going wages to run a locomotive, to blast rocks, or handle dynamite.

Men who are husbands, fathers, and sons will not say this or anything like it. But when the lawyer finds his female competitor by the charms of her beauty and eloquence winning his clients; and the doctor, that the woman physician by her motherly tenderness has seduced his patients; and the minister, that some reverend lady by her superior sanctity has supplanted him in his parish; and all men in all their vocations, high and low, by those toils they had gained bread for their families, are pressed with the competition of those it had been their chief spur to industry and their pride to maintain without the necessity of repulsive work, will not the feeling become universal that men are released from their obligations of duty and support toward the weaker sex?

The naturalists tell us that the human race acquired its strong parental affections by performing the needed offices of care and help which the prolonged infancy of its young—so much longer than among all lower animals—made necessary. We know that the tenderness, affection, and sympathy which are the essential grace and charm of womanhood, as well as the courage, disinterestedness, and chivalric sentiment which form the nobility of manhood, have sprung from that very relation of strong to weak, protector and protected, which have for ages subsisted among all the civilized races. What guarantee can they give us who are seeking to destroy that relation, or at least the cause and reason of its existence, that those cardinal virtues that adorn and dignify both sexes will not be involved in its destruction? For one, I should not dare to vote to drag woman from the high estate in which man honors himself in being her minister and servant, until at least the intelligent majority of women deliberately express their judgment in favor of a social change so consequential.

WOMAN AND THE BALLOT

ALICE B. TWEEDY

June 1896

IF every man considered it a matter of conscience to give voice in his vote to the feminine element in his household, it would put another aspect upon the demand for woman suffrage. If, after a family conclave, the husband, father, or brother quietly pocketed his own conflicting opinion, sallied forth and supported the measures favored by the home majority, what right-minded woman could complain? It would be merely an extension of the main principle of republican government. Only those women without male relatives would be unrepresented, and for them special provision could be made.

This hypothetical condition, however, is so far from fact that it sounds facetious, and the picture of a household wherein a gentle-minded man revises his sentiments to adequately set forth the contrary views of his womankind seems altogether Utopian, yet such a situation is one in which it might be justly claimed that men were the actual political representatives of women.

Some men there are, though *rarissimoe aves*, fair enough to acknowledge that woman ought to be represented in this fashion, or else allowed to deposit a ballot for herself. The proposition of woman suffrage alone does not trouble them, but they stumble over the corollaries of political life and officeholding, and, rightly judging that the trio are logically involved and claimed by suffragists, they demur at the result or reject all together.

Political avocations seem to them utterly alien to the womanly nature, or at least to what they know of it; and since their conception of this elusive quality is undoubtedly founded on the particular instances which have fallen within their experience, it would be useless to oppose it with a flurry of words. One of their number, however, in a paper on The Political Rights and Duties of Woman, is explicit, and furnishes us with several statements which may be debated. To the performance of political functions by women, he holds there is "a serious natural impediment" that "four fifths of the women all the world over, between the ages of twenty and sixty, are occupied with paramount domestic obligations incompatible with public service." "Under this disability of Nature, or closely related to it, all the objections to the exercise of political functions by women may be classed, so that no other objection need be considered."

It is no longer, then, a vaporous theory that confronts us, but an array of questionable facts. The condition of four fifths of the women "all the world over" is certainly beside the issue. We have no reliable statistics regarding them, and we are not at present concerned with their political disabilities. The ballot is demanded only for the women of civilized communities, where the right of suffrage is already possessed by men, and the question is immediately pertinent to those in the United States. Here statistics are available, and in New York State they run as follows:

Women between the ages of nineteen and sixty-five.

Total number 1,707,655
Married women 1,244,291
Mothers[1] 1,238,070
Mothers disqualified for public service 550,252
Eligible women 1,157,403
         = 67 percent of the whole.

Comparing men, we find certain classes among them ineligible to political office by reason of their professional or business duties, yet disfranchisement of their sex on that account has never been considered. Priests and ministers of the gospel, even if devoting some time to politics, could not give to public office "that entirety of energy which an official oath exacts" without disregarding the spiritual welfare of their flocks; and if they are true pastors, it would not be amiss to compare them in the multiplicity of their cares to the mothers of young families. Physicians in active practice can not well be judges or sheriffs without neglecting the vocation for which they are especially fitted. Scientific men engaged in original research are not expected to abandon their laboratories, where they may be on the eve of bringing forth the fruit of lives wedded to patient observation, even if a mistaken populace should nominate them for mayors or Congressmen. Manufacturers and business men have even been known to decline senatorial honors, since these conflicted with the responsibilities of their callings.

If a count could be made of all these men who, for various reasons, will not accept political candidacy, it might be found to equal in number the mothers who are disqualified for office-holding.

It is to be observed that at any given time only a minority of mothers are

---

[1] The general proportion of mothers among married women is ninety-five per cent. Of these, the maximum number disqualified would be fourth ninths.

even thus conditioned. That four fifths of womankind between the ages of twenty and sixty are ineligible for public office proves thus to be an exaggeration.

Planted upon this astounding proposition, our antisuffragist then proceeds to discuss the complications that may arise if women enter upon political life. While they attend committee meetings, the scarlet fever may invade the nursery. If they engage in jury duty, the husband, fretted with financial cares, will fail to find sympathy at home.

It may be presumed that women with young children will not generally accept candidacy for public office; but should they in some cases think best to do so, such contingencies are not unlike those that occur outside of political life. A wife is called to the bedside of a dying mother, one thousand miles away. She leaves her children; the measles breaks out among them, and the father, although an inexperienced man, nurses the flock back to health. Instances are not wanting in which men have wrestled victoriously also with other diseases, so that a great gloom need not settle down upon mankind at the prospect of a mother's occasional attendance upon a committee meeting.

The dearth of sympathy at home is no matter for jesting. No doubt thousands of women, in times of anxiety, have gone entirely unconsoled while their husbands were jurymen. If men have a taste of this experience, where is the injustice?

Not very relevantly our opponent breaks in here with the assertion that "the suffrage is a question of readjusting the occupations of men and women as established by all civilized and uncivilized people." As the occupations of men and women *vary* with the state of civilization and the industrial development of a country, this generalization is valueless. The employments of men and women also depend upon the condition of the nation, whether militant or peaceful, and in regard to certain kinds of work no universal rule can be made. Women act as horse-car conductors in South America; Chinamen prefer the laundry in the United States; while in East Central Africa men insist upon sewing their own and their wives' garments, leaving the women to build the houses and hoe the corn. The modern readjustment of vocation in our midst arises, as it has been pointed out, from the increased leisure afforded women by the introduction of machinery. It is a wonderful evolution for woman, proceeding as noiselessly as the spinning of countless cocoons, liberating many who would have grubbed a hundred years ago to try their wings to-day if they will.

The next statement volleyed at us is very like an explosive used by Mrs. Lynn Linton in one of her harangues against women.[2] "The political disability is one irrevocably connected with that very office and *raison d'être* which called woman into existence."

[2] The Wild Women as Politicians. Mrs. E. Lynn Linton. Nineteenth Century, July, 1891.

Despite our advancement in science it seems next to impossible to extricate some minds from the mire of tradition. Brushing biology and common sense aside, these primitive souls continue to regard woman as the mythical rib of Adam. Those of us who have progressed beyond this dogma look upon it just as flatly contradictory to Nature as the biblical view of the earth as a plane. Woman's sexual life is shorter than that of man, her individual life longer. Therefore, if either was "called into existence" for the office of parenthood, it was obviously the man, not the woman. From a biological point of view the functions of life are two—nutrition and reproduction; and there is as much sense in saying that nutrition is the reason of man's existence as to state that motherhood—if that be "the office" meant—is the *"raison d'être"* for women.

As for us, we frankly confess we do not know anything about "reasons of being" or causes of existence. If Mrs. Lynn Linton, Mr. Talbot, *et al.*, have been taken into the creative confidence, no doubt they have interesting revelations to offer the world!

Our antisuffragist, not being quite content with delving into prehistoric purposes, next hazards a prophecy of the feminine officeholder. As wives and mothers are, according to his premises, ineligible, only "those who have made shipwreck of their domestic ventures," the forlorn and *déclassées*, will pose as nominees. He provides, however, "a contingent disability," that of getting married, which may overtake these. As our prophet waxes eloquent over matrimony he forgets what manner of woman he has pictured as a politician, and tells us "only intelligent and agreeable women will be popular, and only popular women would be candidates and elected." The forlorn and *déclassée* woman is metamorphosed into "the brilliant, educated, and accomplished lady stump speaker," and when she marries, what can be left for the suffragists?

Having thus disposed of the phantasmagoria of his creation, he asks two momentous questions:

1. What wrongs are there affecting society which the women's vote will set right?

2. What oppression does woman suffer at the hands of man which she must rise in her might to redress?

I am not aware that woman suffrage is proposed as a panacea for social evils, or that it will usher in a millennial condition. Man would be disfranchised if such requirement was made of his vote. Legislation does not beget character, and man is not made temperate and pure by law. Stringent laws, however, are needed to prevent various evils and to make certain offenses punishable. Women are quick to recognize vicious tendencies that men with a greed for money-getting often overlook. The work of Mrs. Fawcett in England, and of many earnest women in the United States, shows what good would accrue to society if women helped to frame the laws.

Our opponent does not pause to consider whether woman's vote would be beneficial or not to the community, but spends his full strength in fortifying the second query. "The woman's grievance against man, what is it?"

he asks. "The moment you attempt to inflate its emptiness . . . you are dealing with hysteric fancies rather than hard facts. . . . Woman has no grievance against man. . . . Cruel Nature has committed an offense against woman."

English law is more nearly defined as "a hard fact" than as "a hysteric fancy," and English law contains a long "bill of grievances" which woman may publish against man.[3] True, in this land of boasted freedom most of these laws have been repealed, many others are a dead letter, and still others have been enacted that favor woman. These changes have been brought about by the growth of the sense of justice, but also directly through the efforts of women agitators who have pleaded and written against decrees of oppression. These writings and arguments are a matter of record, and they antedate all betterment of the laws relating to women. Without them we do not know when "man's own sense of equity and right" would have impelled him to annul the obnoxious statutes. Even here in New York we come occasionally upon instances which betray the defects of a masculine code,[4] while in the civilized countries of Europe the laws generally discriminate in man's favor.

Outside of unjust enactments, the former subjection of woman is stamped on our customs, our literature, and our language. It is hardly possible for any one to investigate the origin of many of our conventionalities, titles, terms of obloquy, without coming unexpectedly upon proof of man's injustice to woman. . . .

Our opponent reaches at length his principal tenet: women are a privileged class. Their privileges consist not in the minor courtesies of life, but in various immunities and exemptions which are "a generous attempt on the part of men to make for their mates and yoke-fellows an easier pathway through a rugged world. . . . Having in the right of his strength the opportunity to determine the customs of society, he has exempted his mate from all those vocations that expose to premature death or great physical suffering." An inventory of these exemptions follows:

1. From the perils, wounds, and deaths incident to war.
2. From all kinds of labor dangerous to life or exposing to hardship and privation.
3. From the care of earning her livelihood and that of her offspring.

One is at first sight aghast at this record of masculine arrogance. Women might retort, and say Men have exempted themselves—

1. From the care of their progeny.
2. From the preparation of clothes, food, and household toil.
3. From nursing the sick.

---

[3] John Stuart Mill. *Subjection of Woman*, pp. 56-58.
[4] A woman appointed administratrix refused to pay an exorbitant bill. Her arrest for contempt of court resulted in the death of her babe. The surrogate said it was a case of great injustice, but the code made it mandatory upon him to issue the order (*New York Times*, June 11, 1890).

All these "exemptions" are misnomers. Men have "exempted" women from nothing. They have excluded women in former times, and still exclude them in some degree, from the higher institutions of learning, the professions, and government. These exclusions, however, would form another "bill of grievances." The immunities mentioned are purely imaginary. Man *chooses* to fight, to sail the seas, to dig for gold and iron, to hew wood, and cut his own pathway in the world because he is a man and likes it, not to save any woman nor womankind from such tasks. He has the combative instinct that greets a struggle, the well-knit muscles that crave vigorous action, the adventurous spirit that courts the unknown, and the courage that defies danger. Does a boy wrestle with his playfellow to spare his sister; or run away to sea, or to the gold mines of South Africa, from an altruistic feeling for womankind? . . .

Woman is not only weighted by these gratuitous immunities but, according to this document, her natural delicacy is owing to them, and she is warned that she may part with womanhood if she persists in her unnatural endeavor to change her occupations. The bee is cited as an example that "sex itself may be determined by continuous special regimen or diet."

Now, so far as any naturalist has observed, sex is not *altered*[5] by any regimen or diet; and as the subject of our inquiry is not a neutral bit of protoplasm, but a developed individual, woman, we do not need to study the origin of her differentiation so much as its possible modification. . . .

Thousands of women work in the mines of Belgium, England and Cornwall.[6] In the first-named country they formerly worked from twelve to sixteen hours a day, with no Sunday rest.[7] The linen-thread spinners of New Jersey, according to the report of the Labor Commissioner, are "in one branch of the industry compelled to stand on a stone floor in water the year round, most of the time barefoot, with a spray of water from a revolving cylinder flying constantly against the breast; and the coldest night in winter, as well as the warmest in summer, these poor creatures must go to their homes with water dripping from their underclothing along their path, because there could not be space or a few moments allowed them wherein to change their clothing."[8] Yet women are "exempted" from labor attended by hardship!

Despite these washerwomen, miners, and linen-thread spinners, we are told "it is woman's privilege generally to be exempted from the care of earning her livelihood and that of her offspring."

---

[5] The genesis of sex in certain orders seems to depend upon differing temperature and nutrition.
[6] Census of England and Wales, 1891, vol. cvi, table 6. Miners, female — *coal*, 3,267; *copper, lead, tin*, and *ironstone*, 1,425.
[7] *Vide* Report of Reichstag, 1889, forbidding women to work in the mines of Belgium on Sunday and at night.
[8] Report of Bureau of Labor, State of New Jersey, 1888.

It would seem to be time that this libel upon woman should be scorned by fair-minded men. From all antiquity the majority of women have been faithful workers, rendering a full equivalent in labor for their scanty share of the world's goods. The origin of every industry bears testimony to this. In our own era, while women were still homekeepers, did they not earn their livelihood? What was the weaving, the sewing, the cooking, the doctoring, the nursing, the child-care, "the work that was never done," if it was not *earning* a subsistence? Even in these days, when woman goes forth and receives the reward of her labor as publicly as man, she is no more worthy of her hire.[9] Her ancestress—sweet and saintly soul!—did not dream of recompense.[10] But was it not her due; and shall we refuse to credit it because man was then a self-sufficient ignoramus who deemed himself the only one fit to acquire property?

One by one the old industries have been transplanted from the home, and still man constructs new schemes of enterprise from the little tasks that once rounded out woman's day of toil. In the census of 1890, three hundred and sixty-nine groups of industrial work are enumerated, and in all but nine of these women are employed, the actual ratio of women workers to men being 1 to 4.4. The United States Commissioner of Labor writes: "A careful examination of the actual earnings of women discloses the fact that in many industries their average earnings equal or exceed the earnings of the men. . . . "

This phase of the industrial evolution is unrecognized by our antisuffragist, and he depicts for us how a census-taker would find the sexes relatively employed—the man going to work and the women engaged at home in household supervision and social duties. He admits that now and then women teach, act as clerks, or do literary work, but these are exceptions.[11] "In the healthy normal society the true order seems to be that 'men must work and women must weep,' unless a cheerful temperament converts the weeping into a song." This, which might answer for a poetical view of the lives of the fisher-folk

---

[9]"The never ceasing industry of the women was the principal factor in the development of a manufacture that was probably contributing more directly to the personal prosperity and comfort of the people than any other then in existence in 1790" (*Industrial Evolution in the United States*, p. 20). Carroll D. Wright.

[10]Women colonists rarely worked for wages; . . . they carded the wool, spun the yarn, and wove the cloth for the male members of the family. In many instances they worked on the land, and did their share in every way to enable the family not only to secure a livelihood but to build itself upon stable lines (*Industrial Evolution in the United States*, p. 112).

[11]The percentage of women workers for the United States in 1880 was forty-nine. [Census statistics do not support this claim. See Appendix in Chapter 6. (Ed.)] The number of women employed in mechanical and manufacturing industries for 1890 was 505,712. They received as wages $139,329,719. There are 549,804 women in New York city over fifteen years of age. Those regularly employed number over 250,000.

of whom it was written, was not typical of the general social condition even in Charles Kingsley's time. To-day it is not true of a respectable fraction. The number of women who live in absolute leisure is an insignificant item and is constantly diminishing. . . .

Our antisuffragist again brings forth the bugaboo which is dear to the conservative heart: the threat of unsexing woman.

"The inevitable ultimate result of subjecting the two human sexes to the same labors, the same employments, the same cares will be just the same as when domestic animals have been subjected for long periods to the same conditions. Sexual differences, physical and mental, will tend to disappear, and the two branches of the race will tend to approximate a common type."

We can safely let the matter of sex rest entirely with Nature. It is a fundamental fact of our being, not to be disturbed by any little transformation scenes that we can bring about. We may go for analogies to the domestic animals, birds, or fishes, and in none of them will we find *sexual differences* disappearing or tending to disappear. What are called secondary sexual characteristics are very fickle in their nature, and do for various reasons often desert the sex with which they are identified. These are characters merely associated with one sex but having no essential connection with the sex itself, such as the brilliant plumage of the paecock, or, as Mr. Darwin suggested, the baldness of Englishmen. These in a majority of instances depend upon the preferences of the opposite sex, the last example being a probable exception. So men have only themselves to blame if an undesirable type of woman persists. . . .

Among animals we can, by breeding and training[12] through several generations, increase desirable qualities, such as the pace of horses or flight of pigeons, but it is not claimed by any breeder or zoölogist that the sexes are any nearer each other than they were in protozoan times. It is also an assumption to declare that the "graces of womanhood—affection, tenderness, and sympathy"—have sprung from the relation of the sexes. According to all authorities, the general relation of the sexes in all but recent times has been characterized by anything but "affection, tenderness, and sympathy." So far as we have proof of the origin of these qualities, they have arisen from the offices of motherhood; and just in the degree that we elevate, ennoble, and endow the mother with moral, mental, and political responsibility, do we put it in her power to exercise the wisest affection toward her offspring, the fullest sympathy with her mate.

Our antisuffragist "fears to drag woman from her high estate wherein man is her servant." This has for me the melodramatic ring of "a hysteric fancy." With all the opportunities for progress which recent years have given her, it

[12]In every case of change, breeding is certified to be more potent than mere conditioning; *vide* Alfred Russel Wallace, in *The Popular Science Monthly*, vol. xxxviii, p. 94.

does not appear to me that woman is yet on so high a plane as man. She is, however, climbing step by step, and all unprejudiced men and women will welcome the day when she may stand beside him as his coworker in life. That the ballot, officeholding, or any other right which she can exercise or pursuit which she will undertake can render her less a woman, is a hypothesis without a grain of evidence. No biologist can hold it with any consistency. Over and over again such a result was predicted of education. As she was not educated out of womanhood, so she can not be metamorphosed by politics, and will remain, when acknowledged as an individual, still the counterpart of man.

OCCUPATIONS, PRIVILEGES, AND DUTIES OF WOMAN

GRACE A. LUCE

September 1896

SIR: In consideration of the great interest I felt in an able article in your magazine for May, entitled Political Rights and Duties of Woman, I venture to express some of the thoughts which stirred me upon its perusal.

As I understood them, the writer's objections to the principle of woman suffrage can be classed under three general heads: objections as to the advisability or possibility of certain *occupations* for women; objections on the plea of the *privileges* which they already enjoy; and objections based on the idea of any change in the *character* of woman, as wife or mother.

It is to me a matter of surprise, as it must be to many, that the question of occupation should be considered as having any bearing whatsoever upon the subject. Although irrelevant, in a consideration of it, we must own the magnitude of the subject, as viewed not only in regard to woman but to all human kind. What class can take upon itself the responsibility of dictating to any other class what occupation it is or is not fitted to enter upon? It is easily seen that such a course would inevitably clip the wings of progress, as it is a tenet of its movement that the fittest survive, and the unusually gifted of one generation become in some degree the type of the next. Of one thing we may be sure, that no one performs tasks for which he is incapable, and those succeed who possess the faculties necessary to success. A majority of the walks of life have already been thrown open to women, so that the question of a new occupation opened for them by the right of ballot, narrows itself down to the one of officeseeking and officeholding. We on both sides of the question own, of course, that not all women will desire to enter upon this work, or to take advantage of their political rights, with any more alacrity than do a large share of men. There is one thing of which I may be permitted to feel sure, that if any woman succeeds in wresting office from a masculine candidate, or even in time reaches the White House, it will be by means of abilities which no one can gainsay, for, rather than that votes will be given her because she is a woman, the likelihood will be that she will wrest them from prejudice and conservatism in spite of that fact.

Every year is further proving that sex does not extend to intellect, and those who still hold to that belief will in the course of time have to blind their eyes to a great many facts in order to cherish it. The *grande passion* stirs men as well as women, and has power to inspire or weaken in the same degree.

Women are a privileged class, the paper says. It is true that few of us have any remembrance but of kindness and love from father, husband, and brother, and that very many of us have no great wrongs to bring to light, no troubles for which to claim redress. But it is hardly a privilege we enjoy to be loved, but rather mere justice, for do we not love also, and are we not in the same degree kind? These privileges, if we may call them so, which we mutually enjoy, I hardly think can be weakened by the ballot or by anything less than a sudden change and upheaval in the heart of the universe itself. It is a privilege, we read, that women enjoy in being "exempted from the perils, wounds, and deaths incident to war"; that the ballot now takes the place of the more savage conflict of war, and in this conflict, as in the other, women are exempt. You can't exempt women from fighting; five out of six fight. They fight, as does man, the forces of Nature, time, flesh, and the devil. Woman is in the thick of the world's conflict, whatever and wherever her arena, as are all human creatures, struggling with that friction which is progress, with those forces with which processes of evolution polish the stone for the workman, the soul for its soaring. The potter binds her soul to his wheel, as yours is bound, and what she desires is the same freedom, the same room for her wings.

Certainly, now that the conflict has been removed from the open field of war to the more peaceful one of the ballot, the old and earliest valid reason for dictating to her — her minimum of physical strength — has been removed. In no way of life, except in those old, savage, hand-to-hand struggles, is the race invariably to the physically strongest. Do the athletes, the prize fighters, bestow a privilege upon weaker men when they refrain from knocking them down? The necessary requisite, after all, is not brute strength, but *health*.

The question of character is a very large one, and moved by far too mysterious and wonderful forces to be decided by the ballot. It may be that many men are mistaken in their idea that the qualities of gentleness, amiability, obedience, or a small range of thought, make women better wives and mothers than human beings who are capable of justice, breadth of view, strength of judgment, and wide sympathies and interests. Though we please men and men please us, if we keep pace, it will be rather through our higher qualities of mind, character, and heart, than by our lower nature, weaknesses, and faults; but, Heaven knows, both men and women will ever have a sufficient amount of the latter.

One word more. An often-quoted picture is this: The husband, the wage-earner, from morning until night busied with cares and labors, which leave him little time for culture or the more refining pleasures of life, while the wife and daughters are kept in idleness at home, entertaining themselves with gay or frivolous pastimes, expending the income which was earned at such cost. Would it not be better for custom to break its bonds a little and look about it, and allow those idle women occupation that would assist the father and develop their own dormant faculties? Would the sympathy in that home not be of a deeper and more enduring sort?

For women to be idle is no better than for men, and this waste of life and time, which so many are guilty of, at the cost of some overworked man, is a condition of things which cries to Heaven.

William Jay Youmans (1838–1901) was sole editor of The Popular Science Monthly *when this response to Alice B. Tweedy's "Woman and the Ballot" appeared in 1896. Many Americans shared Youmans' belief that granting women suffrage would represent an acknowledgement of and a further contribution to disharmonious relations in the home. As Youmans put it, "to summon women to the polls would signify an antagonism between their interests and those of men." As it was man's duty to vote not only "his own interest as a male individual, but all the interests, domestic and social, which he represents," women had no just claim to the ballot. Instead, Youmans recommended that they develop more trust in man's sense of familial responsibility.*

WOMEN AND POLITICS

EDITOR'S TABLE

August 1896

THE MONTHLY has lately given place to two articles on the subject of the demand which is now being made by some women on behalf of their sex to be allowed to participate in political life on a footing of perfect equality with men. One of our contributors [Talbot] has tried to show cause why the demand should not be granted, taking the ground that the change would be injurious to society as a whole and particularly injurious to the female sex. The other [Tweedy] treats the arguments of the first with scorn, and, if we are not mistaken, betrays not a little of that "antagonism of the sexes" which nevertheless she declares to be "unnatural and vicious." The question is one which ought to be discussed with complete dispassionateness; and we think that on this score there was no fault to find with the earlier of the two contributions, that by Mr. George F. Talbot, in our May number.

Our second contributor, Miss Alice B. Tweedy, disclaims the idea that "woman suffrage is proposed as a panacea for social evils, or that it will usher in a millennial condition. Man," she adds, "would be disfranchised if such requirement was made of *his* vote." The retort is sharp, but is it logical? Miss Tweedy's main contention is that a suffrage restricted to men is fundamentally insufficient for the best social results; and yet she does not want that complete system of voting which she advocates judged by any higher standard than the present incomplete system. If, however, woman suffrage is not "proposed as a panacea for social evils," what is expected of it? Our contributor says that "stringent laws are needed to prevent various evils, and to make certain offenses punishable"; adding that "women are quick to recognize vicious tendencies that men with a greed for money-getting often overlook." "Men with a greed for money-getting" is a phrase which suggests reflections. What is the chief cause of the greed which men display for money? We do not think we are far wrong in saying that it is the social ambition of the women of their families. It is women far more than men who establish social ideals; and, so far as there is a scramble for money, it is their scramble, to say the least, quite as much as the men's.

This, however, is a side issue: the contention that concerns us is that laws are wanted to make certain offenses punishable that are not punishable now; and that women, being quicker than men to recognize vicious tendencies, would get such laws passed if they only had the suffrage. This is a case in which a few examples would be very serviceable. The proposed laws are either such as would recommend themselves to the approval and support of men, or they are such as would not so recommend themselves. If they are of the former kind, they can get passed now; if they are of the latter kind, it is presuming upon an easy compliance worthy of the immortal Captain Reece, R.N., to ask men to make a constitutional change for the express purpose of defeating their own views and principles. Our contributor acknowledges that in this country "most of the laws (that were unjust to women) have been repealed, that many others are a dead letter, and that still others have been enacted that favor women." We must not, however, thank "man's own sense of equity and right" for these beneficial changes. Why? because they have all been subsequent to certain "writings and arguments" of "women agitators." So that man does not exhibit any "sense of equity and right" when he is influenced by the pleadings of "women agitators." Poor man! He is judged very severely these days. We should like to remark, however, that *post hoc propter hoc* is not a very sound form of argument. Grant that the "writings and arguments are a matter of record," it does not follow that these writings and arguments really determined the changes in legislation referred to. What we know is that the changes were made, and that they were made by men under no actual compulsion.

At the outset of her article Miss Tweedy states that, "if every man considered it a matter of conscience to give voice in his vote to the feminine element in his household, it would put another aspect upon the demand for woman suffrage." How is it now, we feel like inquiring, in this matter? We imagine that the great majority of men who put any conscience into their voting at all do consider, as far as it is possible to do so, the interests of the feminine element in their households. When a man votes, he votes for a certain individual who is seeking a certain office. The cases in which there can be any division of interest in the family as to which candidate should be supported must be exceptional. When, however, a man gives a vote for one side or the other, there is good reason to believe—corrupt motives apart—that he thinks, not solely of his own interest as a male individual, but of all the interests, domestic and social, which he represents. In that sense the average elector's vote is meant to be, and is, representative. Our contributor's idea is that "after a family conclave" the husband, father, or brother should "quietly pocket his own conflicting opinion and support the measures favored by the home majority." The plan is beautifully simple in appearance, but we fear would present difficulties in practice. The man who was earning a living for his family could scarcely be expected to pocket his opinion upon a question, such as protection or free trade, which he believed had an important bearing on his business prospects; but at the same time we are sure that most men would be very glad to have any assistance which the female members of their households could give them in arriving at right conclusions on questions of the day.

If women are to be called upon to vote, it should be for very broad and sufficient reasons. The mere fact that some are demanding it is not a sufficient reason, inasmuch as others, and probably the great majority, not only do not join in the demand but are prepared to oppose it. Let us endeavor to indicate briefly how the matter presents itself to our mind.

In the exercise of the suffrage the individual asserts himself, claiming his share of political power. The vote is given to him for the protection of his political rights against the encroachments of other men. On voting day society is momentarily resolved into its constituent units. As long as men alone do the voting, they are supposed to represent the non-voting sex. Every man has or has had a mother, most have one or more sisters, and a very large proportion have wives. Every man's vote, therefore, we do not hesitate to say, ought to express his consciousness of and respect for the family tie. To summon women to the polls would signify an antagonism between their interests and those of men. It would signify that a man and the women of his household are separate social units in the same sense in which two men are, and that they require protection against one another—that each must be armed with the ballot lest the others encroach. This assumption, in our opinion, is not warranted. Making all deductions for unfortunate instances, the family is in general a unit, and the wife, daughter, or sister has no desire to antagonize the vote of the husband, father, or brother. How about those women, it will be asked, who have no husband, father, or brother to represent them in a satisfactory manner? Our answer is that their case does not appear to us to be one of hardship unless it can be shown that, *considering them as a class apart from those who have male relatives*, they are suffering through lack of political influence. Simply as women they receive whatever benefit accrues to the sex in general through such improvements in the law as are daily taking place, and through the sympathy with woman which characterizes the normal man. To a considerable extent also the same means of influence are open to them as are open to other women. They are not cut off from society: they can speak and write; and how potent "women agitators" can be in procuring changes of the law Miss Tweedy has told us. What is mainly needed, in our opinion, is the deepening of the sense of trusteeship in men, and that fortunately is a process which is realizing itself more and more before our eyes. Far better so than that all trusteeship should be snatched from man with the snappish declaration that henceforth his wife, daughter, and sister will take care of their own interests. A singular time indeed for such a change to be made, when things have so shaped themselves that so earnest a female suffragist as our contributor is hard put to it to say what the disadvantages are under which women labor through man's control of the suffrage, or what laws they want passed which if duly explained and urged they could not now get passed!

There are other views of the question which we have only space to glance at. We can not lose sight of the fact that all law means compulsion — physical compulsion in the last resort; and this to our mind points to the conclusion that the responsibility for making laws should rest with those who could if necessary fight for their enforcement. It has before been pointed out that the situation which would be created if a large majority of women, in combination with a minority of men, passed laws repugnant to a large and effective majority of men, would be a very critical one for social order. Yet if nothing of this kind is going to happen, it is difficult to see where the special influence of woman's vote will come in.

Another point deserving of consideration is that the male sex, when all is said and done, is the progressive sex. Mr. Havelock Ellis's interesting and certainly far from prejudiced book on Man and Woman makes this clear. Broadly speaking, woman shows the statical, man the dynamical, aspect of humanity, and, as the work of legislation is in its nature continuous and progressive, it seems natural that it should be intrusted to that sex which best represents the onward movement of the race. Here, however, we must adjourn the discussion, which is one difficult to confine within narrow limits. Much probably remains to be said on both sides, and we have no doubt the soundest arguments will prevail in the end.

Chapter 6
WORKING FOR WAGES (1875–1915)

UNSEXED

*It was wild rebellious drone
That loudly did complain;
He wished he was a worker bee
With all his might and main.*

*"I want to work," the drone declared.
Quoth they, "The thing you mean
Is that you scorn to be a drone
And long to be a queen.*

*"You long to lay unnumbered eggs,
And rule the waiting throng;
You long to lead our summer flight,
And this is rankly wrong."*

*Cried he, "My life is pitiful!
I only eat and wed,
And in my marriage is the end—
Thereafter I am dead.*

*"I would I were the busy bee
That flits from flower to flower;
I long to share in work and care
And feel the worker's power."*

*Quoth they, "The life you dare to spurn
Is set before you here
As your one great, prescribed, ordained
Divinely ordered sphere!*

*"Without your services as drone,
We should not be alive;
Your modest task, when well fulfilled,
Preserves the busy hive.*

> *"Why underrate your blessed power?*
> *Why leave your rightful throne*
> *To choose a field of life that's made*
> *For working bees alone?"*
>
> *Cried he, "But it is not enough,*
> *My momentary task!*
> *Let me do that and more beside—*
> *To work is all I ask!"*
>
> *Then fiercely rose the workers all,*
> *For sorely were they vexed;*
> *"O wretch!" they cried, "should this betide*
> *You would become* unsexed!*"*
>
> *And yet he had not sighed for eggs,*
> *Nor yet for royal mien;*
> *He longed to be a worker bee,*
> *But not to be a queen.*
>
> Charlotte Perkins Gilman,
> *In This Our World and Other Poems* (San Francisco:
> James H. Barry and John Marble, 1895), pp. 52-53.

Until 1910, no more than 20% of the female population was employed outside the home at any given time.[1] Marriage and motherhood commonly were thought of as full-time occupations, with which paid work would interfere. Most jobs open to women paid so poorly that there was little financial incentive for middle-class women to enter the wage labor force. Moreover, most women considered homemaking and childrearing their primary moral and social responsibilities.

As industry increasingly demanded inexpensive labor during the 1800s, young single women began to occupy unskilled and semi-skilled positions in factories. Later in the century, public schools attracted better educated single women to teaching posts. As households became easier to manage, teenage daughters left home to work in mills, stores, and offices. Thus, the pattern that would prevail far into the twentieth century became established: After finishing their schooling and before marrying, young women worked in low paying, low status positions for short periods of time. In 1890, approximately 70% of all employed women were single; the majority of the rest were widowed or divorced. Married women made up a very small percentage of the total: Never at any moment during the late nineteenth century did more than 5% of all married women work for wages.[2]

In 1870, the first year in which the Census Bureau counted the numbers of women "gainfully occupied" (a term that reflected and perpetuated the belief that women's unpaid labor was "ungainful" or unproductive), 1.9 million women worked for wages, representing 15% of the total work force and 13% of all women, age ten and over, in the country. (See Table 6.1 in Appendix.) Roughly half this number, almost one million women, were employed as domestic servants and laundresses; another 23% (450,000 women) were agricultural laborers and farmers. An additional 15%, or almost 300,000 women, worked in the textile industry as tailors, seamstresses, and milliners. A small percentage of the total, approximately 5% or 85,000 women, were teachers. The remaining 5% worked in the manufacturing, trade, and transport industries.[3]

The 1880s, 1890s, and early 1900s wrought great changes in the types of jobs performed by women. The invention of the typewriter and the development of department stores made available positions that did not exist before. Female office workers and saleswomen, virtually nonexistent in 1870, represented approximately 10% of all employed women in 1910. Concurrently, the ranks of female teachers grew steadily, while the categories of female farm laborers and domestic servants decreased in relative size.[4] But the practice of employing young single women in low paying positions prevailed. Most married women stayed at home. Even as late as 1920, only 9% of married women worked for pay. Social attitudes toward employed women remained remarkably stable. Most Americans continued to view older working women as objects of pity or disdain. Widows and women whose husbands could not support them were pitied; divorced women and spinsters were disdained.

Throughout the late nineteenth and early twentieth centuries there existed an unresolved tension between paid work and womanhood. Americans believed paid labor degraded women and corrupted their moral character. The ideal of the "lady of leisure" was of a woman who did not "work," at least not for money. Women's social class, although essentially defined by their husbands' occupation, was inversely related to how much they earned. It was permissible for women to volunteer their labor, but then their labor was not called work; it was called service or charity or love, depending on who was the beneficiary. Payment for labor demeaned women, reduced the perceived value of their efforts, and lowered their class standing. Female factory workers frequently were thought to be one step away from prostitution. Outside employment for married women was thought to have deleterious effects on family life, since working wives were not at home to look after husbands' needs or to bring up children properly. Although very few married women actually worked outside the home, the working woman nonetheless symbolized the dissolution of the family.

Even as new occupations emerged in the early twentieth century, the division of labor between the sexes continued.[5] Although there was a substantial increase in the number of women engaged in professional work, 75% of women professionals were teachers. Teaching, along with certain other occupations—nursing, librarianship, and social work—were referred to as the "female" professions. The traditional "male" professions—law, medicine, and the clergy—admitted very few women. In offices, women became stenographers and typists and soon commanded the majority of these positions. Office and bank managers, finance and government officials were almost always men. The newly created professions of accounting, dentistry, architecture, engineering, and design also were composed largely of men.[6]

Paid work performed by women was differentiated by marital status, race or ethnic background, and economic class. Jobs that required women leave the home for most of the day were done largely by single women or by immigrant and black women. Only 3% of white married women were counted as "gainfully occupied" in 1900, and these women tended to do work they could do at home: laundering, taking in boarders, sewing, and so on. By comparison, 26% of black married women were "gainfully occupied," the great majority in agriculture or domestic service. Teaching, the most prestigious job for women (but not the best paid), comprised single women who were generally well-educated, middle-class, native-born, and white. Clerical and sales work became the province of less educated, white, middle-class women. Factory work was performed primarily by immigrants, and domestic service, which came at the bottom of the hierarchy in status and wages, was left to immigrants and blacks.[7]

Particularly among the middle class, the "idle" wife or "lady of leisure" was evidence of the family's economic prosperity. In immigrant, black, and working-class families, women's financial contributions were more desperately needed, and thus working women were more common and accepted. In 1900, 43% of adult black women and 19% of immigrant women worked, compared to 15% of white women of native-born parents.[8] But as soon as families grew prosperous enough, married women stopped working to remain at home.

Social injunctions against working women were partially overcome for brief periods, when economic logic prevailed. Early mill owners successfully attracted the daughters of New England farmers in the 1830s and 1840s, in part because farmers needed the extra cash that their daughters could provide. Mill owners offered reassurance that the work environment would be controlled carefully so as to protect the girls' moral characters. Many of the mills made church attendance mandatory and implemented rules concerning visiting hours, hygiene, and behavior. Consequently, for several decades young, white, native-born women worked in the mills to help support families, send brothers to college, and save for dowries. But the stigma of being a "mill girl" never disappeared entirely. By 1850 mill and factory work no longer were considered respectable occupations for middle-class women. The majority of female factory workers in the second half of the century were immigrants.[9]

Since teaching did not raise suspicions about women's moral characters, it was unusual among female occupations. But female teachers suffered another stigma: Their occupation, particularly if they taught past their mid-twenties, characterized them as unattractive and unmarriageable. The stigma is the more ironic because many schools forbade female teachers to marry. Male teachers were not subject to this constraint. In fact, a justification frequently given for paying male teachers higher salaries was that they had families to support. On the other hand, women with families to feed were not allowed to teach.

Thus, regardless of occupation, the idea of a woman working for wages conflicted with social notions regarding the ideal woman, notions that emphasized female purity and motherhood. Many women shared society's view of the ideal woman and were sensitive to the stigma attached to domestic labor and factory work. Leonora M. Barry, a Knights of Labor organizer, implied that this sensitivity was prevalent among women laborers themselves: "If there is one cause more than another that fastens the chains on . . . working women it is their foolish pride, they deeming it a disgrace to have it known that they are engaged in honest toil."[10]

Men often appealed to these ideals of womanhood to justify excluding women from the "male" professions. The claim generally was that professional work would degrade women or taint female purity. A Wisconsin judge barred women from becoming justices of the court because:

> It would be . . . shocking to man's reverence for womanhood and faith in woman . . . that women should be permitted to mix professionally in all the nastiness of the world which finds its way into courts of justice; all the unclean issues . . . of sodomy, incest, rape, seduction, fornication, adultery, pregnancy, bastardy, legitimacy, prostitution, lascivious cohabitation, abortion, infanticide, obscene publications, libel and slander of sex, impotence, divorce.[11]

The judge evidently considered it irrelevant that for most of the "unclean issues" cited, women were involved as actors or victims and already exposed to the evils from which he attempted to protect them. But the argument that women should not work because their highly refined moral natures were susceptible to corrupting worldly influences was used frequently in support of women's exclusion from "men's" work.

Employment was believed to harm women in other ways too. Physicians and gynecologists testified that women's reproductive system made them incapable of withstanding the harsh demands of the work place and argued that they must be shielded from these treacherous conditions, lest their health and fertility suffer. Dr. Ely Van de Warker in "The Relations of Women to the Professions and Skilled Labor" (see pp. 259–270) outlined women's "anatomical unfitness" to industrial labor: Their knees, pelvis, and feet made them unable to stand for long periods; their muscles lacked the necessary dexterity and strength to perform skilled tasks. Along with many other physicians of his era, Van de Warker believed that menstruation impaired women's faculties of reason and judgment. Menstruation was frequently cited as the decisive factor in excluding women from difficult mental or physical work. In 1874 a British physician, Tilt, explained why the London Obstetrical Society decided to bar women: "for the profession felt that the verdict really meant that women were not qualified by nature to make good midwifery practitioners, that they were unfit to bear the physical fatigues and mental anxieties of obstetrical practice, at *menstrual periods*, during pregnancy and puerperality."[12]

Few people challenged the conventional wisdom regarding the debilitating effects of menstruation. An exception was Dr. Mary Putnam Jacobi, who found it significant that 68% of the women reporting severe menstrual pain in her survey nevertheless managed to work. Jacobi also observed that most of these women had "acquired" the habit of pain with the onset of their first period, long before they were employed. In her conclusion, Jacobi stated that the occurrence of menstrual pain correlated inversely with the extent of education and steadiness of outside work, findings that conflicted with the beliefs of many eminent physicians who believed that education and work exacerbated menstrual difficulties.[13] (See Chapter 1.)

Women argued that if they were strong enough to withstand the physical demands of domestic work, they would be able to bear the tribulations of the work place. They also pointed out that many women needed to work to support themselves and their families. Opponents responded that female laborers brought down male wages, making it impossible for men to support families, which forced women into the work force and perpetuated the cycle. But men and women rarely competed for the same positions. Even on the relatively rare occasions when men and women worked in the same factories, certain jobs were reserved for men, others for women and boys. Sexual division of labor was rigidly enforced. When women took over occupations previously filled by men (as happened in school teaching), they received much lower salaries, and the work lost status, it was now "women's work."

What can we understand about women's work experiences in the late nineteenth and early twentieth centuries? Did exposure to the factory, the office, and the school serve to emancipate women, or did the realities of women's work environments, as historian Leslie Woodcock Tentler argues, reinforce rather than diminish women's psychological dependence on the family and the domestic role?[14] There is truth in both statements. On the one hand, work had certain liberating effects for women: Some young women, for the first time, experienced a freer environment than the confines of home. They caught a momentary glimpse of the empowerment and exhilaration of economic independence. Although most young employed women lived at home and depended on their families for room and board, many learned what it meant to have pocket money that they had earned and could spend however they wished, without feeling accountable to fathers, brothers, or husbands.

On the other hand, working conditions were arduous: Hours were long, factory and office work was physically tiring, poorly paid, and boring. Women had very little power or independence in their jobs and were frequently subject to the authority of a male boss. Surrounded by other young and inexperienced women, female laborers looked forward to the moment they could leave a world where they evidently were not valued. Marriage and motherhood seemed far more noble, glorious, rewarding, and enjoyable pursuits, and romance was at least momentarily exciting. Working women fell in love, married, and stopped working.

Most women considered jobs temporary, and for this reason the low wages and unsatisfactory conditions were bearable. There could be no doubt that marriage provided women with a higher and more secure standard of living. Employers saw no reason to pay women higher wages and knew most women did not (indeed could not) depend solely on their wages for their support. It becomes impossible to separate cause from effect in this never-ending cycle that enforced women's subordination to and dependence on men in both the work place and home. The expectation of marriage led women to accept unsatisfactory wages and working conditions, which in turn made marriage a necessity. Because employers treated women as unimportant, easily replaceable parts, women rarely felt themselves an integral part of the work place. Consequently, they assumed little responsibility and showed as little loyalty as was shown them. If women demonstrated little desire to learn and advance, it was equally true that there were very few opportunities for them to do either, and in any case, wages remained low. Historians estimate that women earned from one-quarter to one-half less than men in positions requiring similar skills.[15] Because employers saw no reason to raise women's wages, improve working conditions, or expand opportunities for advancement, women were forced to turn to marriage. And so the cycle continued. In 1909, Elizabeth Beardsley Butler identified these hopelessly entangled phenomena, which were both cause and effect of women's position in the industrial order:

> Expectation of marriage, as a customary means of support, stunts professional ambition among women. This lack of ambition can have no other effect than to limit efficiency, and restricts them to subsidiary, uninteresting, and monotonous occupations. The very character of their work lessens their interest in it. Without interest, they least of all feel themselves integral parts of the industry and in consequence assume no responsibility, affect no loyalty. They do not care to learn; opportunity to learn is not given them; both are causes and both are effects. Women see only a fight for place, and a very uncertain advantage if they gain it; wages are low, again both cause and effect of their dependence on others for their support. They shift around on lower levels of industry from packing room to metal work, from metal work to laundry work; a very few, through unwonted good fortune, unwonted determination, break through the circle and rise.[16]

Most late nineteenth and early twentieth century observers saw simply that women, almost without exception, occupied low paying, low status jobs and took this as proof of what scientists and doctors had been saying all along: Women and men had fundamentally different natures that suited them to different types of work. The success of men in the work place was attributed to their competitive instincts, ambitious natures, physical size, strength and dexterity, and their innate intelligence. Women, it seemed obvious, could not compete successfully with them.

There was a widespread fear that if men and women did the same type of work, whether it be in the factory, office, school, or home, men would become "womanly" and women would become "unsexed." The ultimate fears were that women would become less fertile, that their superior moral capabilities would be threatened, and that their maternal instincts would disappear. To prevent these horrors, most Americans supported separate spheres of activity for men and women. The separation of economic tasks into men's jobs and women's jobs both reflected and reinforced society's belief in unalterable, quintessential sexual differences. Initially, belief in sexual differences led to the structuring of sex-segregated work environments, but the effects of this organization were then taken as proof of the original hypothesis: Only men could do certain tasks, and only women could do others.

APPENDIX

Table 6.1. The Employed Population, 10 Years and Over, of the United States, 1880–1920

YEAR	TOTAL EMPLOYED POPULATION (Thousands)	TOTAL NUMBER OF MEN (Thousands)	MEN GAINFULLY OCCUPIED NUMBER (Thousands)	AS % OF EMPLOYED POPULATION	AS % OF ALL MEN	TOTAL NUMBER OF WOMEN (Thousands)	WOMEN GAINFULLY OCCUPIED NUMBER (Thousands)	AS % OF EMPLOYED POPULATION	AS % OF ALL WOMEN
1880	17,391	18,735	14,744	84.8	78.7	18,025	2,647	15.2	14.7
1890	23,337	24,352	19,312	82.8	79.3	23,060	4,005	17.2	17.4
1900	29,072	29,703	23,753	81.7	80.0	28,246	5,319	18.3	18.8
1910	38,166	37,027	30,091	78.8	81.3	34,552	8,075	21.2	23.4
1920	41,613	42,289	33,064	79.4	78.2	40,449	8,549	20.5	21.2

Source: "Facts About Working Women," *Women's Bureau Bulletin No. 46* (Washington, D.C.: Government Printing Office, 1925), p. 4.

Table 6.2. Employed Women, 16 Years and Over, by Marital Status, Race, and Nativity, 1900

	FEMALE POPULATION 16 YEARS OF AGE AND OVER					
	ALL CLASSES			SINGLE (INCLUDING UNKNOWN)		
		BREADWINNERS			BREADWINNERS	
RACE AND NATIVITY	TOTAL	NUMBER	PER-CENT	TOTAL	NUMBER	PER-CENT
Total women	23,485,559	4,833,630	20.6	6,843,140	3,143,712	45.9
Total men	24,851,013	22,489,425	90.5	9,633,157	8,355,666	86.7
Women who are:						
Native white—both parents native	12,130,161	1,771,966	14.6	3,483,867	1,177,420	33.8
Native white—one or both parents foreign born	4,288,969	1,090,744	25.4	1,802,436	929,852	51.6
Foreign born white	4,403,494	840,011	19.1	832,945	586,173	70.4
Negro	2,589,988	1,119,621	43.2	710,031	447,750	63.1
Indian and Mongolian	72,947	11,288	15.5	13,861	2,517	18.2

Source: Bureau of the Census, *Statistics of Women At Work* (Washington, DC: Government Printing Office, 1907), p. 15.

Table 6.2. (continued)

	FEMALE POPULATION 16 YEARS OF AGE AND OVER							
	MARRIED			WIDOWED			DIVORCED		
	BREADWINNERS			BREADWINNERS			BREADWINNERS	
TOTAL	NUMBER	PER-CENT	TOTAL	NUMBER	PER-CENT	TOTAL	NUMBER	PER-CENT
13,810,057	769,477	5.6	2,717,715	857,005	31.5	114,647	63,436	55.3
13,955,650	13,150,671	94.2	1,177,976	907,855	77.1	84,230	75,233	89.3
7,251,375	217,257	3.0	1,332,334	347,563	26.1	62,585	29,726	47.5
2,212,946	68,976	3.1	256,953	83,107	32.3	16,634	8,809	53.0
2,855,446	102,169	3.6	702,585	145,240	20.7	12,518	6,429	51.4
1,443,817	376,096	26.0	414,107	277,655	67.0	22,033	18,120	82.2
46,473	4,979	10.7	11,736	3,440	29.3	877	352	40.1

Table 6.3. Selected Occupations of Women, 1870–1920

YEAR	NUMBER OF WOMEN	% INCREASE	AS % OF ALL EMPLOYED IN FIELD	AS % OF ALL EMPLOYED WOMEN
Women in Private or Public Housekeeping[a]				
1870	901,954	—	86	47
1880	970,257	8	83	37
1890	1,302,704	34	82	35
1900	1,430,656	10	82	27
1910	1,593,586	11	79	21
1920	1,356,531	−15	76	16
Women in Laundry				
1870	58,683	—	92	3
1880	102,280	86	89	4
1890	218,797	100	87	6
1900	338,635	55	87	6
1910	606,409	79	89	8
1920	478,078	−21	86	6
Women in Teaching[b]				
1870	84,548	—	66	5
1880	153,372	81	68	6
1890	244,467	59	71	6
1900	325,485	33	74	6
1910	479,792	47	79	6
1920	645,181	34	82	8
Women in Agriculture[c] *(Farmers, Managers)*				
1870	24,859	—	1	1
1880	58,680	136	2	1
1890	229,270	291	6	4
1900	311,695	36	6	5
1910	281,617	−10	4	5
1920	281,208	0	3	4
Women in Agriculture[c] *(Laborers, wage & unpaid)*				
1870	430,085	—	22	12
1880	567,169	32	21	14
1890	566,709	0	14	13
1900	696,670	23	13	14
1910	894,722	28	12	17
1920	890,230	−1	10	18
Women in Textile Industry[d]				
1870	104,180	—	43	5
1880	161,283	55	45	6
1890	221,711	38	48	6
1900	280,200	26	49	5
1910	387,062	38	50	5
1920	452,981	17	48	5

Table 6.3. (continued)

YEAR	NUMBER OF WOMEN	% INCREASE	AS % OF ALL EMPLOYED IN FIELD	AS % OF ALL EMPLOYED WOMEN
Women in Office Work[e]				
1870	930	—	3	—
1880	2,315	149	4	.1
1890	45,553	1868	21	1
1900	104,450	129	29	2
1910	386,765	270	37	5
1920	1,038,390	169	50	12

[a] Service work in private households and in "public housekeeping" have always been dominated by women, and almost half the total number of women employed in 1870 worked in this category. Public housekeeping includes charwomen, cleaners, cooks, housekeepers, hostesses, and waitresses.

[b] In teaching, educational developments played a double role: The extension of schooling among the population created more demand for teachers, and the increasing availability of secondary and advanced schooling for women produced an increasing supply of women teachers. Women college presidents, professors, and college instructors were not reported separately until 1920, but they constituted an insignificant fraction of all women teachers. In 1870, only 6.4% of women teachers were married; in 1920, this number was 9.7%.

[c] The Census data for women agricultural laborers greatly understates women's role in agriculture and can be used only to provide a rough indication of trends. The bulk of farm wives were omitted from the Census counts because, although these women constantly performed a significant amount of farm work, they frequently did not consider themselves as farm laborers. They viewed their work in the fields as part of their housework, and thus were not counted. Even so, the number of women farm laborers doubled from 1870 to 1910. After 1910, the advance of industrial farming—the movement from small to large farming operations—slowed the growth of men and women employed in agriculture.

[d] In 1870, 80% of women operatives and laborers in industry were employed in the sewing trades as tailors, dressmakers, and seamstresses working outside of factories and as operatives and laborers in apparel and accessory factories or in cotton manufacturing.

[e] There were very few women office workers in 1870, less than 1,000 in the entire group that included stenographers, typists, secretaries, shipping and receiving clerks, clerical, and kindred workers. The numbers of women office workers burgeoned in the decade from 1880 to 1890, when the numbers of women in these office occupations multiplied nearly 20 times, evidence of the growing need for and acceptance of the typewriter and the trained woman typist.

Source: Janet M. Hooks, "Women's Occupations Through Seven Decades," *Women's Bureau Bulletin No. 218* (Washington, D.C.: Government Printing Office, 1947), pp. 75, 102–104, 139–140, 145–146, 158–159, 161, 191–195.

NOTES

1. The statistics given throughout this chapter are from the United States Census Bureau, but they understate the actual numbers of women who worked for wages because investigators often forgot to inquire about working wives. Also, women sensitive to the social stigma attached to working women may have neglected to mention that they did paid work in addition to being full-time housewives. Moreover, many women took care of gardens and livestock (this was true even for many women living in urban areas), sewed the family's clothes, and took in laundry—labor that is not accounted for in the Census statistics. Lastly, the most significant omission is that the Census did not count the vast numbers of women performing unpaid work in the home until 1930, when a new category, "homemaker," was added.
2. Bureau of the Census. *Statistics of Women At Work* (Washington, D.C.: Government Printing Office, 1907), pp. 14–16. These generalizations do not hold for all groups of women, however. For example, in 1900 26% of all married black women worked for wages, compared to 3% for white women and 3.5% for white immigrants. See Table 6.2 in Appendix.
3. Janet M. Hooks, "Women's Occupations Through Seven Decades," *Women's Bureau Bulletin No. 218* (Washington, D.C.: Government Printing Office, 1947), pp. 34, 103–113, 139, 158, 191, 193. In the interest of brevity, I occasionally use the phrases *employed women* or *working women* to mean women who worked for wages; I acknowledge that these terms are less than ideal in that they are used in contradistinction to women's unpaid labor and so may give unintentional support to the misguided belief that housework, childrearing, and other forms of unpaid labor are not work or are less meaningful than paid work. I certainly do not hold to such tenets, and I hope my statements will not be misconstrued as denigrating to the masses of women who did not work for wages. I regret that I was unable to find or invent phrases that would not reflect these unfortunate, unintended assumptions.
4. For statistics and discussions of woman's engagement in domestic labor, see Hooks, "Women's Occupations Through Seven Decades," pp. 137–146; in agriculture, pp. 190–195; in textiles, pp. 103–116; in teaching, pp. 157–161.
5. The sexual division of labor raises critical questions concerning women's economic and social roles and status in society and is commonly considered by scholars of women's history and feminists to represent the key to women's subjection to men. The sexual division of labor predates capitalism and thus cannot be attributed to industrialization. Some scholars argue that it derives from patriarchy, which Heidi Hartmann has defined as the set of social relations having a material base, in which there are hierarchical relations between and solidarity among men, enabling them to control women. In this formulation, the sexual division of labor is seen as a system devised by men to promote male interests and to further their control of women. For a review of this topic and an introduction to some of the relevant anthropological literature, see Heidi Hartmann, "Capitalism, Patriarchy and Job Segregation by Sex," *Signs: Journal of Women in Culture and Society* 1 (Spring 1976): 137–169.
6. See Julie A. Matthaei, *An Economic History of Women in America: Women's Work, The Sexual Division of Labor, and the Development of Capitalism* (New York: Schocken Books, 1982), pp. 190–191, 206. According to Matthaei, 94% of all nurses, 73% of all librarians, and 75% of all social workers in 1900 were women. By contrast, over 94% of physicians, professors, lawyers, judges, and the clergy were men. In the emerging fields of veterinary science, architecture, surveying and engineering, over 99% were men.
7. Ibid., pp. 187–232.
8. Bureau of the Census, *Statistics of Women At Work* (Washington, D.C.: Government Printing Office, 1907), p. 15.
9. See Benita Eisler, ed., *The Lowell Offering: Writings by New England Mill Women (1840–1845)* (New York: Harper & Row, 1977).
10. Cited in Alice Kessler-Harris, *Women Have Always Worked: A Historical Overview* (Old Westbury: Feminist Press and New York: McGraw-Hill, 1981), p. 63.

11. *Reports of the Supreme Court of Wisconsin*, vol. 39, pp. 245–256. Cited in Robert W. Smuts, *Women and Work in America* (New York: Schocken Books, 1976), p. 116.
12. *British Medical Journal* (January 16, 1875). Cited in Mary Putnam Jacobi, *The Question of Rest for Women During Menstruation* (New York: G. P. Putnam's Sons, 1877), p. 5; emphasis in original.
13. See Jacobi, *The Question of Rest For Women During Menstruation*. For a review of Jacobi's work and her response, see "The Question of Rest For Women During Menstruation," [Book Review] *The Popular Science Monthly* 12 (December 1877): 241–244 and Mary Putnam-Jacobi, "The Question of Rest for Woman," *The Popular Science Monthly* 12 (February 1878): 492–494.
14. See Leslie Woodcock Tentler, *Wage-Earning Women: Industrial Work and Family Life in the United States, 1900–1930* (New York: Oxford University Press, 1979), pp. 3–4, 28, 37–57.
15. See Smuts, *Women and Work in America*, p. 91; Matthaei, *An Economic History of Women in America*, p. 215; Alice Kessler-Harris, *Out to Work: A History of Wage-Earning Women in the United States* (New York: Oxford University Press, 1982), p. 59.
16. Elizabeth Beardsley Butler, *Women and the Trades: Pittsburgh, 1907–1908* (New York: Charities Publication Committee, 1909), p. 373.

ARTICLES ON WOMEN'S EMPLOYMENT IN *THE POPULAR SCIENCE MONTHLY*

(Arranged by Date of Publication)

Ely Van de Warker. "The Relations of Women to the Professions and Skilled Labor," (February 1875) 6: 454–470.
"Women and the Study of Science," (November 1878) 14:119.
H. Carrington Bolton. "The Early Practice of Medicine by Women," (December 1880) 18: 191–202.
"Woman as a Sanitary Reformer," (July 1881) 19: 427–428.
Emily Blackwell. "The Industrial Position of Women," (July 1883) 23: 388–399.
Editor. "Progress and the Home," (July 1883) 23: 412–416.
Abraham Jacobi. "The Historical Development of Modern Nursing," (October 1883) 23: 773–787.
E. LaGrange. "Women in Astronomy," (February 1886) 28: 534–537.
Lucy S. V. King. "Women in Business," (October 1888) 33: 842–843.
Otis Tufton Mason. "Woman as an Inventor and Manufacturer," (May 1895) 47: 92–103.
Henrietta Irving Bolton. "Women in Science," (August 1898) 53: 506–511.
"The Employment of Women," (October 1903) 63: 571–572.
Amanda Carolyn Northrop. "The Successful Women of America," (January 1904) 64: 239–244.
E. S. "Celibate Education To-Day," (November 1908) 73: 423–428.
Cora Sutton Castle. "A Statistical Study of Eminent Women," (June 1913) 82: 593–611.
Elfrieda Hochbaum Pope. "Women Teachers and Equal Pay," (July 1913) 83: 65–72.
D. R. Malcolm Keir. "Women in Industry," (October 1913) 83: 375–380.
Annie Marion Maclean. "Trade Unionism Versus Welfare Work for Women," (July 1915) 87: 50–55.

SUGGESTIONS FOR FURTHER READING

Primary Sources

Ames, Azel. *Sex in Industry; A Plea for the Working-Girl.* Boston: J. R. Osgood, 1875.
Bureau of the Census. *Statistics of Women At Work.* Washington, DC: Government Printing Office, 1907.

Campbell, Helen. *Prisoners of Poverty, Women Wage Workers, Their Trades and Their Lives.* Boston: Robert, 1887.
Dall, Caroline Healey. *Woman's Right to Labor.* Boston: Walker, Wise, 1860.
Gilman, Charlotte Perkins. *Human Work.* New York: McClure, Phillips, 1904.
Hill, Joseph A. "Women in Gainful Occupations," *Census Monographs IX.* Washington, DC: Government Printing Office, 1929.
Hooks, Janet M. "Women's Occupations Through Seven Decades," *Women's Bureau Bulletin No. 218.* Washington, DC: Government Printing Office, 1947.
Meyer, Annie Nathan, ed. *Woman's Work in America.* New York: Henry Holt, 1891.
National Manpower Council, *Womanpower.* New York: Columbia University Press, 1957.
Sumner, Helen Laura (Woodbury). "History of Women in Industry in the United States," *Bureau of Labor Report on Conditions of Woman and Child Wage-Earners in the United States.* Vol. 9. Washington, DC: Government Printing Office, 1910.
Van Kleeck, Mary. "What Industry Means To Women Workers," *Women's Bureau Bulletin.* Washington DC: Government Printing Office, 1923.
Van Vorst, Mrs. John, and Van Vorst, Marie. *The Woman Who Toils.* New York: Doubleday, Page, 1903.
Women's Bureau. "Facts About Working Women: A Graphic Presentation Based on Census Statistics and Studies for the Women's Bureau," *Women's Bureau Bulletin No. 46.* Washington, DC: Government Printing Office, 1925.
Women's Bureau. "The Occupational Progress of Women: An Interpretation of Census Statistics of Women in Gainful Occupations," *Women's Bureau Bulletin No. 27.* Washington, DC: Government Printing Office, 1922.
Wright, Carroll D. *The Industrial Evolution of the United States.* New York: Flood & Vincent Chautauqua Centurn Press, 1895, pp. 200–214.

Secondary Sources

Gordon, Linda; Baxandall, Rosalyn; and Reverby, Susan, eds. *America's Working Women.* New York: Random House, 1976.
Hartmann, Heidi. "Capitalism, Patriarchy and Job Segregation By Sex," *Signs: Journal of Women in Culture and Society* 1 (Spring 1976): 137–169.
Kessler-Harris, Alice. *Out to Work: A History of Wage-Earning Women in the United States.* New York: Oxford University Press, 1982.
Kessler-Harris, Alice. *Women Have Always Worked: A Historical Overview.* Old Westbury: Feminist Press and New York: McGraw-Hill, 1981.
Matthaei, Julie A. *An Economic History of Women in America: Women's Work, The Sexual Division of Labor, and the Development of Capitalism.* New York: Schocken Books, 1982.
Smuts, Robert W. *Women and Work in America.* New York: Schocken Books, 1976; 1st ed., Columbia University Press, 1959.
Tentler, Leslie Woodcock. *Wage-Earning Women: Industrial Work and Family Life in the United States, 1900–1930.* New York: Oxford University Press, 1979.
Walsh, Mary Roth. *"Doctors Wanted: No Women Need Apply": Sexual Barriers in the Medical Profession, 1835–1975.* New Haven: Yale University Press, 1977.
Wertheimer, Barbara Mayer. *We Were There: The Story of Working Women in America.* New York: Pantheon Books, 1977.

Edward Ely Van de Warker (1841–1910) was educated at Troy University in New York and studied medicine at Albany Medical College, where he graduated in 1863. He then served as an assistant surgeon in the Civil War, after which he established a private practice in Syracuse, becoming surgeon to the Central New York and the Women's and Children's hospitals. Van de Warker specialized in women's diseases and was one of the founders and presidents of the American Gynecology Society.

Van de Warker subscribed to a Spencerian view of human development, believing in a fundamental antagonism between "individuation" (physical growth and mental development) and "genesis" (reproduction). This antagonism, Van de Warker argued, accounted for the "marked impairment of fertility" in women of "great mental activity" and also explained the relative scarcity of intellectual women, since women who had advanced furthest in individuation had "greatly inferior chances" of "leaving daughters who will inherit their superior mental vigor." Moreover, while Van de Warker admitted that some women were the mental peers of men and could master the necessary knowledge to practice law, medicine, and theology, he believed that the anatomical differences and physiological processes peculiar to women (in particular menstruation) made it impossible for even these women to attain the degree of excellence achieved by men. The following selection presents a view of women's nature that was prevalent among scientists and physicians of the late nineteenth century. A response to this article was made by Dr. Frances Emily White in Penn Monthly *7 (July 1875): 514–528.*

THE RELATIONS OF WOMEN TO THE PROFESSIONS AND SKILLED LABOR

ELY VAN DE WARKER, M.D.

February 1875

WOMAN is now submitting her fitness to find employment, in the learned professions and skilled labor, to the rigid test of actual trial. Will she succeed, or will those of her sex, who achieve success in these fields of labor, be the exception, rather than the rule, in the future as in the past? In order to answer this question, it is my purpose to study woman in this relation, as a gynaecologist,[1] leaving out of consideration the social aspects of the case. One, who has devoted years to the study of women and their diseases, has a right to be heard upon this vital question. I do so the more readily because I know of no gynaecologist who has devoted his special learning to the

---

[1]Gynaecology, that branch of pathology which treats of the diseases of women.

study of woman's relation to man's work as a means of subsistence and of usefulness. . . .

Whatever may be woman's fitness in the future to become man's peer in the professions and skilled labor, there is this fact against her in the present: she is laboring under the accumulated inherited tendency of countless generations. That which had its origin in common with sister animals in physical and moral subjection to the male, has, in spite of the operation of that intellectual force which we see operating so potently at the present day to cast off this subjugation, continued in full force. I can explain this in no other way except as the result of heredity. This position of woman is as clearly a sexual trait as in lower animals. Darwin says that, "as peculiarities often appear in one sex, and become hereditarily attached to that sex, the same fact probably occurs under nature, and, if so, natural selection will be able to modify one sex in its functional relation to the other sex."[2] Dr. Maudsley, in speaking of one relation of woman to man, says: "Through generations her character has been formed with that chief aim (marriage); it has been made feeble by long habits of dependence; by the circumstances of her position, the sexual life has been undesignedly developed at the expense of the intellectual."[3] Mr. Herbert Spencer insists upon this. "Certain powers which mankind have gained in the course of civilization cannot, I think, be accounted for, without admitting the inheritance of acquired modifications."[4]

The law of sexual selection also comes in as a factor to account for the present relation of the sexes. Since our modern civilization there have ever been women who aimed to relieve their sex of their dependent relation to man. With this moral force ceaselessly antagonizing the natural relation of the sexes and the forces of heredity, why has it not, in a more marked degree, accomplished the noble purpose at which the more intellectual and stronger minds of the sex have aimed? To offer a reasonable explanation of this, I apply the law of sexual selection. Men and women do not appear to wed out of free choice, but in obedience to law which finds its expression in individual preferences. This, in the human family, may be called sexual selection. Mr. Walker states it in this way: "Love from a man toward a masculine woman would be felt by him as an unnatural association with one of his own sex; and an effeminate man is equally repugnant to woman. In the vital system, the dry seek the humid; the meagre, the plump; the hard, the softer; the rough, the smoother; the warmer, the colder; the dark, the fairer, etc., upon the same principles; and so, also, if here any of the more usual sexual qualities are reversed, the opposite ones will be accepted or sought for."[5] Dr. Ryan, in

---

[2]"The Origin of Species," p. 83.
[3]"The Physiology and Pathology of the Mind," p. 203.
[4]"Principles of Biology," vol. ii, p. 249.
[5]"Intermarriage," by Alexander Walker, 1839, p. 116.

speaking of selection in relation to marriage, uses nearly identical language.[6]

The annals of literature show that the most eminent of the sex either are unmarried, or are married late in life, and are thus often without issue. The women who intellectually leave their impress upon the age in which they live are the very class to which this law of sexual selection applies. The chances of this order of women leaving daughters who will inherit their superior mental vigor are greatly inferior to those of the average woman. The woman of the average, her mind and ambition being of the measure of the ordinary matters of life, not only seeks a husband by the force of education, but is sought by men. Thus, married early in life, she becomes the source from which the population is recruited. This, in my judgment, is not only a potent cause of the present relation of the sexes, but will serve to explain the chances of women becoming prominent in the professions of the future.

There is another set of laws which apply to this part of the subject. These are the phenomena which are observed in studying human increase, and are called the laws of population. The forces engaged in the evolution of nervous and mental (cerebral) structure are opposed to those necessary to reproduction. Mr. Herbert Spencer expresses it as an antagonism between Individuation and Genesis; and that this antagonism is more marked "where the nervous system is concerned, because of the costliness of nervous structure and function."[7] There is no part of individuation so costly as that of cerebral growth. The more solid expansion of mind is accomplished after general structural development has ceased. If mental growth be unduly forced before structural completion, structural and sexual genesis are retarded or impaired. Professional training in women must, therefore, fall within the child-bearing period. In the case of men of great mental activity there is marked impairment of fertility.[8] The cost of reproduction to males is greatly less than to females, and it therefore follows clearly that the prolonged and intensified mental growth, the result of professional training, is to be deducted in full from the sum of the forces necessary to reproduction. M. Quetelet cannot doubt the influence of professional life upon fertility.[9] Mr. Herbert Spencer says, "that absolute or relative infertility is generally produced in women by mental labor carried to excess, is more clearly shown. . . . This diminution of reproductive power is not shown only by the greater frequency of absolute sterility, nor is it shown only in the earlier cessation of childbearing, but it is also shown in the very frequent inability of such women to suckle their infants. In its full sense, the reproductive power means the power to bear a well-developed in-

---

[6]"The Philosophy of Marriage," 1873, p. 70.
[7]"Principles of Biology," ii., p. 502.
[8]"Principles of Biology," ii., p. 487.
[9]A Treatise on "Man and the Development of his Faculties." Translation. Edinburg, 1842, p. 21.

fant, and to supply that infant with the natural food for the natural period."[10]

Even were it not that absolute and relative infertility are against the woman undergoing severe mental discipline having children to inherit her improved cerebral evolution, and in favor of the average or inferior woman, still the very condition of this mental discipline, if the woman is preparing for the professions or skilled labor, involves a postponement of marriage to a period when, in the mass of wives, fecundity has received a permanent check. The average individual wife shows a degree of fecundity which, at the age of twenty-five, diminishes,[11] and this is the period at which the professional woman is prepared to enter upon her business career. The opinion of Mr. Sadler, that delayed marriages developed a degree of fertility in women which compensated for the loss of fecundity consequent upon the delay, is completely overthrown by the tables of Dr. Duncan.[12] If women are to enter the learned professions and skilled labors, they must be devoting themselves to training at a period of their lives when the mass of women are wives—mothers. I think that it must be conceded as a fact that, to contract matrimony during this period of mental and bodily training, would totally defeat the selected lifework of the woman. The desire to become the co-worker with man upon the highest level of man's work belongs only to superior women; if, in addition to this innate superiority, we add that acquired from increased cerebral development, the law of heredity would tend to continually enlarge the number of women fitted by inherited traits to occupy this advanced field. But we have shown that the laws of sexual selection and of population are entirely opposed to the increase of women thus favored, and in favor of the average woman by a large per centum.

I am inclined to regard all forces which have hitherto acted, and yet continue to act, upon the great mass of humanity in the creation of sentiments common to the majority upon a given subject, as acting with the force of a law. I conceive, therefore, that there is yet another law which explains, and tends to perpetuate, the present relation of women to the other sex and to society. This is the law of public opinion. The exponents of public opinion upon this subject are the women themselves. I do not think any one will controvert me when I assert that a vast majority of women are opposed to their own sex entering the professions. One would naturally suppose that, in the matter of religion, a woman's opinion is as good as a man's; that, with equal learning and experience, a woman is as competent to discharge pastoral duties as a man (I am assuming the physical equality of the sexes); and yet you may count upon the fingers of one hand the number of pulpits filled successfully

---

[10]*Loc. cit.*, p. 486.
[11]Dr. Matthews Duncan, "Fecundity, Fertility, Sterility, and Allied Topics," Edinburgh, p. 43.
[12]Sadler, "The Law of Population," ii., p. 279.

by women in this great country. In this country women are free to enter the medical profession; but, with about as many exceptions as that of women filling pulpits, they are gaining but a precarious and scanty support. Now, in both the professions named, women are retarded by the force of opinion of their own sex. In all social questions, women wield a great influence. In these matters they are the throne, and the power behind the throne. In Protestant congregations, if women were a unit in favor of women preaching, women would preach in a fair proportion of church organizations. If a woman made a free choice of a physician of her own sex, there is scarce a household in which she would be denied her choice. Women seem to lack confidence in their own sex in this position. In the desperate diseases peculiar to women, the sorely afflicted ones seek the medical man instead of the medical woman. The future has yet to produce the anomaly of the female ovariotomist. In the literature of medicine there has been but one Boivin, and but one La Chapelle. The reliance upon man in moments of bodily peril is easily explained; it is an inherited trait, strengthened by education.

I have said enough to explain philosophically the present relation of women to the other sex, and to society. It is this relation which has, in the past, regulated woman's admission into the professions and skilled labor. But we have now to accept the fact that women have entered the professions to contend with man in the struggle for existence. In this struggle, I presume, women expect no favors. In this new field of contest all they ought to ask for is a fair chance to win—the same chances man must take. But, in view of her present relation, and the radical physical differences between the sexes, have they a fair chance, and can they take the chances of man and reach his level in his professions? This is the question I shall endeavor honestly to solve.

Mentally, I believe woman to be the peer of man; that there is nothing about law, medicine, or theology, that woman cannot learn as well as he; that her mental difference is a sexual difference, just as her bodily differences are sexual. To the question of her mental fitness for this work, other than sexual, I shall give no attention. The physical difference between the sexes will form the first part of our study of woman's future relation to the professions, and brings us naturally to the discussion of the law of anatomical development as modified by sex. . . . Now, there are certain skilled labors which belong to man by virtue of his superior strength. The anatomical peculiarities of woman do not need to be contrasted with man's in reference to this class of labor, and women in the lower walks of life have demonstrated their ability for severe bodily toil. But the intellectual work—to which I am mainly directing my attention—to which women are reaching, implies that the candidates possess the delicate structural development, the inherited result of civilizing forces. We can draw no inference, therefore, from that fact that women, less exposed to these forces, and more nearly approximating man in physical strength, fully equal him in the value of their labor. If we examine some of the lighter and more delicate forms of skilled labor, such as we would naturally conclude were peculiarly fitted for the delicate and nice touch of women, we find them in the hands of men almost exclusively. The question of mental fitness must be excluded, for mentally they are as competent as man is to acquire and practise these arts. I think it can be shown that anatomical unfitness, aside from her inclination, is the obstacle. In the manufacture of instruments involving great delicacy, and, until the introduction of machinery, in the manufacture of watches, as a class women were excluded. While not involving any great muscular outlay, these samples of skilled labor demand great delicacy of educated touch. While we must make great allowances, as an anatomical factor, for the advantage to men so employed of the inheritance of mechanical taste and skill from fathers, oftentimes so employed for generations, yet it is a common error to suppose that employments which involve delicacy of manipulation do not require strength, as great in degree as, but differing in kind from, that demanded by more hardy labor. Educated, coördinate muscular movements depend more than any other upon strength and certainty of muscular contraction. In no particular, aside from sexual differences, is the male skeleton so greatly different from the female as in the irregularities and asperities of the bones for the attachment of the muscles. While in man they form a marked feature of his bony structure, in well-formed females they present but a comparatively scanty development. It is true that both muscles and osseous irregularities, for their origin and insertion, may be developed by training, yet woman, as at present related to the other sex, has not only to acquire his strength, by a course of extra training, but she must equal him in skill, if she is to prove a successful competitor for his place. These comparisons between the physical strength of the sexes would be altogether unfair, were it not for the fact that they are invited by the position women have elected for themselves, and are essential in giving an opinion of woman's chances of success.

The fact that those employments are chosen by women which permit a sitting position is significant in this relation. Woman is badly constructed for the purpose of standing eight or ten hours upon her feet. I do not intend to bring into evidence the peculiar position and nature of the organs contained within the pelvis, but to call attention to the peculiar structure of the knee, and the shallowness of the pelvis, and the delicate nature of the foot as part of a sustaining column. . . . Comparatively, the foot is less able to sustain weight than that of man, owing to its shortness and the more delicate structure of the tarsus and metatarsus. I do not think there can be any doubt that women have instinctively avoided some of the skilled labors on anatomical peculiarities. . . .

If we examine carefully the mental action of women, we perceive in it an undercurrent of sex. As there are organs which characterize sex, so also is there a sexual cerebration. We know from experience that this unconscious dominance of sex in cerebration in no way interferes with high culture, and the exercise of the best qualities of mind. It is a normal condition of mental action in women, but its existence implies conditions which may at any moment render mental action abnormal. Take the emotions, for instance, the undue exercise of which are so liable to assume morbid proportions, as in hysteria. Here sex, when it asserts itself unduly, obtrudes inharmoniously into what otherwise would be healthy mental action. It is in this class of mental actions, termed the emotions, that the mind of woman forms part of the sexual cycle. Some of these actions are so elementary that they are called instincts. The maternal affection, and also love, partakes of this instinctive character. The exercise of the sympathies is more general and active in women than in men. This is one of the features which give such beauty to the character of women, and is not the result of education. . . . I will give one instance, which is a type of character, and shows how sympathy and natural feeling may interfere with professional advancement.

The wife of a practising physician, being of a scientific tendency of mind, acquired a theoretical knowledge of her husband's profession. The husband died, and left the widow poor and with several children, some of them so young as to demand much of her time and thought. She continued the study of medicine with the design of making it the means of support for herself and children. To this end she attended lectures at a woman's medical college. Before she obtained her diploma, an old, superannuated Presbyterian clergyman excited her sympathy by his forlornness. She gave him a home in a very womanly way—she made him her husband. Here was a double burden—an old man, and little children. This physician, although laden with her great, womanly heart, was prosperous in a small way. She secured the position of house-physician in the hospital connected with the college, with a small salary, and with sufficient time to attend to private patients. Her pecuniary prospects were better than those of young physicians of the other sex. The husband soon died. At this point in the history occurred an incident which seems to me to be phenomenal, and yet is typical. A second old clergyman, equally forlorn and wretched as the first, accepted the charity of this woman by becoming her husband. Her practice slowly increased; her children were well clad and well educated. A daughter married, and moved, with her husband, to a distant city. A son studied medicine, and the last husband died. The next act in this singular history reveals an intensity of maternal feeling entirely opposed to a business success in a difficult profession. Gifted with a fine mind, as thoroughly educated in her profession as the majority of medical men, with good health, and having reached that time of life when she was functionally at rest, and with every encouragement to remain at her post, yet she made a better mother than doctor. She resigned her position in the hospital; abandoned her private practice; and moved to the city in which her daughter resided, in order to be near her child and grandchildren—and there, in a strange community, recommenced the difficult occupation of a female physician. This history is truly a physiological study, and reveals the intensity of feeling which may exist in all women upon subjects which lie near the heart. . . .

There is one fact in woman's functional life which is of vast importance to the subject of this paper, and which I refer to with great reluctance. This fact is ovulation. The mental reaction of this function is oftentimes of such a character as, for the time, to totally incapacitate for professional or other mental work. As this paper is written solely with the view of arriving at the truth in a matter of great practical importance, I must let this serve as my apology for referring plainly to this subject; and this importance requires that I let others, who are acknowledged authorities in gynaecology, speak for me.

Dr. Robert Barnes, of London, the author of the latest work upon gynaecology, uses the following unequivocal language: "The mind is always more or less disturbed. Perception, or at least the faculty of rightly interpreting perceptions, is disordered. Excitement to the point of passing delirium is not uncommon. Irritability of temper, disposition to distort the most ordinary and best-meaning acts or words of surrounding persons, afflict the patient, who is conscious of her unreason, and perplex her friends, until they have learned to understand these recurring outbursts. . . . Not even the best-educated women are all free from these mental disorders. Indeed, the more preponderant the nervous element, the greater is the liability to the invasion. Women of coarser mould, who labor with their hands, especially in out-door occupations, are far less subject to these nervous complications. If they are less frequently observed, if they less frequently drive refined women to acts of flagrant extravagance, it is because education lends strength to the innate sense of decorum, and enables them to control their dangerous thoughts, or to conceal them until they have passed away."[13] Another of the accidents attendant upon ovulation is hysteria. Dr. Tilt defines it as a disease peculiar to women during the reproductive period of life, and is often known to return at each period of ovulation.[14] This function is constantly liable to accidents. Speaking of the mental effects of aemenorrhoea, a disease to which every woman is liable who follows an intellectually rather than a physically active life, Sir J. Y. Simpson says that she becomes "subject to fits of excitement which come on most frequently at a menstrual period, and which usually assume an hysterical form, but are, at times, almost maniacal in character."[15] I shall make but one other quotation, and I am glad to say that it bears directly and practically upon this matter. Dr. H. R. Storer, of Boston, is reported to have spoken as follows in a debate at the Gynaecological Society of Boston, May 1870: "In the present excited state of public opinion, it were foolish, and

---

[13] "A Clinical History of the Medical and Surgical Diseases of Women," p. 162.
[14] "Diseases of Menstruation and Ovarian Inflammation," p. 129.
[15] "Diseases of Women," p. 617.

at the same time unkind, to object to female physicians upon any untenable grounds; and he frankly stated that the arguments that physicians had usually employed, when discussing this subject, were, almost without exception, untenable. Some of the women who were desirous of practising physic and surgery were just as well educated for the work, had just as much inclination for it, and were as unflinching in the presence of suffering, or at the sight of blood, as were many male practitioners. They had a right to demand an acknowledgment that, in these respects, they were as competent to practise as are a large proportion of ourselves. There is, however, one point, and it is upon this that the whole question must turn, that has till now almost wholly been lost sight of: and this is the fact that, like the rest of their sex, lady doctors, until they are practically old women, regularly menstruate and are therefore subject to those alternations of mental condition, observable in every woman under these circumstances, which so universally affect, temporarily, their faculties of reason and judgment. That these faculties are thus affected at the times referred to is universally acknowledged. . . . "[16]

We have been studying woman, in her relation to the subject of this paper, as a sexual being; and, if we continue the study in the same direction, we must arrive at the conclusion that marriage is not an optional matter with her. On the contrary, it is a prime necessity to her normal, physical, and intellectual life. There is an undercurrent of impulse impelling every healthy woman to marry. That this is a law of her sexual being we know by the positive evidence of medical men and others. We also know that the married woman exerts a more marked influence upon men, and society in general, than the celibate. There is also, among married women, a more perfect equilibrium between the intellectual, physical, and sexual forces; and yet, necessary as marriage is for woman, in the present relation of the sexes, it must in every way impair her prospects of success in professional work.

The effect of celibacy upon women has often elicited the remarks of gynaecologists. Dr. Tilt says of marriage: "It is easier to prove the benefits of marriage than to measure accurately the evils of celibacy, which I believe to be a fruitful source of uterine disease. The sexual instinct is a healthy impulse, claiming satisfaction as a natural right."[17] Again: "An enlarged field of observation convinces me that the profession has not in any wise exaggerated the influence of marriage on women, and that its dangers are infinitesimal as compared with those of celibacy."[18] Nearly every treatise upon gynaecology may be quoted to establish the same fact. It is upon the mind of woman that the defeated sexuality acts reflexly in a morbid manner. Dr. Maudsley, who has had abundant opportunities for observation, says: "The sexual passion is

---

[16]"The Journal of the Gynaecological Society of Boston," vol. ii, p. 267.
[17]"Uterine Therapeutics," p. 224.
[18]*Loc. cit.*, p. 127.

one of the strongest in Nature, and as soon as it comes into activity it declares its influence on every pulse of organic life, revolutionizing the entire nature, conscious and unconscious; when, therefore, the means of its gratification entirely fail, and when there is no vicarious outlet for its energy, the whole system feels the effects, and exhibits them in restlessness and irritability, in a morbid self-feeling taking a variety of forms."[19] While it is true that the engrossing cares of professional life, or of a skilled labor, will serve as a partial "vicarious outlet for its energy," in contrast to an idle life, yet this will in no manner act as a substitute for the natural expression of this physiological want. Its constant suppression will tinge the thought and manner of the woman. This is not an unreasonable statement, when we reflect that bodily derangements, not at all serious, will often account for changes in the mind and manner, as well as for the entire mental habit of men otherwise strong. If we contrast her with man in this respect, the chances of infinitely against woman in professional life. The penalty of sex is an episode in man's life. The tribute to his sexuality once paid, he is practically unsexed, and the trained intellectual man moves among women and men with scarcely more than a consciousness of his reproductive faculty. But sex in woman is a living presence. From the age of fifteen to that of forty-five, her life is crowded with startling physiological acts. Ovulation, impregnation, conception, gestation, parturition, lactation, and the menopause, contend with each other for supremacy—each act a mystery; each attended with its peculiar peril; and most of them evoking in its behalf the highest efforts of which her physical organization is capable. It will demand genius indeed to enable woman to rival man in the field of labor, and, at the same time, contend with the *inexorable law of reproduction*. . . .

The end and aim of woman's sexual life is perfected by maternity. It broadens and elevates her intellectually and physically. The influence over society reached by wives-mothers is a natural outcome of the stimulus of maternity. The maternal instinct, which lies dormant in the nature of every woman, awakens her mental being into increased activity the moment it is called into life. I think that it is for this reason that frail women, with no knowledge of life, when widowed, often succeed in keeping their families together and providing for them. With the woman who is constantly liable to the demands of a profession, or skilled labor, the maternal affection, anxiety, or care, may intrude at moments when her occupation will demand her highest mental efforts. The manual labor of rearing children the professional woman may delegate to others, but the ceaseless love, care, and forethought, so beautiful in a mother's love, the true woman must assume herself. Physically, children are necessary to the married woman. The sterile wife is constantly exposed to diseases that the fecund wife is comparatively exempt from. The

---

[19]"The Physiology and Pathology of the Mind," p. 203.

sterile wife is not a normal woman, and sooner or later this physical abnormality finds expression in intellectual peculiarities. Not upon the mind alone, but upon the body as well, does motherhood have a maturing influence. Gestation is nearly the completion of the sexual function. The process involves increase in the size of the heart, and in the volume and strength of nearly all the muscles of the body.[20] It is evident from this that gestation is not only a functional completion, but it is necessary to structural maturity, and to me it seems a natural corollary that it has an equal effect in increasing mental vigor. Having shown that marriage is in obedience to a physiological law, and that maternity is necessary to insure mental and bodily health in the mass of women, it is proper for us to ascertain if the last of these conditions—gestation—is not of itself, physically and mentally, an obstacle to professional life in women. The physical incapacity is too evident to need any comment.

Mentally, the changes undergone are most singular and multiform, and operate upon the cultivated and ignorant alike. Dr. Montgomery, speaking of the nervous irritability of pregnancy, says: "It displays itself under a great variety of forms and circumstances, rendering the female much more excitable and more easily affected by external agencies; especially those which suddenly produce strong mental or moral emotions. Hence the importance of preventing, as far as possible, pregnant women from being exposed to causes likely to distress, or otherwise strongly impress their minds."[21] These objective mental conditions described by the author must not be regarded as exceptional; on the contrary, they are classed among the usual symptoms of that condition. Still more marked mental disturbances may occur and are not rare, as in the following quotation from Dr. Storer: "Strange appetites, or longings, as they are called, and antipathies, are well known as frequent attendants on pregnancy in many persons."[22] And further: "The evidence that I have now presented proves that the state of pregnancy is one subject to grave mental and physical derangements, giving rise to serious anxieties, and requiring judicious treatment."[23] These mental effects are of minor importance in the relation we are studying, when we consider the fact that absolute insanity may be an accompaniment of either gestation, or follow parturition. Dr. Maudsley refers to this as explaining the excess of female insane over that of the other sex. . . .[24]

We have but one other sexual accident to consider which may act as a bar to woman's progress in the professions. These accidents are incident to the

---

[20]Dr. Alfred Wiltshire, "On the Influence of Childbearing on the Muscular Development of Women." Transactions of the Edinburgh Obstetrical Society, vol. ii, p. 237.
[21]"Signs and Symptoms of Pregnancy," p. 17.
[22]"The Causation, Course, and Treatment of Reflex Insanity in Women," p. 139.
[23]Ibid., p. 148.
[24]*Op. cit.*, p. 207.

climacteric period of life. This period includes the years between forty and fifty, and, judging from men, a professional woman ought then to be most actively engaged in her occupation. It is during the functional changes then taking place that women are exposed again to the dangers which attend the advent of puberty. It is the second and last crisis in the functional life of woman. We will let the mere bodily diseases of this period pass unnoticed, and refer to those of cerebral origin, as mind forms the working organ of the professional woman. Dr. Bedford regards the varieties of nervous irritation peculiar to this period as "beyond calculation."[25] In fact, it is upon the nervous system principally that the cessation of ovarian function acts reflexly in an abnormal manner. Thus, 500 women divided among them 1,261 forms of cerebral disease, confirming the general belief in the frequency of cerebral diseases at the change of life.[26] The liability to insanity at this period is greater in women than in men. Leaving out of consideration such an extreme result as insanity, yet the lighter shades of nervous derangement which would entirely unfit a woman for healthy mental work are so multiform, and to which men are in no way exposed, that it is evident that at this period woman would encounter some of the most stubborn barriers to her success in professional life. The professions, in giving undue employment to the mind, would greatly predispose a woman so employed to nervous disease at the change of life. Her very employment, to which many are working their way so bravely, is almost sure to entail suffering and danger at a period when educated and refined women, more than any others, require mental and bodily repose, and which the nature of their employment forbids. . . .

[25] "Clinical Lectures on Diseases of Women and Children," p. 374.
[26] Dr. Tilt, "The Change of Life in Health and Disease," pp. 164, 185.

Emily Blackwell (1826–1910) was born in Bristol, England and was the younger sister of Elizabeth Blackwell, the first woman to receive a medical degree in the United States. Emily, deciding to follow Elizabeth into medicine, began her medical schooling in 1848 by studying under Dr. John Davis, demonstrator in anatomy at the Medical College of Cincinnati. In 1852, Emily was accepted by the Rush Medical College in Chicago, but the State Medical Society censored Rush for admitting a woman, and she was barred from attending after the first year. In the fall of 1853, she entered the medical college of Western Reserve University in Cleveland and graduated with honors in 1854.

In the 1850s and 1860s, the Blackwells established and enlarged the New York Infirmary for Women and Children, an institution that gave women the opportunity to consult physicians of their own sex and provided clinical experience for women physicians. In 1858, the Blackwells began a course to train nurses, and in 1868, they established a medical school to provide women with a level of training unavailable to them elsewhere. For the next 30 years, Emily served as dean and professor of obstetrics and gynecology in the New York Infirmary's Woman's Medical College. In 1898, when the new Cornell University Medical College accepted women on equal terms, the Blackwells arranged for the transfer of their students and closed their school. In the 31 years that the Woman's Medical College existed, it graduated 364 women physicians. The New York Infirmary for Women and Children continued and still exists.

Emily Blackwell's work brought her in touch with women of all classes. Her experiences, along with her medical knowledge, led her to believe that it was not women's constitution that disqualified them from employment, but "conventional requirements and prohibitions." The following selection contains Blackwell's analysis of women's relation to domestic life and paid employment.

THE INDUSTRIAL POSITION OF WOMEN

EMILY BLACKWELL, M.D.

July 1883

AMONG all the questions affecting women, and society through women, there is none more vital than that of their industrial position. It is conceded that women should work, but there is a great difference of opinion as to what their work is, and how they should do it. This difference of public opinion is not merely a matter of theory; it leads to very positive practical results, for the support of public opinion is necessary to make work in any special direction possible. . . .

Upon this subject there are two views, the holders of which are endeavoring to enlist on their side this final arbiter of the question, the force of public opinion. On one side it is held that women urgently need greater facilities for work; a wider range of occupations, in order to give them greater power of self-support; that many grave social evils result from this want. It is maintained that the claims accompanying this effort, for equal general and special education, for participation in any kind of work which women feel that they can do, for employment in any occupation for which they have fitted themselves, are just; that the movement is in the direction of progress, and that it is the interest of society to support it. On the other side it is urged that woman has her own peculiar sphere, that of domestic life and work. This, well understood and followed, is sufficient for her. She is unfitted by her physical and mental constitution for the occupations carried on by men. Success in the effort she is making in this direction is impossible. The attempt is leading her to do violence to her own organization, to abandon or slight domestic life, and to become an inferior competitor instead of a companion to man. Progress is to be sought, not by favoring the effort, but by promoting such an extension of home-life as shall render it unnecessary.

Both parties are agreed as to the paramount importance of domestic life. This being admitted, the objection to non-domestic work for women is based upon the implied supposition that, were domestic life as universal as it should be, the domestic work connected with it would be sufficient to absorb the great body of women-workers.

To estimate the force of this objection, let us consider what is meant by the terms domestic life and domestic work. There are two elements in the domestic position of women: first, their personal relation to the family as wives and mothers; secondly, the work which necessarily devolves upon them in the fulfillment of the duties of these relations. The first, the personal relation, is a fixed and constant element. It grows out of the constitution of human beings, and exists under every form of society. It attains its highest expression wherever the union of one man and one woman is the foundation of the family. Here we find most marked the personal affection, the intimate companionship, the community of interests, the common responsibility and care for the children, which are the characteristics of the family. The related life of the group so formed constitutes domestic life. But if the personal relations of woman to the family are thus fixed and enduring, her industrial relation to it is by no means so unchanging. The work which she must do for it varies according to external conditions. There is no one kind of work which absolutely belongs to domestic life; there is hardly any kind of work that has not in some phase of society been considered to belong to it.

In the savage state, women built the wigwam, raised the corn, prepared the clothes, carried on in its rudest and most elementary form all the work which is to-day the object of modern industry. But since these simple forms of labor have developed into architecture and agriculture and manufactures, it is held that women can not follow their old occupations under their new forms, under penalty of personal deterioration and social disaster. It is the conditions under which work is done, apparently, which constitute it domestic work, rather than the nature of the work itself. Weaving was domestic work when done at home, but ceased to be so when done in a factory.

Domestic work, therefore, is all work for the family which, under our present arrangements, must be done at home, upon a small scale, by individual workers, free from the organization and competition of business. Precisely in the degree that outside occupations partake of any of these characteristics of domestic work, are they considered appropriate to women. On the other hand, how feminine soever the nature of a work, as soon as it is seized upon by the modern system of outside labor, begins to be carried on upon a large scale, and to be subject to the laws of business competition, it thenceforth ceases to belong to women. It has departed from the conditions of domestic work.

If this definition of domestic work be correct, two questions naturally arise in connection with it: 1. Will the work now done at home always continue to be so done? 2. If the conditions of domestic work are those most favorable to the well-being of women, what is the reason of their growing distaste for domestic service? The answers to these two questions are very closely connected with each other, and with the main question of the industrial position of women.

When we consider, on one hand, how pressing and increasing an evil is the lack of skillful and reliable servants, how severely the want of efficient service weighs upon the mothers of families, and, on the other hand, how liberal is the compensation and how certain the employment for women having even a moderate degree of skill in housework, there seems, at first sight, some truth in the assertion that the difficulty with women is not the want of work, but the inclination to shirk their own work in order to invade that of men. The complaint of the difficulty incident to finding well-paid work does not come from our domestic servants. The Irish girl finds work from the day of her landing and begins almost immediately to send remittances of money to her friends at home. The American girl, thrown upon her own resources, struggles miserably to keep soul and body together upon the scanty wages of the shop or the factory. Yet so decided is the disinclination to domestic service, the largest and most profitable field for women to work, that American women have virtually abandoned it. The Irish girls gradually absorb the same distaste, and are less available as they become Americanized. We already hear the suggestion in favor of the Chinese, that they are needed to supplant the Irish servants, as the Irish have taken the place of Americans.

Is not the cause of this dislike to be found in the servile nature of domestic service, which renders it necessary to bring in a constant succession of servile labor to fill it? Is it not just in proportion as women rise above the servile tone of feeling that they become restive in the position, and will sacrifice comfort and pecuniary advantage to escape from it? Almost every feature of domestic service partakes of this repellent character. On entering it, the woman, like the slave, drops the surname which marks her as a member of a family of a social connection, for the personal name which sinks her at once into a rank below that in which social connection is recognized. Reversing the natural order of things, the woman addresses the children and young men of the family by terms of respect implying superiority, while they address her by the familiar name implying inferiority. She abandons family life, having no daily intercourse with her relatives as do out-door workers living in their own homes. She loses her personal freedom, for she is always under the authority of her employers. She can never leave the house without permission; there is no hour of the day in which she is not at the bidding of her mistress; there is no time in her life, except the few stated seasons of absence, for which she may not be called to account. Though her accommodations are probably far better than she would have at home, their relative inferiority renders them less acceptable than the poorer quarters in which she shares freely the best there is to have. Every distinction of dress which is a badge of domestic service is universally felt to be derogatory. It is creditable to a man to refuse any domestic position that entails the wearing of a livery, while the uniform of even the lowest ranks of the public service—of the policeman or the postman—is assumed with satisfaction. The white cap and apron that become almost a uniform when worn by the graduates of the training-schools for nurses, as the mark of a superior class, are assumed with reluctance as an accompaniment of domestic service. . . .

In all such matters feeling is quicker than reason. Every woman instinctively feels that, in exchanging the position of an outside worker for that of domestic service, she descends one step in the social scale, and approaches one degree nearer to personal servitude. Upon what does this servile nature of domestic service depend? It is not due simply to difference of wealth and social standing; that difference exists everywhere between the employer and the employed. It is due to the conditions under which the work is done in the house, each servant dependent upon the mistress in the details of her personal life, doing work more or less undefined in its nature, amount, and time of doing. These conditions imply a direct, perpetual, personal subordination, necessarily servile. It is the absence of these conditions that renders non-domestic work independent, instead of servile. The limitation of the work within certain hours, outside of which all subordination or accountability to the employer ceases, the freedom of personal life thus gained, the more defined nature of the work, its larger scale, the numerous workers engaged in it—all these characteristics render the relation between employer and employed a business, not a personal one. . . .

There is no reason why what is now done by domestic service should always continue to be so done. As weaving and tailoring have gone, so the making of women's and children's clothing is now going. There is no reason inherent in the nature of things why washing, cooking, mending, etc., should not go also, and be done by business organizations from outside, instead of by domestic service. Thus domestic work will be reduced to the minimum, to that part most intimately connected with the personal life of the family. The need of domestic service will diminish in the same proportion, and the problem it presents will be solved by its diminution, or gradual disappearance; while domestic life will be more and more freed from the necessity of carrying on a variety of domestic work.

The obstacles to be overcome in bringing about this result do not differ in kind from those which are disappearing elsewhere before the ingenuity and perseverance of business enterprise. The difficulties in the way of supplying cooked instead of raw food are very similar to those being now overcome in the transport of delicate and perishable food, and in the preserving such food in perfection through the whole year. There is no reason why bakers should necessarily supply inferior bread, or why cooking done on a large scale should always be inferior to that done at home. That the work which remains to be so dealt with is the most difficult to be thus treated is the reason it has remained to the last. . . .

It may be objected that so radical a change in the conditions of household work must imply the destruction of the home as we at present understand it. But why should this be the result of the changes to come, any more than of the equally great changes that have been already accomplished? . . .

It may be objected that the failure to marry is the reason so many women are seeking employment; and that, were marriage sufficiently universal, the immense majority of women would be occupied in their own homes. Facts do not seem to bear out this view. The proportion of persons who pass through life unmarried is comparatively small. The mass of working-women is composed not of middle-aged single women, to whom alone the criticism could refer that they have preferred other work to marriage. The great bulk is composed of young women under twenty-five, whose families can not afford to support them for the sake of their domestic work, and the majority of these will probably eventually marry. There is also a considerable number of married women who, by the death or inability of the husband, are thrown back upon the necessity of self-support. This last is a much larger class than is usually supposed. It would probably at least equal the number of single women of corresponding age — that is, of women who have remained single to middle life.

As a matter of fact, support through marriage can not be co-extensive with the need of support for women. It does not cover the whole period of working-life, and it fails to be a support in a considerable proportion of cases. . . .

We believe, therefore, that a careful consideration of the movements which have gone on and are going on in social life leads to the following conclusions:

1. There is no necessary connection between domestic life and domestic work.
2. Domestic life means the united life of the members of a family, and is a constant social element.

3. Most of the work now done as domestic work is only so done because it has not yet been brought within the grasp of business organization. The range of such work is constantly growing narrower.
4. Our method of doing it by domestic service is imperfect, because domestic service involves a servile relation that does not exist in nondomestic labor.

But if all work tends thus irresistibly to become organized into departments of business, the question of the future industrial position of women is settled. They must follow their work under its new forms, or cease to work at all. Extremes meet, and the organization of industry must end by giving back to women what it began by taking from them, a place in the varied work of the world.

So far from the perfection of domestic life being imperiled by the gradual substitution of non-domestic for domestic labor, many advantages would be thereby gained:

1. It would help to free marriage from any but personal considerations. The question as to the capacity of a woman for house-work would become as foreign to that of her desirability as a wife as is now her ability as a tailor. It would be a wife only, not also a domestic, that the young man would need to seek.

Still more important a change would it be were marriage, to women, only the entrance into a wider and happier social state, and need never be regarded as the only recognized business opening.

2. It would bring more varied ability to the service of domestic life.

Despite the many kinds of work which have been gradually taken out of the housekeeper's hands, her position still calls for a variety of faculties rarely combined in one woman; and household life is in most families correspondingly imperfect. The business ability that makes a good housekeeper, in the sense of a good provider for material needs, of a capacity to use money to advantage, and to secure order and perfection of work, is one thing. To be a good educator, to possess the faculty of understanding and training children, is another. Neither of these qualifications is necessarily connected with the gifts and tastes which are required to make the home a social center, to bring its inmates into the friendly and easy relations to other families upon which its social standing depends, and which, under the present state of things, are so essential to the welfare of its young people as they approach the age for marriage. The mother of a family, whether rich or poor, must be a sort of "Jack of all trades," and often goes through life with the discouraging sense that, in one or other of these important departments, her good intentions will never supply the lack of natural faculty. The less complicated and extensive the work that necessarily devolves upon a woman in her household, the more chance for its successful accomplishment. The more she can call upon skillful help, the less likely the family will be to suffer from her deficiency in any direction. . . .

3. Another great advantage that would come from a general recognition that the occupation of women in non-domestic work tends inevitably to increase, would be the impulse it would give to the industrial training of girls. Parents do not think it worth while to educate their daughters for any pursuit, because they consider industrial occupation for a girl an undesirable exception, not to be provided for. An immense amount of misery would be avoided did custom require that every girl should be taught some paying work. It should be considered more obligatory in the case of girls than of boys, thus to guarantee them the possibility of independence, both because they are less able to make opportunities for themselves when unexpectedly called upon to do so, and because of the greater dangers to which helplessness exposes them. There is no greater source of suffering and vice among women than the fallacy of taking for granted that they will not need to support themselves.

4. The wider the range of occupations for women, the more numerous will be the points at which the lives of men and women touch. One of the objects to be accomplished by advancing civilization is the bringing of men and women into easy and natural companionship. Under existing circumstances almost the only meeting-ground for young men and women is in society. Those who can not take an active part in this are almost shut off from acquaintance with the opposite sex. Numbers of girls educated at girls' schools, and afterward living at home in narrow circumstances, or going into work conducted by and among women, remain single because they pass the age for marriage without sufficient opportunity for meeting men of their own standing, or make unsatisfactory marriages, because they do not choose from knowledge, but accept the only opportunity that offers. The same is true of young men not in society. Their life is passed almost exclusively among men from their school-days upward. Their acquaintance with women of their own age is extremely limited and superficial. The more complete the separation of men and women in work, the more must this division in life be the result. The more numerous the common interests and occupations in which they meet in recognized and honorable companionship, the more numerous the chances for suitable and happy marriage. So far, therefore, from deploring the encroachments of business organization on domestic work as a danger to the happiness of domestic life, we should see in them an agency which will lead to its higher development.

But if, as we have shown, it be in the natural course of things for women to take part in industrial pursuits, what is the meaning of the warning notes that attend their steps in that direction? We are told that women break down under the strain of college education; that their health gives way under the requirements of book-keeping, telegraphy, factory-work, every kind of business; that their work is poor and unreliable, and will command only starvation wages, etc., and these discouraging reports come not only from illiberal opponents, but from sincere friends and well-wishers. The most important of these objections is based upon the assumption that the physical constitution of women unfits them for safely bearing the strain of brain-work or business.

It is true that the health of women is not what it should be; but the cause of this lies neither in their peculiar organization nor in their efforts in new directions. It is to be found in the influences surrounding them from infancy, which prevent our girls from acquiring the physical vigor which should accompany maturity. This defective health is nowhere shown more conspicuously than in domestic life. Nowhere do women break down more frequently and completely than in the bearing and rearing of children, under the strain of maternity, and the wear and tear of domestic duty.

It is not only, nor chiefly, our college graduates and industrial workers who crowd the offices of specialists, and the same department in our charities. The girls who stay at home and are subjected to no educational strain, the wives and mothers who are pursuing the most natural of avocations, are quite as fully represented. We must believe that it is the physical education, not the organization, of women that is at fault, unless we accept the conclusion that the special constitution that is supposed to disqualify them for other work disqualifies them also for its own ends. . . .

Many of the difficulties which now embarrass women in work are such as belong to a transition state. They will disappear as the presence of women in these fields is accepted and provided for. The fewness of the occupations open to women and their consequent overcrowding; the difficulty, often the impossibility, of acquiring special education for occupations in which special skill is required; the opposition of the workers already in the field — these are only a few of the obstacles which are due to the novelty of the effort. Business has been arranged to suit men; and women, upon entering any new branch of labor, are required to accept its existing conditions. There are many kinds of work which women could do perfectly well if they could modify these conditions. But if, without this, they fail to do it as well as men, or suffer more in doing it, it is taken for granted that the work is unfit for them, that the remedy is to exclude them from it, not to adopt the mode of doing it to their requirements.

As an illustration of the different effect of the same work according to the circumstances under which it is done, take agricultural labor. Nothing is more frequently quoted as an exemplification of the brutalizing effect of masculine work upon women, than the results of field-labor as it is carried on by them in some parts of Germany and other places, where women are considered, and treated, as mere drudges. Contrast with these reports the following statements in regard to the effect of field-work upon women in the north of England, extracted from the "First Report of the Commissioners on the Employment of Children, Young Persons, and Women in Agriculture," presented to Parliament in 1868. In it Mr. Henley states that the women who work in the fields of Northumberland are "physically a splendid race." The same witness says: "There are many who consider field-work degrading; I should be glad if they would visit these women in their own homes, after they have become wives and mothers. They would be received with a natural courtesy and good manners that would astonish them. . . . The visitor will leave the cottage with the conviction that field-work has no degrading effect, but that he has been in the presence of a thoughtful, contented, unselfish woman. . . . The very appearance of the habitual workers is sufficient to prove the healthiness of their mode of life; and the medical testimony is overwhelming as to the absence of disease and the usual complaints attendant on debility."

Mr. John Grey testifies of the same women: "The healthful and cheerful appearance of the girls in the hay or turnip fields of the north, and their substantial dress, would compare favorably with those of any class of female operatives in the kingdom," etc. Here we have the same kind of work, destructive in one case, beneficial in the other. And this is due to the different conditions under which it is done. So in other work, it is not necessary that women should do every part of it precisely as men do it. The question is, Is there not in most kinds of work a place which women can fill to advantage under suitable conditions?

Women are much more fettered than men by conventional requirements and prohibitions. They come to any new occupation hampered by the restraints and burdens so imposed. Their dress is modeled upon fashions adopted by women in society, to whom dress is a profession, occupying a great part of their time, strength, and intelligence; yet custom forbids any material modification of it to suit the requirements of work. Equally liable to misrepresentation is any assumption of unconventional freedom in going about, ways of living, etc. Women are hindered at every turn by endless restraint in endless minor details of habit, custom, etc., which, often trivial in themselves, by their number and perpetual action often trammel them as effectually as the threads of his Lilliputian adversaries did Gulliver. In these respects we might apply to men and women the common French saying in respect to English and French law, viz., that "to one everything is permitted that is not expressly forbidden, to the other everything is forbidden which is not expressly allowed." Most women who have been engaged in any new departure would testify that the difficulties of the undertaking lay far more in these artificial hindrances and burdens than in their own health, or in the nature of the work itself.

Finally, is not much of the objection that work is destructive to the workers applicable to all work and all workers—to men as well as to women—to domestic as well as non-domestic work? . . .

It is only too evident that we have not yet solved the most fundamental problems in regard to labor, when we see such glaring contradictions as produce spoiling in the fields because there is no market for it, and mills stopping work because the market is over-supplied, when at the same time thousands are suffering from want of food and clothes. So long as the relations between workers and work are so imperfect, the hardships thus entailed must fall upon women as well as upon men. . . .

In answer to Blackwell's "The Industrial Position of Women," (see previous selection), Edward Livingston Youmans (1821–1887) and William Jay Youmans (1838–1901) published the following editorial. In it, they presented an ideal of the family as not merely a "union of the sexes in marriage," but a "union of interests which made their [men's and women's] respective spheres of occupation supplementary to each other." For the security and protection of women as well as for the maintenance of social order, the Youmans argued, it was necessary to foster "the identification of her [woman's] interests with those of man." However, after the elder brother's death, William Jay softened his position considerably. Compare this editorial with the two later ones, "Individuality for Woman" and '"The New Woman' and the Problems of the Day," which appear in Chapter 7.

PROGRESS AND THE HOME

EDITOR'S TABLE

July 1883

DR. EMILY BLACKWELL discusses the industrial position of woman in a way that appears to us especially significant at the present time. We said not long ago, "If there is one thing that pervades and characterizes what is called the 'woman's movement' it is the spirit of revolt against the home, and the determination to escape from it into the outer spheres of activity that will bring her into direct and open competition with men." This statement has been criticised as unjust; but we certainly did not mean to intimate that there may not be many women thoroughly enlisted in the "woman's movement," and who, nevertheless, retain a strong home interest. Our statement was general, and simply affirmed a widespread tendency, the unmistakable drift of which, we think, the article on "The Industrial Position of Women" decisively illustrates.

It will be seen that Dr. Blackwell writes as a student of social tendencies. She appeals to the primitive condition of society, falls back upon the law of progress, and forecasts the results of its future working upon domestic life. The industrial progress of mankind, as is well known, has been carried forward by the division of labor, in which, through greater proficiency of specialized work, improved machinery, and efficient organization, the productive capacities of society have been much diversified and augmented. Dr. Blackwell's argument is that this great social tendency had taken effect upon the domestic sphere, and must take much further effect by removing those forms of domestic labor with which women have been so long burdened, to the outside sphere of business organization. She maintains that woman must follow out these industries into the outer field of competition, or be left without the means of subsistence; while, by thus getting rid of all work hitherto called domestic, she will achieve her liberation from that home bondage of which she has so long been the victim. The social movement here referred to has two effects—the enlargement of external competition for woman, and a corresponding diminution of the internal sphere of home occupation. We must very briefly object to Dr. Blackwell's views upon both points.

As to the industrial tendencies of social evaluation invoked by Dr. Blackwell, she seems to have left out the most important, and, indeed, in this case, an all-determining consideration. While the common differentiations of industry are a result of progress, that between the sexes is not a result of progress. The division of labor between the sexes is primordial—older and deeper than all social development, and a fundamental condition to it. Any one who will consult the comprehensive "Cyclopaedia of Descriptive Sociology," by Herbert Spencer, and refer to the operative division of his tabular summaries, will find superabundant proofs that in the very lowest stages of all savage societies there was a fundamental and universal separation in the active spheres of the sexes, so that "no division of labor except that between the sexes" becomes almost a stereotyped formula. Men devoted themselves to hunting, fishing, and war, for the maintenance of the life of the tribe, while women cooked the food, made the clothes, took care of the children, and occupied themselves chiefly with the drudgeries of the rude home. Thus, before industries began to take any separate shape, there was already a division of occupations so broad and clear as to be evidently grounded in the nature of things, and all the subsequent progress of mankind has been achieved in subordination to it. The first great specialization of human activities is, therefore, not a product of social evolution. We have here to do with a fact of exceptional import, deeply grounded in the constitution of things, and not to be studied as an effect of social progress. And in its essential quality, moreover, this separation of the spheres of action of men and women is totally different from the ordinary differentiations of industry. The historic relation of the sexes, in regard to their distinctive spheres of action, is a non-competitive relation. The family arose not merely by a union of the sexes in marriage, but by a union of interests which made their respective spheres of occupation supplementary to each other. There is here no industrial rivalry, but the common ambition centers in the prosperity of the home. This is the fixed order observed equally in all stages of progress. As men fished, hunted, and fought in the pre-industrial stages of society, while women were occupied with the domestic cares, so the men still labor without, struggling with their fellows in the arena of business, and earning wealth which it is their pleasure and pride to expend upon the home and for the advantage of the family, while wives and mothers co-operate in the household sphere, contributing their indispensable and co-equal share to the common domestic welfare. But the relation of woman throughout has not only been non-competitive, but the fundamental fact in the case is that it has been a relation of protection. Not more in the predatory life of the savage than in the highest civilized life, woman has been the protected sex. Her security, and through it the maintenance of the social order, and the progress that has resulted, have not arisen from the independent competitive struggles of woman, but from the identification of her interests with those of man, through the division of their spheres of action. We await the reasons which are to convince us that these deeply-grounded relations must not continue throughout the future of humanity. The precipitation of woman into the outer world of conflicts, where the strongest have their way, would involve a dissolution of human society, and is not even possible as an experiment. Granting that the protection of woman has always been, and is still, very imperfect, progress must consist in making it more perfect, and not in subverting the order of which it is a natural and necessary part.

And now let us note the correlative effect upon the home of Dr. Blackwell's thorough-going social reform, and point out its radical error. She says: "There is no reason why what is now done by domestic service should always continue to be so done. As weaving and tailoring have gone, so the making of women's and children's clothing is now going. There is no reason inherent in the nature of things why washing, cooking, mending, etc., should not go also, and be done by business organizations from outside, instead of by domestic service. Thus domestic work will be reduced to the minimum, to that part most intimately connected with the personal life of the family. The need of domestic service will diminish in the same proportion, and the problem it presents will be solved by its diminution or gradual disappearance." But this process of the "gradual disappearance" of domestic activities and their relegation to outside organization is to be carried still further, so as to remove from the home the nurture of even "very young children." "There is nothing which would seem more absolutely dependent upon the mother than the care and training of very young children. Yet the careful study of the best modes of training these early years, which has come in with the Kindergarten, shows how far the nursery alone is from meeting their needs; how early and how much skilled teachers, other children, a variety of apparatus—that is, outside help—are desirable for the best interests of the child, as well as for the assistance of the mother."

It is not surprising that Dr. Blackwell should anticipate the very natural objection "that so radical a change in the conditions of household work must imply the destruction of the home as we at present understand it." She intimates that the dread is illusive, but she by no means replies to the objection. And that the logic of her position leads inevitably to this result is undeniable. For, when the process of removing all that can be removed from the domestic sphere, and handing it over to outside organization, is completed, not a remnant of family life "as we at present understand it" can remain. As spinning and weaving, brewing and drying fruits, tailoring and knitting, butter and cheese making, have gone, so washing, mending, sewing, and cooking can go also in the same way, and the sick can be sent out to the hospitals. And when the Kindergarten becomes a state institution, and "compulsory education" takes away the "very young children" to be cared for by outside arrangements, and all kinds of domestic occupation are thus eliminated, it might appear that the home has been reduced to its minimum as a place for the bare organic processes of gestation and lactation. But will progress, under such heroic interpretation, leave even this shred of a domestic sphere? Is not the stirpi-culturist abroad with his lamentations over the evils of the unregulated multiplication of human beings, and is he not predicting the time when human perfection shall be attained through the total disappearance of present domestic relations, and their better discharge under the control of outside organization? Dr. Blackwell declares that the personal relations in the family are "a fixed and constant element"; what is her warrant for the declaration in the light of the slow-working progress she invokes.

Dr. Blackwell's argument rigidly carried out would sweep the family and the home out of existence, and merge it in the outside life of society, where all regulation falls within the province of the state. Her reasoning goes to the most chimerical lengths through a failure to recognize that there is a permanent sphere of legitimate distinctive womanly work. Her affirmations that "there is no one kind of work which absolutely belongs to domestic life," and that "there is no necessary connection between domestic life and domestic work" can not for a moment be accepted as true. They are no more true than would be the proposition that there is no necessary connection between life and work at all. There are plenty of people who live and never work, but it remains true that human life is inexorably conditioned upon work. There are women who never do domestic work, who abandon the home, and live in hotels; but it is still true that domestic work is a condition and necessity of home-life—so true that, if domestic work disappears, the home is impossible. If there is a house, there must be housekeeping; if there are children, they must be cared for; if there are invalids, they must be nursed; if there is food, it must be prepared, and all these things involve work as a simple practical necessity. Because there has been a great deal of foreign and unfeminine work carried on in the household is no reason for asserting that there is no such thing as proper feminine domestic work. The home has, of course, been burdened by these industries, and women made drudges to them, and we all bid Godspeed to their exodus. But for what reason? That woman may be released from exhausting, unfeminine occupations, to give more strength to the proper performance of her legitimate duties as wife, mother, and household administrator. Weaving, cheese-making, and domestic manufactures stand in no relation to the essential nature and characteristic duties of woman. Such occupations have robbed her of leisure for self-improvement, and want of suitable culture has hitherto prevented the mass of women from properly performing the duties which lie in the very heart of home. Every step of progress from the primitive state to the present has been in the direction of woman's emancipation from the hardships of physical labor, and coincident with this relief there has been an improvement in her nature, the gentler virtues appear and the finer qualities of the feminine mind are developed. But the ideal of womanhood toward which such considerable progress has been made is not the fine lady, idle of hand and brain, the gadding and gossiping woman of leisure and society, who evades or discharges with wretched incompetence the cares and responsibilities of domestic life. Womanly talent and cultivation are demanded in the line of strictly feminine occupations, that the home shall become more and more instead of less and less in the social life of the future.

We have no space here even to enumerate the varied forms of womanly activity involved in the home, when all its extrinsic burdens are removed. That which progress must bring us is not exemption from them, but their more intelligent and congenial performance. The adequate education of woman for the home sphere we have never had, and it is now resisted with all the power of traditional habit and all the influence of the old educational ideals and the organized systems of study. Men are educated by the newer colleges for their special work in life; women never! The prejudice against studying things domestic, although the problems opened are many and of the deepest intellectual interest, abides with a strange inveteracy. Dr. Blackwell recognizes no amelioration of the home through inte ent preparation for it. Though education is now the standard solvent of all the difficulties in our civilization, she concedes to it no potency in renovating and developing home-life. She asserts, indeed, that women must have technical training for the sphere of outside competition, but nothing is said of its need as a preparation for domestic activities. As long as the home endures, it is to continue the stronghold of servility and degradation. Progress is to do wonders, but the home must remain the asphyxiating Black-Hole of menial ignorance and stupidity as lasting as may be the vestige of the institution. There are perhaps not many who will go to this visionary extreme, but in so far as the "woman's movement" exemplifies the feeling it merits unsparing condemnation.

Elfrieda Hochbaum Pope (1877–1962) was a writer living in Ithaca, New York. Her published works include an article, "What the University Loses by Underpaying its Instructors," Educational Review 31 (January 1906): 55–66; a travelogue, Passion and Pageant (1933); and a biographical novel, published posthumously, Burning Arrows (1963). In the selection below, Pope made a cogent argument for paying men and women equal wages on the basis of "comparable worth." The article is as apropos today as it was in 1913.

WOMEN TEACHERS AND EQUAL PAY

ELFRIEDA HOCHBAUM POPE
July 1913

WHAT is the clear and natural justice of paying women teachers equally with men? Two persons are expending an equal amount of energy in rendering services of equal value. In exchange a return energy is given in the form of financial reward. There is no reason why the return energy should diminish in quantity, the moment the recipient is a woman, but retain its normal volume if the recipient happens to be a man. Is it not an ancient principle of justice that the laborer is worthy of his hire?

The immediate reply to this will be: It is just that a man receive more, because he has to support a family. And Mr. Perry, whose argument rests on the ethics of not violating the principle of the "market-value" of teachers, unmindful of the principles of the bargain counter, says:

> The fact that the great majority of men have families to support has led to an economic balance whereby men's wages expressed in terms of money are such as enable a man to support his family.

This is plainly an economic fallacy, since wages and salaries are not a result of a nice adjustment to personal and family needs. A man supports his family in accordance with his wages; he does not receive wages in accordance with his family. And does the man who has no family receive less and the woman who has a family receive more? Is it the custom to arrange salaries on a sliding scale in accordance with celibacy or marriage among men? Why is it that late marriages are so common? Is it not because the incomes earned are thought not to be sufficient for the support of a family? Does any one know of a scheme like the following? An instructor in a university receives a salary of $1,000 a year, and manages to be fairly comfortable on it. He marries, and the trustees grant him an additional $100. He has a child, and his income is increased again by $100, and again for every succeeding child. We leave it to the trustees to estimate the proper value of a child on the basis of a full professor's salary. Now the wife dies, and $100 are subtracted from his salary, and his children become self-supporting the salary is reduced in proper measure, leaving him, when all his children have departed in the *status quo* with his original salary, for as a single person he requires no more. No doubt such a sliding scale would be most acceptable to college instructors so long as it went up and would encourage early marriage. The only bitter pill would be to have the scale slide down. In the actual world, however, the bachelor does not receive less because he has no family and the married man does not receive more because he has. The woman teacher still generally receives even as high as fifty per cent. less than a man, whether she has a family to support or not.

But, it is replied, the single man expects to have a family in the future for which he must lay a financial foundation now. Is then the young woman not expected to have a family? Will her savings be less of a help to the future family because they are feminine? Or will they go farther for the same reason and do they therefore not need to be so great? Is not the family the ultimate loser by this principle of stinting women, since the family funds are derived from one source alone? After marriage, if both father and mother are capable of earning, is the family not the gainer if the earnings of the mother are at the same rate as those of the father? If the father dies, is the family not the gainer by having full support instead of two thirds or thereabout? . . .

It has been stated that where men and women teachers receive equal wages the men will vanish. This assertion, however, seems to rest on the implication that the wages are low, for when it has been suggested that women receive the high wages of men, opening the way for a natural competition irrespective of sex, the answer has frequently been that the natural result of this competition would be that the men would be chosen and the women would be left. Now certain qualities excellent in a teacher have been conceded to women. For instance:

> Taking it altogether the fine women who as a whole make up our teaching force exert a healthful influence over their boys and are successful disciplinarians.
>
> The woman is quite as apt as the man to establish that connection between her mind and the child's which is the foundation of instruction.

Even a woman's knowledge is apt to be sufficient, at least for the high school. But it is possible that these virtues exist only at a low rate of wages and take wings at an equal high rate. It is one of the characteristics of arguments springing from the traditional view of women that quite opposite assertions are made to fit the same theory. Thus women are better than men and they haven't so pronounced a moral sense; they have no time for professions, and they waste time in frivolity; they are thrifty, and they are extravagant; they are physically weak, and do the physical work of the household; in the case of women teachers, they drive men out of the profession of teaching, and they can not compete with men; and again: they are not worth so much because they leave teaching to marry, and they are not worth so much because they do not marry. Perhaps it would be safest to adopt the high and equal rate of salaries even if it leaves man, as the superior teacher, victor in the field, since in education we are concerned with the best results obtainable. We sincerely trust and believe, however, that even at an equal high rate of pay it will be realized that men and women are needed in the schools as in the home. Woman is as much a factor in human life as man, and her interpretation of life and knowledge is just as necessary for a complete view. If there is no difference between the masculine and the feminine viewpoint surely there is no reason for discrimination. But the very possibility of a difference of conception is of immense potential value educationally, and forbids a lessening in value because of sex. . . .But as long as we regard education as a thing to be provided, in large measure to be sure, but at as low an expense as possible, we shall encourage the cheap labor of women teachers and the proportion of men and women will not be normal. As soon as we realize that education is an investment whose returns are to be measured in quality and diversity of knowledge and of character, we shall be glad to invest capital in that enterprise, though for intangible and indirect returns, and we shall recognize that woman's share in the product is as important as man's.

As far as the theory goes, held by Mr. Perry, that a budget permitting the expenditure of only a certain fixed amount for teachers' salaries forces a reduction of the men's salaries because of the necessary averaging, it is a matter of fact that a budget can always be increased where the need is felt to be actual. Even the "practical administration difficulties" of a huge system like that of New York City should not prevent meeting actual needs. In smaller systems, the budget can always bear an increase when a man teacher is needed who will not come at the salary allotted to women. In women's colleges the principle of exclusive femininity is inevitably disregarded through the crying need of some masculinity, possible naturally only on the teaching staff. A certain proportion of men is felt to be absolutely necessary either for the sanity of the educational process or for the protection of the masculine teachers themselves from the danger of feminization. When that proportion is threatened the budget does not stand in the way. And thus it is possible for a new-fledged doctor of philosophy, untried and without teaching experience, simply because he is a man, to obtain a salary 25 to 50 per cent. higher than that of a woman professor who may have greater knowledge, greater experience and every requisite for a successful teacher. Is it a wonder under such conditions that we do not find women stimulated to do more productive work? It is quite possible, too, that the youth has no obligations whatever, and that the woman has financial and family obligations. It is possible that the case of obligations may be *vice versa*. It is certainly wrong for us to assume that the man has always the financial burdens to bear, the woman never. Can we, who have taught in college and high school, not name numerous cases of that kind? Do we not also know of men teaching in the high schools who were merely taking advantage of the relatively good salaries they could obtain as teachers until they could get a foothold in another profession, thereupon to leave the teacher's calling forever? And have we not seen in the same schools women teaching at lower salaries, some of whom were supporting relatives, sending brother or sister to college, and some even husbands? . . .

The same principle and view of life that reduces the pay of the woman teacher reduces the pay of women in whatever field. It forces girls in factories and department stores into lives of shame, and gives the washerwoman who supports a family (and who ever hears of a washerwoman who has not a family and sometimes a husband, too, to support?) for labor that is by no means unskilled $1.25 a day, while the unmarried Italian who digs ditches gets $1.75 or $2.00.

There are other causes, to be sure, besides sex discrimination, which have encouraged unequal wages. Foremost among these is that women have not realized their own worth, have not demanded equal wages, have not been able to do so, in fact, through lack of organization. Moreover, in the past, women were crowded into a very few callings, among which teaching was a very prominent one, and thus they competed with each other. In the past, too, when women left the home to work, it was because they were forced to. Any addition, however meager, to the family income was welcomed. In the higher walks of life, again, women were content to earn the luxuries, depending on their families for the home and necessities. The parents, meanwhile, took pride in the fact that their daughters did not "have to" work. The effect on the worker, on the profession and on the family was bad. You got cheap labor, poor and half-hearted labor, and the family was out something, too. With modern times has come the realization that labor and self-support are necessary for the dignity, the character and the development of women, and that the welfare of society and of the family demands that she become a contributor of wealth rather than a mere consumer. But we shall not have the best efforts from women in professions until professional rewards are open to them. That increase of salary, with advance in position, based on merit alone is a necessary stimulus no one can deny. It will be well when all women realize the harm that is done, not only to their sisters, but to their profession, when they permit themselves to be stamped as cheap labor.

And as for the man, we fear that it is not chivalry that fails to recognize the equal value of woman's labor with his own. We fear that it is not chivalry that frowns upon the married woman teacher. And so we hope that he will be moved to a more generous spirit when he realizes that woman's loss is his own. The modern marriage is a halving of resources, whereas the colonial family was a doubling of resources. What the wife formerly actually produced by the labor of her hands in the way of food, clothing and household supplies, in a personal field of industry, quite free from competition either with her own sex or with the other, she must now produce in the form of the wherewithal to buy the food, clothing and household supplies. Her field has become wider, she must compete with others, but her capabilities have also grown wider, and must increasingly grow as she, with her husband, progresses farther and farther from that rude and simple life that was enclosed by four walls and called forth only a few of the manifold potential powers of hand and mind. Men must come to an insight of the economic waste of an unproductive life for their women, or of production without fair returns. But perhaps they will also begin to realize forms of waste that are not so material. When women, through motherhood, have that insight into the growing mind that no one else can possess, we prefer to have them withdraw from the profession of teaching. Is there no sense of a tremendous pedagogical loss? Because women alone can be mothers is that a reason that they should be nothing else? Shall their souls and their minds be refused their proper occupations even after motherhood is past, and shall they be condemned to atrophy because of a great though not exclusive function? Is there no insight into this spiritual waste? Does not society suffer from all these forms of waste? When we demand that a woman sacrifice her talents and ambitions, in other words, her natural powers, in order to become a wife and mother, we must not close our eyes to the fact that it is a sacrifice, and that sacrifice means waste. Our marriages rest upon a wasteful basis, and must become increasingly wasteful as civilization takes away more and more woman's former productivity in the home, unless she is granted a free field for her energies outside of the home. The young woman teacher must look forward, then, to contributing her share to the establishment of the home by her earning powers before she is married and afterwards when she can. The more she earns, the better. With this earnest view of the necessity of contributing to the family support, her profession will become something more than a means of occupying durance vile. It follows as the night the day that early marriage will be encouraged where two are contributing, and when marriage does not mean sacrifice and dependence on one side, and sacrifice and a heavy burden on the other side. Men, while necessarily bearing the financial burden alone for some of the time of married life will yet not be sacrificing more of their individual energy to the family than the mother who is giving of her life substance. For the woman a life of development and service will be added to motherhood, as a life of development and service are added to the fatherhood of the man. For we conceive of fatherhood as something more and nobler than the occupying of all one's time and energies with earning money for the children. Will there not be more time for fatherhood when the pressure of financial responsibility is lessened? And who knows what rich rewards of womanly forces future society will reap from allowing women to develop according to the divine promptings from within rather than by rule of man. For the full honors and rewards of effort, whether in the household or in scientific academies, have never yet been granted to women. They have never yet been permitted to drink freely of the cup of life. Let the men who openly or covertly regard women as their inferiors consider this, and for the sake of the future give her an equal chance. It was Schopenhauer who said, in quite a different connection, we may be sure, "First they bind our arms, and then they sneer at us because we are impotent." And it was Wilhelm von Humboldt who wrote of "the absolute and essential importance of human development in its richest diversity." If women are to develop humanly, they must not be arbitrarily cut off from the inspirations and the rewards that stimulate the growth of human mind and character.

A discussion of this general nature seemed necessary, because it was felt that prejudice against remunerating women teachers equally with men was mere prejudice based on a failure to grasp the wide bearing of the forces of work in the natural and historic evolution of women. We have still to consider the fact that there will always be some women in the profession of teaching who no longer look forward to marriage, though the *terminus ad quem* of this hope is nowadays very problematical, and who have no dependents whatever. This will be true, even after women have become large factors in all the professions, in most of which they already are represented, and after they have invented some new ones. But their number will not be very much greater than that of the single men in like circumstances unless women preponderate immeasurably in the population. If there were an injustice in giving them the full return for their labor, it would yet be less than the sum total of injustice of the old system. Moreover, dare we not hope, with the special penchant of women for charity and philanthropy, with the noble roll of "old maids" who are milestones in the progress of civilization, Frances Willard, Florence Nightingale, Clara Barton, Jane Addams, that the surplus energy earned will go for the improvement of society? Society needs the development of all its latent energies for its own purification and advance. Who dares, unless he was present when the foundations of the earth were laid, brand woman's energies as inferior because proceeding from a woman, and say to her, "Hitherto shalt thou come, but no further"?

Robert Malcolm Keir (1887–1964) was born in Buffalo, New York and attended Wesleyan University for two years (1905–1907) before dropping out to work in order to help support his family. Later he continued his education at the Wharton School of the University of Pennsylvania, receiving a bachelor's degree in 1911, and was then invited to teach at Wharton while working on his graduate degrees. He earned a master's in 1913, a doctorate in 1917, and went to work for the War Department as an assistant to Dartmouth's president, Ernest Hopkins. In 1918, Keir joined Dartmouth's faculty as a professor of economics, where he remained until his retirement in 1956. Keir became a nationally known expert in labor economics and published numerous books on American industry as well as more than 100 articles. The following selection was written very early in his career, while he was an instructor at the University of Pennsylvania.

WOMEN IN INDUSTRY

D. R. MALCOLM KEIR

October 1913

The Physical Effects of Wage Work

WOMEN'S efforts to obtain a vote have directed attention to other problems which confront members of their sex. Long hours of work in factories and stores and the evils of the sweat shop have been investigated, but little has been written upon the effect that working may have on women's ability to bear children.

It is said that the hue and cry over the work of women in industry is misplaced and overemphasized. Women have always been employed at the very same things for which they now draw wages. Since history has been recorded they have woven cloth, prepared food and borne burdens. The only difference between former times and the present is that most of this work was once done individually in the home, whereas now it is carried on collectively in a factory. Women are not doing men's work. They can not, for they are smaller, less agile, less strong. Rather it is true that men in spinning, weaving and sewing are invading women's sphere and crowding out the women. It is claimed that the work in mills is for no longer hours, nor under worse sanitary and hygienic conditions, than women's tasks have always been. A parallel argument is that scarlet fever is not a dangerous disease because it is no worse than smallpox. If it is true that there are 156 women sick for every 100 sick men in the cotton mills; if the sick-insurance societies of England, Switzerland, Germany, Austria and France report that women are ill oftener and for a longer duration than men; if medical authorities report that 40 per cent. of married women who have been factory girls are treated for pelvic disorders before they are thirty years old; then it must also be true that factory work has in it something that is more injurious to health than similar employment at home. When the labor is performed away from the domestic hearth new elements enter into it that make it dangerous. In the home the woman prepared the raw material for spinning, twisted it into thread and then wove the cloth. Each of these operations called for a change of position. In the factory the whole task has been so subdivided that each woman does only a very small share of it, and so she must stand or sit continually in one place. Such intense specialization permits no variety in the motions of the work, thus producing a monotony that is deadening. Furthermore, the number of machines to which a woman must attend, and the speed at which the machines are driven are constantly being increased. Coupled with piece-work wages, the "speeding up" results in a nervous tension and strain almost wholly lacking in the domestic system.

Although a woman might work for long hours at home, she could stop when it was necessary to attend to her natural bodily needs. In the factory she has not that freedom, and the result is a whole train of ills.

As a quadruped the female suffered little handicap because of the functions peculiar to sex, except when actually carrying or nursing the young. But after mankind had learned to stand erect, her support was far from ideal. The bones of the ankle and feet are too small to sustain great weight. A woman's knee is not so well adapted as a man's to form part of a sustaining column. The muscles of the leg, too, have a shorter purchase than a man's, hence the leverage between the trunk and the extremities is less. The strain of support is transferred to the back. Thus any work which requires long standing for a woman is injurious. All the pressure of the body's weight is brought to bear upon a portion where the sex organs and others are crowded together, and produces a dragging feeling above and about the hips. Women performing such work are especially liable to congestion of all the organs enclosed by the hip bones, because standing and the habit of resting on one leg only, causes a narrowing of the hips. This narrowing is especially apt to occur because the greater proportion of women workers are too young to have become securely and permanently established physiologically before going to work. The average age for men at work is between 25 and 30, whereas the average age for women is between 16 and 20. In 1900 49.3 per cent. of the women were under 25 years of age. In the silk, knitting and hosiery mills there are as many girls between 16 and 20 years as all women over 21 years. By far the greater number of girls do not break down while they are at work, but after leaving the work for matrimony the deformities caused by the work become apparent. Specifically the uterus is very apt to be crowded out of place, or to be congested. Menstruation is made irregular and difficult. Factory women frequently stand at their work to within a few hours before giving birth to a child, with the result of premature labor. Miscarriages occur oftener among factory wives than in the general population. It is more frequently necessary to use instruments in childbirth among such women.

The mill hands are not the only women who suffer from long standing. The girl clerks in department stores are subject to the same conditions. Although 37 states require seats to be placed for clerks, there is no law enforcing their use. Many stores have a rule that clerks must stand at all times, because they look less alert when seated. Clerks on the first floor are seldom allowed to sit down. When sitting is permitted at all, the number of seats is inadequate for the sales force. In addition to standing, clerks suffer from lack of space behind counters, which increases the strain of lifting and puts a further burden upon the pelvic organs. The secondary effects of long standing, among which are broken arches in the feet and enlarged veins in the legs, is to add to the nerve strain, and indirectly affects other functions.

Sitting in one position has an action similar to long-continued standing. Lack of exercise reduces the capacity of the lungs, and so they do not eliminate certain poisons from the body. Because the lungs fail to act the kidneys are forced to do extra work, adding to the congestion of the abdominal organs. Sitting augments constipation, a minor ailment in itself, but one which breeds more bodily ills than any other single cause that might be mentioned. This condition is very prevalent among working women because of their lack of careful personal attention. In mills and stores the toilets are often too few in number, unsanitary in condition and inconveniently placed. In many cases there is no separation for the sexes. In some stores and factories no employee can leave her work for more than five minutes. When in many-storied building the toilets are not on the same floor with the worker this rule amounts to a prohibition. In other places girls must ask permission of men foremen or floor walkers to leave their work, a thing which many hesitate to do. When a store closes at 6 o'clock clerks frequently may not be absent from their posts after 4:30 P.M. Such conditions cause a partial paralysis of the alimentary canal, and abnormalities in the secretions, which puts an undue and constant strain upon the whole body. In women this strain is most apparent in functional abnormalities, hysteria and general anemic conditions. In addition to the restraint of sitting the indirect pressure against the abdominal organs by leaning over a sewing machine or against a desk augments the tendency to chronic inflammatory disease in the pelvis. The total result of long standing, or sitting in one attitude, is either absolute sterility or such organic disturbances as make child-bearing dangerous.

The second new element in modern industry is the monotony of the work, the unending recurrence of unavailing effort. It is difficult to trace any direct effect of monotony upon the more vital organs of the body. Monotony is a mental rather than a physical phenomenon. Modern factory work demands no feeling, no personal interest, no responsibility, nor inventive genius on the part of the worker. She does one thing endlessly, automatically. Work which demands nothing of the intelligence costs the intelligence more than work which demands too much. When only one brain center is employed the brain is more fatigued than if all the centers were worked harder. The result is either a stunting of mentality or an inordinate craving for excitement. The intimate association of the nervous system with the other functions of the body insures the reflection of injuries to the brain centers in the disturbance of all other organs.

The monotony of work is linked to the strain under which it is carried on. In the knitting industry a girl now has to watch from two to ten needles instead of one. In sewing shops the needles make 4,400 stitches a minute. The operator can tell when a needle or a thread is broken or a stitch misplaced only by a variation in a beam of light thrown on the needle. Constant attention to so minute a detail puts a fearful strain on the eyes and nerves. In textile mills the number of machines has so increased that the operator is kept always at the highest rate of speed. In a large publishing house girls who bind the magazine must handle 25,000 copies, each weighing three fourths of a pound, in ten hours. Piece work aggravates the evil of keeping up with a machine. In the millinery trade "rush work" is of a similar character. This speed coupled with the monotony of doing the same operation repeatedly brings about nervous exhaustion. The monotony of the work exercises only little patches of the nervous system. Mental and physical fatigue are closely bound together. A muscle in contracting uses nitrogen and liberates a poison or toxin. Under normal conditions this toxin is carried out of the body by way of the kidneys and lungs, and is neutralized by an antitoxin. If the muscle is exercised too frequently the toxins multiply faster than the ability to eradicate them. The poison accumulates and the muscle becomes fatigued. Further work is performed at the expense of the will, which puts a drain on the nervous system. Fatigue may go to the point of exhaustion, and result in death by chemical self-poisoning. Normally the tired body throws off the toxins during sleep and is then ready for another full day's work. But if the body does not get rest, the fatigued muscles on the second day can do only half the normal amount of work before again becoming fatigued. At the beginning the overwork may pass unnoticed, but since fatigue is accumulative it eventually results in a complete nervous breakdown, because fatigue really weakens the brain centers that control the muscles, although the feeling of being tired is primarily felt in the muscles themselves. Women are predisposed to nervous trouble, and their nerves are weakened by the various sex functions. Nervous tension exaggerates any bad tendencies already present. The industrial woman works to the point of over-fatigue and then goes home to do housework, or seeks excitement in dances and shows, thus adding nerve strain to nerve strain. Sleeplessness and loss of appetite follow; succeeding days of work pile up fatigue until the brain cells and nerves collapse. A usual accompaniment of nerve exhaustion is menstrual irregularities and poverty of the blood. The constant vibration in a mill may help bring about organic troubles, particularly if the organs have been weakened by other causes. The vibrations plus noise act on the nerves as a continual light tapping does on steel. Both steel and nerves disintegrate. One girl said that when her machine stopped at night she always felt like screaming, which proved that her nervous energy was being too greatly sapped by the day's work.

The effect of the strain of industry then is to add mental to physical fatigue, destroying the recuperative power of the body. Since the sexual organs and the nervous system both take the same food elements from the blood and are delicately adjusted to each other, the toll industry takes of the nerves is sooner or later reflected in organic maladjustments.

As with monotonous work, so with industrial diseases no direct result on the fecundity of women can be pointed out. The harm comes indirectly through a lowering of general vitality and nerve strain. Lead poisoning seems to attack women more readily than men. It is a most potent producer of abortion, for it is rare for a woman working in lead fumes to give birth to a healthy child at term. Often the poisoning results in sterility. At first, the odor of carbon bisulphide in a rubber factory makes girls excitable, but it is followed by headache and nervous lassitude, with a loathing for food. As with morphine and cocaine, the cause has the semblance of a cure, a feeling of normality only when drugged. This produces the vicious circle, of poisoning, lassitude and repoisoning. The excitement causes undue fatigue, while malnutrition culminates in poverty of the blood, general debility and organic disturbances. The eating of the hands by acids in pickling factories, bleacheries and soap works tortures the workers and exhausts the nerves. Dust dries the throat. The effort to cough produces asthma or an inflammation which is a good seeding ground for tuberculosis. A hot, damp workroom weakens the body by excessive perspiration, and renders it liable to rheumatism, bronchitis and tuberculosis. Lifting heavy weights or running foot-power machines so injure the sex organs as to induce sterility. This list might be lengthened, but enough has already been written to make the point.

Malnutrition plays a part in lessening the vitality of working women. When a mother has to prepare a breakfast for a family before hurrying away to a shop, that breakfast is to be commended for the speed of its preparation rather than for its adherence to principles of proper diet. Bread and butter, coffee or tea and greasy meat, with the addition of a pickle as a stimulant, is the usual bill of fare, varied but little in the three meals of the day. There is not enough of cooked food and vegetables are lacking. The mother's body is not sufficiently nourished to withstand the double tax of factory work and house work, and if a child comes, the mother is sometimes too impoverished physically to nourish it. A baby should be fed every two hours, but a factory mother sees her infant once in six or ten hours only. Drugs are given to children only two weeks old to keep them quiet while the mother is away from home. Interference with lactation is injurious to the mother and fatal to the baby. The survivors of this heroic treatment grow up never having had sufficient nourishment. When it comes their turn to go to work, they do so not equipped with full vigor to meet the increasing stress of such work, but in a weakened condition, and are susceptible to all the ills before mentioned. Lifelong malnutrition added to pelvic deformities acquired by work is a serious drawback to the motherhood of working women.

Because children are a handicap, in calling for time and attention that is needed for other work, they are unwelcome or impossible to certain groups of women workers. A woman reporter or school teacher can not afford to have the demands of a child interfere with the requirements of her work, so in some branches of industry it can not be stated whether the women engaged in it are physically incapacitated for bearing children, or whether they have none because they do not wish them. All the social and economic factors which are causing a world-wide decline in the birth-rate operate in the various groups of working women and complicate any general conclusions that might be drawn as to the injury done by the work itself.

A hundred and fifty years ago the man who ventured to predict that no women would be allowed to work in mines, bar rooms, buffing or polishing metals or as public messengers would have been laughed at in scorn, but we to-day look upon such occupations for women with horror. May not our children's children think of a time when women worked in factories as a barbarous age? Lawmakers may some day forbid such labor on the grounds of the injury to the future race. We may not have to wait for such laws, however. Working conditions are usually the worst in the smaller concerns. The country is tending toward large-scale production, therefore conditions are slowly improving. But with large-scale production machines are larger, work heavier, speed and number of machines greater, so much so that men must be employed to do the work formerly done by women. The task a woman performs is largely of a mechanical nature, hence with large-scale production unreliable labor is replaced by a reliable automatic machine. The development of the loom from hand power to power driven, and from female to male attendance, and to final automatic action is an illustration in point. The more efficient machine may drive the less efficient woman out of the shop. In stores, offices and schools they must be more adequately protected by law, for it is evident that working, sooner or later, is reflected in fecundity.

Chapter 7
THE "NEW WOMAN" (1890-1915)

No one who understands the feminist movement, or who knows the soul of a real new woman would make the mistake of supposing that the modern woman is fighting for the vote, for education, and for economic freedom, because she wants to be a man. That idea is the invention of masculine intelligence. Woman is fighting today, as she has all the way up through the ages, for freedom to be a woman.

Anne B. Hamman, "Professor Beyer and the Woman Question," *Educational Review* 47 (March 1914): 296.

Beginning in the 1890s, Americans spoke of a "new woman" who differed from past ideals of American women in several ways. First, she was highly educated. She had gone (or aspired to go) to college and had proven her ability to master areas of learning once considered beyond her intelligence. Second, she embraced a special social mission. She desired to use all her capacities—intelligence, energy, grace, gentleness, moral sensibility, and "womanly instincts"—to improve the material and moral conditions of society, particularly the urban poor. Lastly, the new woman aspired to individuality and autonomy, claiming the right and ability to decide how to best employ her talents in living her own life. She was thought both capable of and entitled to self-expression.

Those who praised the ideal, however, were careful to point out that this new woman was still essentially a wife and mother and would not neglect her domestic responsibilities. Education, social awareness, and individuality were of importance because these would help her become a better companion to her husband, teacher to her children, and guardian of the home. Thus, the new woman had reached "the point where she blushes for shame if health be not the normal condition of her family." She had grown more conscientious about rearing and educating children, having learned "that between the kindergarten and the college graduation there is the unbroken link of an immortal soul being trained to live." Moreover, to apply her moral consciousness and domestic knowledge to society at large did not indicate neglect of the family, but an increased commitment to improving domestic life: "The new woman has learned that if she would have a clean house she must have a clean street . . . a clean neighborhood . . . a clean city, town, community." For this reason, the new woman "would be ashamed not to know something of the administration of the city, state or Nation." For she prized "good citizenship for what it gives to her home and maintains for it."[1]

Despite reassurances that the new woman was dedicated to wifehood and motherhood, many Americans found the ideal alarming, in part because the emphasis on autonomy and individuality threatened familial relations, in which women were supposed to subordinate their personal interests to those of husbands and children. Critics argued that in practice individuality translated into selfishness. Women, it seemed to them, were much less willing than before to assume conventional familial roles, as evidenced by their decision to postpone marriage or not marry at all. Worse yet, women seemed increasingly more willing to leave husbands who did not satisfy them, and the quickly rising divorce rate deeply troubled many Americans (See Chapter 3). Those who disapproved of or feared the new woman described her in what were then considered to be the most unattractive terms possible—as an inferior, incomplete, lonely, "mannish" woman, emphasizing those traits acceptable in men, but "unnatural" in a woman. The new woman was supposedly loud and coarse; she smoked, drank, and behaved improperly in public. In speaking of the "Bachelor Maid" (one form of the new woman), Winnifred Harper Cooley wrote:

> It is an anomalous term, Bachelor Maid, used so romantically for authors and artists—which may include weary teachers, "dainty lady typewriters," women in stores, Government departments, and all of the numerous professions . . . [as well as] the conspicuous minority, the heiresses . . . who have not found . . . that ideal man who is their dreamed of life partner.
> Certainly, these women are to be pitied. . . . Failing to find the inestimable delight of a sympathetic sharer in her joys of luxury and travel . . . she is the envied of the poor, yet is of all women perhaps, the most lonely.[2]

Defenders of the new woman tried to refute this image of her as a lonely single woman and represented her as someone longing to "strengthen the cause of right and justice" by raising "her own sex to the highest level it can attain."[3] The new woman, her proponents argued, did not disapprove of marriage, but desired to rectify the evils existing in marriage. Similarly, she did not dispute the social ideal of motherhood, but desired to make the actual condition of motherhood approach that ideal. As one writer put it, the new woman wished "to make marriage no longer an auction of sale to the highest bidder, or an exercise of tyranny on one side and subjection on the other, but a covenant of mutual help and service; and motherhood, not dreaded, despised, and a hindrance to self-fulfilment [sic] and rights of citizenship, but a state of recognition and honor."[4]

Thus, the debate over the new woman centered on whether more independent, better educated women would make acceptable wives and mothers. Some people—and these included single, educated career women as well as politically conservative and socially traditional men—portrayed the new woman as an irreconcilable alternative to the conventional mother and denied that her new activities and ambitions could be incorporated into a conception of socially acceptable motherhood. Throughout the late nineteenth and early twentieth centuries, combining motherhood with full-time paid employment outside the home was not acceptable practice. Motherhood prevailed as the primary and sole occupation of most women, even highly educated ones. As I discussed in Chapters 2 and 3, the majority of college women married (on the order of 60%), and most of these (approximately 80%) had children. Child-rearing made it very difficult, even for highly educated women, to work outside the home, and it should be recalled that most college women left outside employment after marrying. For a small group of women, however, celibacy offered a different life pattern: Reformers such as Susan B. Anthony and Jane Addams became prototypes of women who forfeited having families in order to devote themselves to social reform. This is not to say that women had to remain single in order to work for social change. There were many examples of female reformers (Elizabeth Cady Stanton, for example) who married and had children.

While the ideal of the new woman emerged in response to the experiences and aspirations of young college women, many Americans quickly attempted to reconcile the ideal with the traditional domestic responsibilities of married women. (See "The Mother as a Power for Woman's Advancement," pp. 307–311.) They did not succeed. By 1910, the term, new woman, had such connotations that one woman writer felt she could not use the phrase and have her arguments taken seriously.[5] The social derision brought upon the new woman evidently made it difficult to discuss the real and serious obstacles to women's economic independence, equal political rights, and social recognition as autonomous individuals.

One reason why the new woman evoked such antagonism and scorn was that the ideal attempted to close the biological and social gaps between men and women. The new woman was just as intelligent as a man, she could and did go to college, and she even demanded the right to work and vote. In the face of these developments, it seemed to J. C. Fernald that what was being forgotten was that "men and women constitute different classes of human beings, naturally, originally, and inevitably different." In an ironic vein, Fernald went on: "Why not, then, dispense with the word woman, and say simply 'human rights,' 'human education,' 'the human sphere,' 'the human mission,'—'a convention of human beings will meet to consider the subject of human suffrage?'" For Fernald, the answer to "Why not?" was plain: "'A human being,' without limitation of age or sex does not exist. . . . The only conception of human nature in the abstract that will be good for anything must be formed by knowing persons that actually exist, and men and women in real life will be found to possess such very different characteristics as must affect their respective relations to all life's activities."[6]

It is interesting to consider how much the ideal of the new woman emerging in the 1890s differed from an earlier ideal—that of the "true woman," identified by Barbara Welter as pertaining to women from 1820 to 1860.[7] Two elements of true womanhood—piety and purity—were upheld by proponents of the new woman, although critics accused her of lacking these qualities. A third element, domesticity, was retained but in a somewhat altered form. Domesticity took on a larger meaning: It no longer necessarily confined women to the home, since they were permitted to employ their special homemaking skills in certain outside charitable endeavors. Critics claimed, however, that these women neglected their domestic responsibilities by not marrying, marrying late, or not having children. The fourth element identified by Welter, submission, is more difficult to assess. On the one hand, supporters encouraged the new woman to expand her interests and actions beyond the domestic sphere, and to some extent this represented intentional defiance of men's absolute authority in public activities. Moreover, some women began to deny openly husbands' excessive or unpleasant sexual demands. On the other hand, supporters of the new woman did not envision the ideal as a revolution against basic social conventions, nor did they desire that women alter their roles as wives or mothers. The majority of women, even many of the so-called new women, upheld the doctrine of separate spheres and considered domesticity the quintessential part of their lives. It was in women's interest to retain and enhance their power in the family, for so little was available to them in any other domain.

The similarity between the new woman of the 1890s and the young career woman of the 1980s also deserves comment. Just as there was an upsurge in the numbers of women attending colleges in the 1890s, so there has been a dramatic increase in the number of women attending graduate and professional schools in the 1970s and 1980s.[8] Many college women of the 1890s felt a new sense of emancipation, believing that they were free of the social restrictions that had constrained their mothers. They had greater opportunities for employment; a very few even could become doctors and ministers. Similarly, young women, today, just out of law, business, and medical schools seemingly find themselves at no disadvantage with men in their competition for the most prestigious and lucrative positions in their fields. Both ideals look to education (or educational credentials) as the means of achieving power and status and posit that educated women are different—different from their predecessors and from most of their contemporaries—in having access to the skills and opportunities that will enable them to attain equal economic and political power.

The woman movement in the nineteenth century, however, largely believed that women's special feminine qualities—rooted in their biology—made them better suited to childrearing than men, and then used the responsibilities of motherhood to demand broader educational opportunities and political rights. This limited tactic was effective. Beginning in the 1820s and 1830s, educational reformers gained support for women's education on the grounds that education would make women better homemakers and childrearers, while by the end of the century, the suffrage movement attracted proponents with the argument that politically active and informed women would make better mothers. Throughout the century, women were encouraged to work in occupations—notably teaching—that would better prepare them for homemaking and motherhood.

Today we are reconsidering the meaning of motherhood and reevaluating what is necessary for women to do in order to be "good" mothers. We no longer insist categorically that women with young children stay at home, although there remains an abundance of academic, scientific, and social opinion that claims that women who do not remain at home with small children risk endangering their children's emotional and physical health. The women's movement today is attempting to assist women in the work place by reducing some of the traditional familial responsibilities women have borne: by calling upon men to help with housework, by providing affordable childcare services, by requesting that employers offer more liberal maternity and paternity leaves. And some women are exploring nonconventional familial arrangements, raising children on their own or with other single women. But so long as social practices and the organizational structure of our work lives prevent men from sharing equally in the day-to-day responsibilities of childcare, women alone will continue to face the practical and psychological conflicts that put them at a disadvantage in the work place. The balancing of childcare and career will continue to involve sacrifice, compromise, and frustration for women, and for men as well, so long as family well-being and individual professional success remain irreconcilable, so long as economic support and childrearing remain at odds. If the family is to continue as a viable institution and a source of love and comfort for all its members, the organizational structure of our work lives must be changed. Both parents must be able to decide together how to coordinate professional aspirations with their children's needs. The woman question cannot be resolved until it is also treated as a man question, and its resolution will have to satisfy both working mothers and working fathers.

NOTES

1. Lillian W. Betts, "The New Woman," *The Outlook: A Family Paper* 52 (October 12, 1895): 587.
2. Winnifred Harper Cooley, *The New Womanhood* (New York: Broadway Publishing Company, 1904), pp. 9–10.

3. Nat Arling, "What is the Role of the 'New Woman?'" *Westminster Review* 150 (November 1898): 576.
4. Ibid., p. 576.
5. Margaret Deland, "The Change in the Feminine Ideal," *Atlantic Monthly* 105 (March 1910): 289.
6. J. C. Fernald, *The New Womanhood* (New York: Funk and Wagnalls, 1891), pp. 18–19.
7. See Barbara Welter, "The Cult of True Womanhood: 1820–1860," *American Quarterly* 18 (Summer 1966): 151–174. Also reprinted in Michael Gordon, ed. *The American Family in Social-Historical Perspective* (New York: St. Martin's Press, 1973), pp. 224–250.
8. For example, 230 law degrees were granted to women in 1960 compared to 801 in 1970 and 10,754 in 1980. In 1960, 387 medical degrees were granted to women, with 699 in 1970, and 3,486 in 1980. There were 24,000 women who received master's and second level professional degrees in 1960; 83,000 in 1970 and 147,000 in 1980. Similarly, only 1,000 women received doctorates in 1960, compared with 4,000 in 1970 and 10,000 in 1980. Bureau of the Census, *Statistical Abstract of the United States 1982–1983* (Washington, D.C.: Government Printing Office, 1983), pp. 166, 168.

ARTICLES ON THE NEW WOMAN APPEARING IN *THE POPULAR SCIENCE MONTHLY*

(Arranged by Date of Publication)

Editor, "Individuality for Woman," (September 1891): 39: 696–697.
M. C. De Varigny. "The American Woman," (July 1893) 43: 383–388.
Mrs. Burton Smith. "The Mother as a Power for Woman's Advancement," (March 1895) 46: 622–626.
Clare de Graffenried. "The 'New Woman' and Her Debts," (September 1896) 49: 664–672.
Editor. "'The New Woman' and the Problems of the Day," (November 1896) 50: 120–123.
Ellen Coit Elliott. "Let Us Therewith Be Content," (July 1897) 51: 341–348.
Henrietta Irving Bolton. "Women in Science," (August 1898) 53: 506–511.
Earl Barnes. "The Celibate Women of Today," (June 1915) 86: 550–556.

SUGGESTIONS FOR FURTHER READING

Primary Sources

Arling, Nat. "What is the Role of the 'New Woman?'" *Westminster Review* 150 (November 1898): 576–587.
Betts, Lillian W. "The New Woman," *The Outlook: A Family Paper* 52 (October 12, 1895): 587.
Bisland, Elizabeth. "The Modern Woman and Marriage," *North American Review* 160 (June 1895): 753–755.
Cooley, Winnifred Harper. *The New Womanhood.* New York: Broadway Publishing Company, 1904.
Deland, Margaret. "The Change in the Feminine Ideal," *Atlantic Monthly* 105 (March 1910): 289–302.
Eliot, Charles W. "The Normal American Woman," *Ladies' Home Journal* 25 (January 1908): 15.
Fernald, J. C. *The New Womanhood.* New York: Funk and Wagnalls, 1891.
Gibbons, Cardinal. "Pure Womanhood," *Cosmopolitan* 39 (September 1905): 559–561.
Henry, Josephine K. "The New Woman of the New South," *Arena* 11 (February 1895): 353–362.
Hewitt, Emma Churchman. "The 'New Woman' in Her Relation to the 'New Man,'" *Westminster Review* 147 (March 1897): 335–337.

Humphreys, Mary Gay. "Women Bachelors in New York," *Scribner's Magazine* 20 (November 1896): 626–635.
Ouida. "The New Woman," *North American Review* 158 (May 1894): 610–619.
"New Woman Under Fire," *Review of Reviews* 10 (December 1894): 656–657.
"The American Ideal Woman," *Putnam's Monthly Magazine* 2 (November 1853): 527–531.
"The New Woman," *Cornhill Magazine* 23 (October 1894): 365–368.
Wells, Kate Gannett. "The Transitional American Woman," *Atlantic Monthly* 46 (December 1880): 817–823.
Winchester, Boyd. "The New Woman," *Arena* 27 (April 1902): 367–373.
Winston, Ella W. "Foibles of the New Woman," *Forum* 21 (April 1896): 186–192.

Secondary Sources

Freedman, Estelle B. "The New Woman: Changing Views of Women in the 1920s," *Journal of American History* 61 (September 1974): 372–393.
Forrey, Carolyn. "The New Woman Revisited," *Women's Studies* 2 (1974): 37–56.
Reeves, Nancy. *Womankind: Beyond the Stereotypes*. Chicago: Aldine, Atherton, 1971.
Welter, Barbara. "The Cult of True Womanhood: 1820–1860," *American Quarterly* 18 (Summer 1966): 151–174. Reprinted in Gordon, Michael, ed. *The American Family in Social-Historical Perspective*. New York: St. Martin's Press, 1973, pp. 224–250.

Upon his brother's death in 1887, William Jay Youmans (1838–1901) assumed full responsibility for editing The Popular Science Monthly. *This remarkable editorial contains Youmans' most generous and liberal assessment of women's position, recognizing the legitimacy of women's demand for individuality. This was not Youmans' last word on the subject, however, and should be read along with his later editorial, "'The New Woman' and the Problems of the Day," also reprinted in this chapter.*

INDIVIDUALITY FOR WOMAN

EDITOR'S TABLE

September 1891

AS a general thing, when the importance of individuality has been insisted on, the individuality in view is that of man. It is he who has been exhorted to assert himself, to be true to his opinions, to live his own life; the exhortation has not been to any great extent addressed to his wife or his sisters. Enough for them if they can be so fortunate as to minister not unworthily to some grand male individuality. Women, however, though not particularly invited to the lecture, have been listening to it, and—what people do not always do with lectures or sermons—are applying it to themselves. The best of them are now aspiring also to be individuals. They want to think, to feel, to know, to do something as of themselves, and, if possible, to think clearly, to feel truly, to know surely, and to do efficiently. St. Paul said that a woman should not be suffered to teach: what would he say if he could attend an annual meeting of our National Educational Association, and see to what an extent woman has become the teacher of the youth of the nation? He said that if a woman wanted any information on doctrinal or religious matters she should go home and ask her husband. The husband of to-day knows more about business than he does of theology; and few wives, indeed, would think of consulting their husbands on the latter subject. In any case the conditions have totally changed since these dicta were uttered. Woman has access now to something wider than domestic teaching. The world of science and literature is open to her, and the need of depending solely upon her male relatives in intellectual matters is not very often felt. Among all the changes that mark our modern time we consider this one of the most important. The elevation of woman means the elevation of man. Many persons have distressed themselves over the thought of men and women competing for work, and doubtless such competition has already given rise to some unpleasant results. But, strictly speaking, competition *for* work is a feature of an imperfect social system, and therefore, as we may trust, an evil that is destined to disappear;

while competition *in* work will remain as a powerful spring of progress. On the other hand, man will be roused by the rise of woman to a competition not so much with her as with himself. If he wishes to win her respect, to say nothing of conquering her love, he will have to be something better on the average than he has been in the past. Heretofore man has, consciously or unconsciously, counted too much on the power of instinct for his influence over woman; while she in turn has regarded him as a creature to be captivated mainly by appeals to the senses and by an appearance of subservience to his wishes. In the future the primitive attraction between man and woman will remain, but it will be so modified by intellectual and moral influences that it will not exercise the same mastery that it has done in the past, nor be so determining an influence in conjugal unions. It is vain to represent to women that it is their *duty* to marry; their first duty is to themselves, and only when marriage can give fuller scope to their individuality will the best women of the now rising generation care to commit themselves to it. In some ways this may seem to bode evil, seeing that the less advanced will be as ready as ever to marry on the old terms; but, on the whole, we can not doubt that the reflex action on men will carry with it a large surplus of advantage to the world. We want *individual* men — that has long been recognized; but we want also *individual* women — that has only lately been recognized: when once woman becomes an individual in the truest and highest sense, civilization will have reached the threshold of its most glorious period.

Frances Gordon Smith (1870–?) was born into a prestigious southern family. Her father, General John B. Gordon, had been a distinguished confederate commander. She was educated in Atlanta and Washington D.C. and married Burton Smith in Atlanta (1888), becoming a mother of two children. Mrs. Burton Smith participated in community affairs, lectured on home economics, and joined several women's clubs. She was vice-chairman of the Committee on Home Economics of the General Federation of Women's Clubs (1908–1909) and an active member of the Colonial Dames, Daughters of the Revolution, and the Daughters of the Confederacy. Her published writings include a sketch of her father and several articles in Good Housekeeping *and* The Popular Science Monthly.

In the following selection, Mrs. Burton Smith elaborated upon the ideal of motherhood prevailing in the late nineteenth century, speaking for the majority of American women, who saw their role in society as defined by their relationships to their husband and children.

THE MOTHER AS A POWER FOR WOMAN'S ADVANCEMENT

MRS. BURTON SMITH

March 1895

THERE are still thoughtful, liberal-minded men and women who persistingly declare that there should be no woman question; that women have now all the rights and opportunities which should be theirs, and that a just appreciation of what they have already would leave no time nor desire for further demands. There is a great deal of truth and justice in this position, as there is generally in any honest view of any really serious question; but the unalterable fact remains that there is a woman question, and that a discussion which has had the earnest attention and advocacy of so many high-minded, well-balanced men and women must have had its origin in the real needs of some portion of humanity. Surely, no matter what the point of view, the cause of woman's advancement on the best and broadest lines, whatever may be its highest expression to the individual, will have at least sympathy from every thoughtful human being. In this cause, with all its wide-reaching consequences, in all its breadth and fullness, motherhood has just now a peculiar call for effort.

In all great questions which set the world thinking and listening, which touch men's hearts and stir their brains, there is a necessary tendency to extremes. The very force of conviction and power of feeling which go to make the prophets and leaders, carry them away from lines of moderation. But when thought and agitation have developed into real activity, conservatism, as much as enthusiasm, is needed in any movement for reform.

Just at present the woman question is a most convincing illustration of these truths. The ardor of each side has carried its advocates to extremes, which have probably never been equaled in sociological discussion. There are women who affirm that there is no intellectual, social, or professional advancement for woman except as she asserts her independence of man, and arrays herself against him as the enemy of her sex; there are others who declare all marriage slavery, all married life under the existing state of things mere bondage. Such women are as far from the truth as the novelist who has recently attempted to illustrate in her heroine the "soul-destroying" influences of the higher education for women; or the woman who declares, "With the new school of thought, and the new class of woman it has bred, we have lost both the grace and the sweetness, both the delicacy and the virtues, of the real womanly ideal." Such rash generalizing on either side simply balances against rash generalizing on the other; and the result, as far as their power is concerned, is a standstill, frequently followed by positive retrogression. Those whose work or sympathy might otherwise be enlisted in some branch of woman's development, simply look on such extravagances with amusement or pity, and await the next edition of feminine fantastics. A little more conservatism is needed to tip the balance in favor of sure and steady progress. There is no longer need for the agitator, when the question, in its different phases, is being discussed in legislative halls and by the fireside, by thoughtful men and women the world over; but there is great need for the conservative moderator, and in just that capacity should the mothers of the land make their power felt. They occupy a position, by its very nature, powerful beyond the possibilities of any other position on earth—powerful with God-given rights, which admit of no question and need no acknowledgment. They are burdened with responsibility, it is true, but any responsibility rightly met is a power in itself. There is no class of women who stand upon such vantage ground, who can so well exemplify all that is essentially feminine, and at the same time demand, by their rights and responsibilities, any outside aid, whether it be of higher education or suffrage, or whatever it be. There is no class of women who know so well the delights of all the dear feminine prerogatives, the power of those exquisite qualities, grace, delicacy, and sweetness, and at the same time who feel more deeply the need of any and all means of enlightenment and advancement. There are no women better fitted to temper the present discussion; none who can better offer sympathy, yet counsel moderation, to those restless sisters whose demands so often grow out of bitter personal experience and too often rise to a discordant clamor. Of course, this view has been of mothers as a class. There are, alas! pitiful exceptions—women who do not admit the responsibilities of motherhood, and women who dare not demand the rights which motherhood gives them. Such women present problems which can not be dealt with here. Certainly these remarks may apply to every mother who will exercise a certain just self-appreciation, who will devote a little time and attention to the consideration of this question, and her own duties and responsibilities in relation to it. Is it not possible for such women to show that womanliness does not mean weakness—that the very life of all lives the most womanly needs for its right living not goodness only, but wisdom, knowledge, and freedom? On the other hand, ought they not to demonstrate that in this womanliness essentially, in the clinging to it and emphasizing it, they will gain a peculiar power which nothing else can give? It is surely a strength and freedom, not to be left behind in the march onward to new strength and new freedom. It is a quality which must be cultivated and emphasized in this "new era." It is an emotional superiority, a God-given essence, which we can not afford to lose, in our new grasp upon the intellectual forces within us. If every intelligent mother in this land could bring herself to an accurate realization of these truths—a realization of the power for broad yet conservative advancement which lies merely in her position in the plan of society—what an immediate uplifting of womankind there would be! And beyond this, too, reaching away off into the future, is the influence she exerts upon her children, and through them upon an ever-widening circle. She has great power for good in this never-ending, ever-expanding influence, which must go out to the world from her, through her children, as well as in the strong and right expression of her individuality.

Mutual understanding and sympathy, both so potent in the relation of parent and child, must be established before the woman, as mother, can, through her children, do her part in this progressive age. With that much accomplished (it is the first step, a difficult but a necessary one), let us, then, in our strength, as mothers, push on to this important expression of our work for woman's advancement—the emancipation of our daughters from the slavery of half-developed bodies and unhealthful clothing. These exists to-day a painfully small number of women who have the physical endurance necessary for the right living of any life, whether domestic or professional. All women who have felt the hampering influence of weak bodies would cry out if it would help them, "Give us strong backs and good circulation, and we can do the rest ourselves." Whatever life we contemplate for our girls, whether in college halls or kitchen—whether as lawyers, teachers, doctors, or mothers—in every work, they need physical endurance, and with us, their mothers, rest the opportunity and ability to give them great help or hindrance. It is indisputable that a good circulation and fine digestion have much to do with a normal, healthful, mental development; and no one will deny that a well-developed body, with all its possibilities of symmetry and beauty, with all its suggestions of noble appropriateness, can, and frequently does, have a material effect on the character. The buoyancy, the feeling of mastery over all problems, the exaltation mental and spiritual, which come with perfect health, are not only helps but inspirations in any work. And even if we can not attain perfection, is not an approximation worth striving for? It is a rare case where the watchful care of a mother can not do much, by prenatal as well as postnatal influence, to counteract inherited weakness, cultivate desirable qualities, and bring her child to a full fruition of its physical possibilities. This branch of the mother's work, including as it does the development of a just appreciation of what is appropriate and healthful in dress, deserves separate and careful consideration. It is only possible here to outline those powers for the good of humanity, and of womankind especially, which have always belonged to the mother, and to emphasize the necessity for her use of them just now, when there seems to be a call which she is peculiarly fitted to answer.

Following upon a fine physical development, which is the first object to be obtained, we may expect a truer and more natural expression of tastes and tendencies, and just in that expression we must look for our guide posts and follow to some extent certainly, in the education of our daughters, the roads toward which they point. To repress all evil inclinations, whether inherited or acquired, is an accepted duty to both sons and daughters; but the careful study of capabilities, the consideration and cultivation of special talents, are privileges accorded, as a rule, only to our sons. There is no work which can not be better done with education and special cultivation than without it; and in woman's work especially, from cooking, all through the literary and professional scale, up to motherhood, the greatest and most important work of all, there is necessity for the high development — for the information, skill, discernment, and wisdom — which such advantages bring. With the possible responsibilities of the suffrage, too, either open to women as a privilege or thrust upon them as a duty — according to the individual view of the matter — the daughters of to-day have need of an education not only thorough in its details but broad in its scope. Mrs. Elaine Goodale Eastman wrote recently of a young woman who had attained "real distinction in the sciences." In writing a letter to a friend on the birth of a child, this exponent of the higher education for women uses these significant words: "Your letter brings news which never fails to thrill me. I am sure that any woman would rather hold her own child in her arms than attain to any degree of eminence in science or learning." So long as such an expression of the poet's ideal of woman can come from one who has attained "real distinction in the sciences," we need not fear the consequences of a higher intellectual development for women. The danger lies elsewhere in the derision brought upon the advancement of women by the extravagancies of some unwise enthusiasts, and in the encouragement of a spirit of antagonism between man and woman — a spirit contrary to all the laws of God, and death to the best development of mankind. This last danger was forcibly illustrated in a recent magazine article which demanded that every father should share with the mother the responsibility of the mental and physical training of his children, and entitled Modern

Woman *versus* Modern Man—a most excellent subject, the very central thought of which proves the necessity for *working together*. It was well developed too; but oh! the spirit indicated by that title. That the mother, as a member of society and the guide of future generations, can do more than any other woman to meet these dangers and counteract them, is the conviction which I believe will be born of a just valuation of her powers. In considering the growth of opportunity for women, it is natural that we should give special attention to the needs of our daughters and to the development of which they are capable. But our sons are no less important "seed fields." Even viewing woman's higher development as it affects herself individually, there is need for an influence upon the man of the future, which will awaken in him a spirit of helpful sympathy with the earnest woman who is trying to dignify and broaden her life and work. And considering this increasing earnestness in woman in its wide-reaching effect upon all mankind, it is evident that, without a kindly fellowship and encouragement from men, which will make the working together possible, the future will not bring the great results which are hoped for. Would it not be well to infuse some of this spirit into our sons while their natures are still plastic material?

In writing recently of woman's work, Miss Agnes Repplier said with admirable force, "Now as in the past character is the base upon which all true advancement rests secure," a truth which must commend itself especially to every conscientious mother. It is through a better physical and mental development, it is true, but mainly through them as leading up to a growth in character, that we must look for the best results. If there is to be a "new woman," let us have her by evolution, not revolution.

Let us free our daughters from the unwholesome physical restraints which unnecessary conventionalities would impose, and educate them as *human beings*, with all ordinary possibilities latent, besides those womanly qualities which set them apart. Let us cultivate in them all that is strongest and most forceful, all that is sweetest and best and most womanly; and then, with the realization that neither marriage nor a career is the essential, "the destined end," there will come to them a growth in strength and goodness which will enable them to do any work in life better than they have done it in the past. It is certainly not incredible that such women should be able to counteract every retarding influence, and hand in hand with broad-minded men as husbands, brothers, or co-workers, demonstrate the beauty and strength of united force.

Is it too much to hope that in the near future there will arise in the minds and hearts of mothers a whole army of thoughts and inspirations with which they may do battle for that high development, that noble expansion, which we are pleased to call "the advancement of woman"?

By 1896, the term "new woman" already held pejorative connotations of a woman who viewed man as the enemy of her sex and shunned marriage and motherhood. In this editorial, William Jay Youmans (1838–1901) reaffirmed the prime importance of a harmonious marital and domestic life—seen as the responsibility of women, thus qualifying the more liberal position he adopted five years earlier. Compare this selection with "individuality for Woman," pp. 305–306.

'THE NEW WOMAN' AND THE PROBLEMS OF THE DAY

EDITOR'S TABLE

November 1896

AS there is a new everything in these days, we suppose it was inevitable that there should be a "new woman"; though why a new woman more than a new man it might not be easy to explain. For our part we believe but faintly in "new" woman; we believe in woman. We believe in progress; we believe that new times call for new measures; we believe that these are new times, and that it behooves both men and women to prepare themselves to meet the demands which the age is making on them.

What is really new in the world is knowledge. We see the practical outcome of the new knowledge in the transformation that has taken place in the arrangements under which the life of society to-day is carried on. With the new knowledge there has come a vast enlargement of human power in all directions and a vast development of human individuality. Custom, though still powerful, is no longer such a ruler of men's lives as it used to be. Men and women everywhere have been roused, we might almost say stung, into a sense of individual existence; and, looking round on their changing environment, they are asking a thousand questions to which as yet no very certain answers can be vouchsafed. Woman is awake because man is awake; the keenness of the times has roused them both; and from both we seem to hear the inquiry made by the jailer at Philippi, when startled from slumber by the trembling of the earth and the flashing of a strange light: "What must I do to be saved?" The difference between the so-called "new woman" and woman without that qualification is that the latter would wish to be saved with man and the former apparently without him. The new variety emphasizes the fact that she is a woman, and in that capacity is going to do wonderful things: whereas woman without the "new" is content to know herself a woman and to feel that with her it rests to accomplish her equal part in all the best work of the future.

The great change, as we have said, is that there is more knowledge in the world and that the rule of custom is to a large extent broken. Things that once had all the authority that convention and routine could give them are now open to every one's criticism. Morality no longer rests in absolute security upon dogma. The time has come which Voltaire predicted would be the end of all things, when *the people* have taken to reasoning. Fortunately, there is no need to agree with Voltaire; but it is necessary to recognize that something is needed to give wise direction to the emancipated thought and action of our time. The dogmatic morality of the past was in the main sound; and the problem of to-day is to secure a sufficient sanction for whatever rules of conduct are necessary to the well-being of individuals and of society. That much in the way of wise counsel and true inspiration may be expected from the increased reflectiveness of women we most gladly recognize; but we do not feel disposed to call a woman who thus responds to the needs of the time a "new" woman, seeing that for generations past, and particularly in times of emergency, women have more or less fulfilled the same rôle.

The two principal questions which to-day confront society relate to the future relations of men and woman and the education of the rising generation. The allegation is freely made in many quarters that marriage is a failure; and no doubt frequently it is. None the less, however, is it the case that no scheme that has ever been proposed as a substitute for marriage merits a moment's consideration. It is easy to provide theoretically for the gratification of passion and impulse, but not so easy by any means to show how by any union less solemn and abiding than marriage the higher natures of men and women can be duly developed and their lower propensities kept in check. We do not look to any new woman for light on this question; but we do look to the best women of to-day, those who to purity and soundness of instinct add a trained capacity for independent and intelligent judgment, to join with the best men in indicating the higher path which the generations of the future may tread. We may be sure of this, that the path is one not of less but of greater self-control, and that redemption from the miseries which attach, in too many cases, to marriage as it is will be found in an elevation and purification of the whole idea of marriage. Not that the idea has not been held in its highest purity by many in different ages; not that the world has ever lacked examples of ideal marriage, but that there has never been a sufficiently wide recognition of its true nature and possibilities. There is a gospel on the subject which has to be preached and, so far as individual action can do it, enforced — the gospel that the true happiness of a man and woman united in marriage bonds consists in learning, as years go on, to love and respect one another more and more, and in aiding and stimulating one another more and more to right and noble action, each gaining strength through the other, each finding in the other the means of achieving a true individual completeness. The true gospel is that there is *more* in marriage than for the most part poets have sung or romancers dreamed, and that the failures of which we hear so much have been, in the main, failures to grasp the true conception of it and to make a right preparation for the duties which it involves.

Does not all this mean, it may be asked, that many are unfit through defect of character, and others through ignorance and general inferiority of thought and sentiment, to make the best of marriage? It certainly does, and here the no less important problem of education comes in. . . . If the rising generation is to be adequately educated, the best men and women of the day must come together and consider how it is to be done—how the work of the state is to be supplemented by individual endeavor, so that growth in character may keep pace with growth in knowledge and intelligence. There are two main ways in which, at first sight, it seems possible this might be done, or at least more or less hopefully attempted: first, by an improvement of the home, and, secondly, by the action of a higher public opinion on the schools. We quoted, some months ago, an eminent French writer of our own day as saying that it was necessary to put more "soul" in the public schools. That is precisely what they want, as all the best teachers are fully aware. But you can not make an appropriation for "soul." It is not quoted in the catalogues of school supplies; it is not among the prescribed subjects in teachers' examinations. It is a very real if not a very tangible thing; and it is a communicable thing. There are those who have it and can impart it; in deed, those who have it can hardly fail to impart it. If there is enough of it outside the schools, it will leak in; and our hope is that the best men and the best women of the day will so join forces as to create, especially around the public schools, an atmosphere of higher sentiment that shall affect for good the working of the state machine, and greatly strengthen the hands of all who, within the schools, have set for themselves a certain standard of spiritual as distinct from merely intellectual accomplishment.

Then as to the home. Here is where we want women with new knowledge, but not—we speak with all due fear and trembling—"new" women. The "new woman" would set every one discussing rights; but the *true* woman with adequate knowledge would see what the best women have always seen, that the home requires a principle of unity and not a system of scientific frontiers or an elaborately arranged balance of power. Home life and home influence have, we fear, been suffering in our day through a variety of causes; but the home, like marriage, is an institution which only needs to have its possibilities developed in order to stand forth more than justified. Without entering into the question as to whether the wisest methods are being followed to-day in the education of women, it is beyond all doubt that women have gained a vast enlargement of their intellectual horizon, and that in many cases women are not only the peers but the superiors of men in the same station in life as themselves in knowledge and culture. Such knowledge and culture can nowhere be better employed than in the home, where the physical, mental, and moral development of children has to be watched over. The question is, How far will it be employed in this way, and how far made a means of mere personal self-assertion? The true woman will use it for the good of others, and if possible, will make it available for the improvement of the home; while others—the new type—will use it to make themselves conspicuous in the world, and, as they vainly fancy, add glory to the female sex.

The hope of the future lies mainly in well-ordered homes—homes in which children are trained to be just, reasonable, and humane, in which they are taught to look with an intelligent eye upon the phenomena alike of Nature and of society, in which they learn lessons of industry and self-reliance, of honor, purity, and self-respect, and are guarded against the vulgar worship of wealth and worldly success. It is for the wise and noble women of our time to help to make such homes, and it is for men to see to it that they are worthy of partnership in so sacred a cause. It is no time for any silly rivalry or futile opposition between men and women, who are as necessary to one another now as at any previous age in the world's history—nay, more necessary. On the contrary, it is a time for earnest counsel and vigorous co-operation on the part of all who have the interest of the present and future generations at heart; and the less we hear of the separate and conflicting claims of men and women the better. There is ample scope to-day for the efforts of all, and if any stand idle in the vineyard it must be from lack of will, not from lack of opportunity.

In "Let Us Therewith Be Content," Ellen Coit Elliott analyzed why a large number of American women had serious reservations about woman suffrage. The following article illumines the importance of self-sacrifice and altruism in women's conceptions of themselves and reveals how the American ideal of womanhood conflicted with the new woman's demand for the vote.

LET US THEREWITH BE CONTENT

ELLEN COIT ELLIOTT

July 1897

THE men of America have met the suffrage agitation with an admirable gallantry. Aspersed to their faces from the rostrum as masculine creatures of unfathomable iniquity, they return only a deprecating smile. Assured by the "new woman" that the ever feminine leadeth them on, and that politics will clarify as soon as the superior purity and integrity of the sex are brought to bear upon them, they appreciate her splendid confidence, applaud, and cry her on. There are those who, ever suspicious of the masculine character, take umbrage at this favor, looking upon it as an impertinent condescension. But surely we may grant that the slow partner of our humanity, admiring our victorious advance, and bewildered by our swift onslaughts from all points at once, wishes by his expressions of good will to placate our wrath and further our desires. Stupid and mannish he may be, but after all he is rather good-natured.

American women, however, are taking toward the question at issue a curious attitude. One large and picturesque division, when exhorted that they "ought" to desire a finger in the political pie, if not for the sake of the pie at least for the sake of the finger, show a sweet resignation, and, definitely premising that they do not wish the ballot, cry meekly that if it be the will of God to give it to them they will do their best to make a proper use of it. Others express a frank impatience with our prophets and saviors. Others, still, recognizing that the vantage ground upon which American women stand today is not entirely the result of democracy, give due gratitude and appreciation to those who through hard battles have helped to win the position. "But," they exclaim, "stay in your ministrations of deliverance! Forebear to impose upon us the added responsibility of the suffrage!" And, worst of all, masses of these shackled citizens show an unalterable apathy toward the injustice they are suffering, and indifference to the hands reached out to help them. Surely never did enthusiasts have to deal with more refractory and exasperating material. The suffrage leaders have proved in their own persons the angelic quality of womankind in not giving up long ago the attempt to free such inveterate slaves.

What is the significance of this general reluctance? To give her the suffrage is to add another to the long list of her opportunities for exercising power and influence outside of the home, and the question becomes, Do American women desire this, and if not, why not? The answer is bound up with the hackneyed subject of "woman's sphere," and, as all our philosophy is nowadays biologized, it rests back upon the great physical fact that women for all time must be prepared to bear and rear the children of the race. Granting that much of her physical disability is due to various sorts of foolishness and may be removed, it remains undeniable that in even the most normal of women the reproductive system is by nature so constituted that it requires a much larger proportion of her vitality than is the case with man. Hence, leaving out of account all other possible variations between the sexes, this difference alone is a definite handicap to all women who "compete" with men. For married women there is the further fact that childbearing and the care of children add a new and very serious handicap in any "competition" with men.

If, then, woman is physically at so great a disadvantage in many occupations, shall she not consider that these occupations are, for her, but secondary issues? For her specialty shall she not look along the line of least resistance? Instead of denying her physical constitution, shall she not exalt it by a consistent allegiance to its fundamental significance? Notwithstanding the present apotheosis of the physical sciences, woman will not rest satisfied in a purely physical explanation of her destiny. Bitter rebellion is inevitable whenever she is confronted by her physical limitations and possesses not the spiritual key to their meaning. But a spiritual significance in the life of woman has been more or less felt in all times, and in the present it is not only tacitly conceded by society in general, but it has received definite scientific formulation. From their physical constitution women more than men must inevitably sacrifice themselves for the progress of the race. Unconscious and unwilling though they may have been, necessity and habit have so trained countless generations of women in the practice of self-denial that they have grown to be in the world the special witnesses and exemplifiers of the altruistic principle. So true is it that motherhood and the love and self-sacrifice which it involves, is a woman's pecular contribution to evolution and progress, that, as has been keenly pointed out, "the woman question is not solved until it is solved by mothers." In other words, a woman can not solve her life problem on a purely individual basis except at the price of her influence on the race. A man may lead a life largely self-centered and still transmit his qualities to his children, but the self-centered woman can not pass on her qualities, for she will have no children to inherit them. If she would, in any large way, save her life, she must lose it.

The actual facts bear out this conception of a woman's function. It is not that women are wholly altruistic. Though loath to own it, we are but mortal. Nor will any (except the suffrage leaders) contend that every woman is more unselfish than every man. On the contrary, it is only too easy to point out cases where feminine selfishness is shown again and again in petty ways to which men, as a rule, do not stoop. Yet it remains in general true that the practical life of women the world over calls for a more constant exercise of self-sacrifice than that of men, and that everywhere women have learned in the main to make their sacrifices cheerfully because lovingly, and even to court a life which brings them. That this acquiescence should be often considered an indication of tameness, if not inferiority, is but natural in a civilization which has even now only half realized the dignity of the altruistic ideal. In the affairs of life intellect has enjoyed a long prestige. Character, which, according to the highest conceptions of the race, depends at its best upon altruism, is but slowly growing into an equal recognition. In a rough, general way, men have been the apostles of the one and women of the other. It is true that the ideal of humanity is one. Women have gained in intellect and men in character, and this must go on; but it has not come about, and it will not come about, by a direct exchange of their activities.

These considerations lead to the good old dictum that "home is woman's sphere." It seems well-nigh superfluous to enumerate the obvious qualifications of this general statement. Surely no *fin-de-siècle* person would understand it to mean that woman should look upon marriage in itself as the sole desideratum of her existence, or that, failing to marry, she should devote herself to pets and fancy work, and live upon the charity of her male relatives. Surely at this stage of proceedings no one would attempt or desire to limit woman to purely domestic pursuits. It has been reiterated and most abundantly proved that she need not be circumscribed in freedom or opportunity for the sake of binding her to the home: it is not necessary, for Nature will take care of itself; and it is not expedient, for the more she is allowed to be in herself the greater the gift she can and will bring to the race. Moreover, no one will contend that every woman ought to be a mother, or that an indefinite number of offspring is a wife's chief duty. In a word, marriage, and the bearing and not bearing of children, are individual accidents dependent upon a thousand private considerations. To fulfill the law of womanhood one need not be a mother, but only to be motherly; one need not be a wife, but only to be loyal to the unselfish principle of wifehood; one need not eschew the paths of business or professional life, so only that she recognize hers as the exceptional feminine career, the more normal and significant one lying within the walls of the home.

Consciously sometimes, but perhaps more often with unconscious instinct, a woman does thus stand by her colors. Why this eager activity in the matter of temperance rather than the tariff? Because intemperance menaces the home. Why this quick sympathy with organized or unorganized charities, as opposed to the average apathy over finance? Because charity touches people whom she can love and homes which she can transfigure. And—if one may be pardoned a notion somewhat transcendental—is not her oft-observed lack of creative ability, together with her equally notable power of appreciation, due to the fact that with her an idea is not worked out so readily in purely intellectual formulations as in the material of character? The laws of mechanics as such she does not readily apprehend, but the truths of rectitude which are their moral counterpart she grasps with special illumination. The masterpieces of formal art she does not create, but she, more naturally than man, can live a life which may properly be called a poem or a picture.

And why this respect for womankind deeply rooted in the best of men? The individual character of woman is not, unfortunately, so much loftier than that of man as to compel it, and that she is the "weaker sex" hardly accounts for so large a fact. Nor does it look like a merely left-over remnant of mediaeval chivalry. Is it not, at bottom, that sound and sensible men recognize and reverence the altruistic ideal; which, however faltering her loyalty, it is a woman's special privilege to perpetuate? . . .

Homes must be made, and the masculine half of us, as they make haste to proclaim with amusing emphasis, have neither the inclination nor the ability to assume the task. Says one of them, naively, "If marriage meant to a man what it does to a woman in the way of suffering, labor, and social status, I am convinced that not one man in fifty would marry." It is impossible not to be reminded of the similar disclaimer—

Oh, then I can't marry you, my pretty maid!

and the milkmaid's retort—

Nobody asked you to, sir, she said—

seems singularly appropriate, did we wish to be so impolite as to use it. But, strange as it may look to the masculine mind, women in general do choose to marry. They are not driven to it by the conditions of society, nor impelled by a blind sexual instinct, nor misled by the enthusiasm of the martyr. They know perfectly well what it will mean in their career. And they need not be looked upon as fools for so doing, being in fact possessed of the average degree of common sense of the race. They choose it because they want it, and they want it because, in spite of its restrictions, it brings the most satisfactory fulfillment of their aspirations and development of their powers.

The same masculine thinker is firmly convinced that "women wish to be men, but men do not wish to be women." Both parts of this proposition are delightfully characteristic of the sex which has never been backward in claiming its superiority, and the last clause, by the same sign, is doubtless unquestionable. But the first is as unjust to woman's ideals as it is derogatory to her mission. If she give up social pleasures, literary activity, pecuniary independence, or a hundred other personal ambitions, to minister to the interests of one modest home, and the career of one average husband; if she turn from the gratification of public recognition to years of the unapplauded cares of the nursery; if she drop out of the onward march of purely intellectual progress, and spend her life marking time in the ranks of the housekeeper—it is not because she is the poor-spirited victim of circumstances. It is not that one half the race is, by some mischance of destiny, doomed to a life of tragedy. The bird with one wing broken droops in its flight, and humanity thus hampered would have sorely lagged in its onward sweep. On the contrary, she chooses these things because law and the satisfaction of her life are not that of individual ambition or attainment, but the law of love and service— "unto the Jews a stumbling-block and unto the Greeks foolishness."

Women, it is true, do not always feel or admit this. Many of them have a taste for pity, and they pet and pity themselves and each other. Yet the more sincere own willingly that everything has its price, and that they have paid none too dear for that which they have gained by their sacrifice. The strongest scorn to pose as martyrs, because they see clearly that in life as it runs, a woman exactly as a man, gets what she pays for, and must pay for what she gets. And they conceive of no more just equality of the sexes than this.

As to the women of America, to begin with, they are not, as some would have us think, downtrodden drudges, manacled slaves, or what not, after the same order. Rejoicing in the most perfect social freedom the world has seen, proud in a position and influence quite equal to those of men, they can afford to laugh at such tirades. With the exceptions that must always accompany general statements, woman in America may do whatever she wishes to do. She may run the typewriter in an office instead of a sewing machine at home. She may carry on a farm or a business. She may teach, write, preach, lecture, practice law or medicine. Journalism and *belles-lettres* are her happy hunting grounds. She may marry or remain unmarried with equal honor, and no one dictates in her choice of a husband. She may wear bloomers and ride a wheel. She may carry on public agitations to an unlimited extent. The most serious drawbacks to her complete freedom result from flaws in her own standards and traditions, and are in no wise imposed upon her from without.

American men are neither tyrannical nor condescending toward women. From childhood up they have been in the habit of seeing their sisters walk beside them with independence and privilege equal to their own. Their attitude is one of frank comradery based upon a respect which on both sides is unconsciously taken for granted. They have, besides, a genial tendency to be proud of their women and to applaud rather than discourage their ambitions. If women wish to vote, these men will not deny them. In fact, many an American household presents the edifying spectacle of a husband more ready to vote the suffrage to his wife than she to accept it.

Notwithstanding this freedom — perhaps because of it — one need only obtain an unaffected expression of their feeling to find that, maid and matron alike, the women of the country are, as a rule, content in marriage as a career. They wish for children, and gladly make the prolonged sacrifices necessary to their care and education. One day a young woman — exactly such a one as may be met with any day anywhere in the country — went "in fun" to consult a fortune-teller. But she returned in tears, and confided to her girl friend that she wept because the seer had told her she would never have children.

It can not of course be said that among women there is no discontent, no restlessness. The age is full of discontent of a certain kind, and restlessness is in the blood. Women do not escape these general influences of the time. Moreover, there is, at least among college women, a special dissatisfaction with the drudgery attendant upon home-making. With the increase of individuality which the higher education can not fail to bring, comes the need of a new sort of home; and the conflict and adjustment of old with new ideals, old with new duties, old with new purposes, brings confusion and sadness into the problem of many a modern woman's life. Notwithstanding this, the college woman is found in general to be no more ready than her uneducated sister to go back upon the womanhood which means self-denial, and the career which means self-sacrifice.

When these American women, full of the complicated interests and duties of the American home and its dependent sociological activities, are confronted with the prospect of exercising the suffrage, their instinct seems to be to draw back. Ask the women, one after another, in a representative community, if they wish to vote, and again and again will come the answers, "I haven't time," "My hands are overfull now," "How can I undertake a duty which means that I must inform myself upon all the public questions of the day?" Naturally, many of them, especially those who are temperance workers, or those whose property interests are not represented under existing conditions, desire the ballot. But the great majority are content to occupy themselves with the multitude of interests which are already theirs, and to leave the formal affairs of state to men. The great majority, when they speak sincerely, will say that home-making and its allied interests is their chosen life, and that its demands are so exacting that they must leave the work of government to other hands.

This attitude is certainly open to criticism. Perhaps it is true that the sons could be better educated by mothers who voted, that homes could be better made and protected by wives who held the power of the ballot, that the welfare of schools and charities would be furthered if women who are interested in them had a share in the making of the laws. Yet it would seem that if woman possessed by nature any great aptitude for political life, she would be eager to exercise it. It has been said that "the men are not what they are because they vote, but they vote because they are what they are." They make politics, and they are interested in the work of their hands. Women do not make it and (always in general) are not interested in it. If woman alone were to govern the state, how radically different would be her methods! And how can oil and water mix? Until she can disfranchise man and establish a rule of her own peculiar sort, woman may perhaps be expected to show indifference to political affairs. Furthermore, she might evince more alacrity for reaching out for the august power of the ballot if she observed that the men who exercise it thereby get what they want. But to her puzzled query, "If you want this reform or that measure, why don't you put it through?" the conclusive reply is that "you can't get at it," on account of the "primaries," or "the bosses," or "the spoils system," or the "rings," or the wheels within wheels of whatever other complications interfere to muddle the brain and thwart the will of the sovereign American people. A woman answered thus, and reflecting upon the suffrage, is apt to wonder, in her silly, feminine way, if the game is worth the candle.

Perhaps it is worth the candle. Many a wise man thinks so, and having the suffrage himself, a man should be able to estimate its value. However that question may be finally settled, women will be women. The practical conviction that this is after all what they most wish to be must have an important bearing upon their particular aspirations, and it is this conviction which, to say the least, suggests misgivings and compels reserve in the minds of a very large number of average American women whose voices are not heard in the land.

Earl Barnes (1861–1935) was educated at the Oswego (N.Y.) State Normal School, founded by the progressive educator Edward Austen Sheldon to train farm children to teach in country schools. Barnes was deeply influenced by one of his teachers, Mary Downing Sheldon (1850–1898), the daughter of the school's principal. Mary Sheldon attended Michigan University in the early 1870s, taught history at Wellesley (1876–1880), and studied at Newnham College in Cambridge (1880–1882) before returning to Oswego to teach history and literature. In 1884, Barnes graduated from Oswego, and in 1885, he married Mary Sheldon. The two moved to Hoboken, New Jersey, where Barnes taught school, then to Ithaca so that he could enter Cornell University as an undergraduate. While still a student, he accepted an offer of a professorship in European history from David Starr Jordan, who was then president of Indiana University. (See "The Higher Education of Women," pp. 96–104.) Barnes completed his bachelor's degree at Indiana and earned his master's from Cornell (1891). By this time, Jordan had become the first president of Stanford University and had appointed both Earl and Mary to the faculty; Barnes became head of the Department of Education (1891–1896), and Mary Sheldon became an associate professor of history (1892–1896). In 1897, the Barnes resigned their positions to travel and study in Europe, where Mary died after a recurrence of a chronic illness, followed by an unsuccessful operation. In 1901 Barnes married Anna Kohler, a graduate of Stanford and a high school teacher of history. After spending several years in England and finishing his two volume Studies in Education Devoted to Child Study (1902), Barnes settled in Philadelphia, where he spent the remainder of his life giving lectures at numerous teachers' institutes, ethical societies, women's clubs, and public schools.

While Barnes evidently believed that marriage offered both men and women the most complete life (he seems to have been happily married both times), he realized that greater freedom and opportunities were available to single women. In the following article, he presented a sympathetic view of the difficulties and compensations experienced by unmarried women in the early twentieth century.

THE CELIBATE WOMEN OF TODAY

EARL BARNES

June 1915

THERE were in the United States, in 1910, 8,924,056 women, over fifteen years of age, who were neither married, widowed nor divorced. These single women represent twenty-nine and seven tenths per cent. of all the women over fifteen years of age in the United States at that time. Many of these women have since married or will marry; but at every age there remains a large number of unmarried women facing a life of celibacy. Thus of native-born white women between the ages of twenty-five and thirty-four, thirty and six tenths per cent. are unmarried, while of the same class, between the ages of thirty-four and forty-four, seventeen and eight tenths per cent. are still single. The public school teachers of America, alone, number nearly 400,000 mature women, hardly any of whom are married. Why do they not marry, and what compensations can a life of celibacy bring them?

In the first place, it is a mistake to imagine, as most people do, that the emancipation of women since 1870 has tended to discourage marriage. The contrary is true. In 1890, but sixty-eight and one tenth per cent. of American women of fifteen years of age were married, or had been married; in 1900, this proportion had risen to sixty-eight and six tenths per cent.; and in 1910, it was seventy and three tenths per cent. Higher education, industrial independence and increasing participation in social and political life have apparently increased the tendency of women to marry. But why, since they have so rapidly taken possession of nearly everything else, have they not more generally taken possession of husbands?

The widespread belief that there are not enough men to go around is another old superstition. The men have outnumbered the women in the United States at every decennial census since 1820. In 1910, there were in the whole country 2,692,288 more males than females. Of course, this is partly due to the fact that more male than female immigrants come here to work.

But even among the native-born whites, there were in 1910, 922,502 more males than females. In that year there were 102.7 white native born males to each 100 native-born females. Only in Massachusetts, Maryland and North and South Carolina was there an excess of females. Counting only the people over twenty-one years of age, there were 110 males to each 100 females in 1910; and even among the native-born whites there were 106.8 males to each 100 females over twenty-one years of age.

There have always been women living apart from family life, at least during the historic period. For obvious reasons, men have almost universally considered female celibacy a matter of reproach, and they have even invented such makeshifts as child marriage in India and sealing among the Mormons that unfortunate or undesirable women might be spared the disgrace of dying unmarried. At times, a lack of dowry has condemned the least attractive women to live alone; and the offices of religion have imposed celibacy for at least a part of life upon groups like the Vestal Virgins in Rome and for the entire lifetime upon nuns and other Christian recluses. In none of these cases, however, did women choose to live alone because they hoped thereby to realize a fuller life than they could find in the married state. In the religious orders, they dedicated their virginity to the service of the Deity, and at most hoped to profit in the life to come for their loss on earth.

This is the great difference between celibate women of the past and those of the present time. With our enormous number of unattached men, it would be foolish to imagine that the great majority of single women in America could not marry if they wanted to do so. Man proposes, but woman dictates when he shall do it. Why do so many women elect to walk through life alone?

Doubtless the growth in democratic ideals, which has been steadily working among women since 1870, has had much to do with it. Women have ceased to be merely "the sex"; they have become individuals. Under simpler conditions of life, such as prevailed in our colonial period, if a woman found a man of her race, religion and social position, who was personally agreeable to her, little more was necessary to insure a happy marriage. But now a woman seeks fulfillment not only for her personal liking, but for all the qualities of her varied personal life. She has not only racial, religious and social interests, but she has an intelligent attitude towards the whole of life; she has musical, dramatic or literary tastes; she is interested in social justice or in the vested interests of caste; she cares for travel or she desires a quiet home; and in a hundred other directions she is an individual. Such a complex individuality does not easily find its complement. A person who merely likes music can generally find it; a person with a cultivated musical taste must search for music that suits him.

All this uncertainty is emphasized by the examples of marital unhappiness that intrude themselves on every hand. We have in our midst nearly a million divorced people. The deserted wife and mother is one of the greatest and most common problems that confront social workers. Our funny papers find the majority of their humor in deceived and deceiving wives and husbands. The drama and the novel burden us with sex problems. Yellow journalism lives on the tragic side of married infelicity. Of course, there are millions of happy marriages, but such happiness, like good health, is inarticulate and does not advertise itself in the market place.

Every self-supporting girl must also be deeply impressed with the difference between her own economic position and that of her mother. For years, as a girl, she has seen her father and mother working together as life partners; and she has seen all the net income recognized as her father's personal property. Her mother's relation to the family purse has generally been that of a medieval serf to his lord's estate. Even the house furnishing and the mother's own clothes have often been secured by stealth and indirection. Now the girl has her own pay envelope, and in possessing it absolutely she holds the key to life as she desires it. It is not to be wondered at that a young woman in full possession of youth and health finds it difficult to give up a salary which, even if small, is absolutely her own, to accept a feudal relation to some man's salary, often not much larger than the one she has earned, knowing it must suffice for two and probably for more.

Of course, this feeling fails to recognize the danger of the passing years. With the blindness of youth, it over-emphasizes the value of present liberty. Later, she may see the day when she will realize that she has sold her birthright for a pay envelope. But it is not so much the amount of the income that really troubles the modern woman as it is her personal relation to it. Surely some means might be devised by which a woman could be related to the family income so as to preserve her independence and self-respect as well as that of her husband.

Another difficulty that confronts the young woman of to-day in her search for the altar is her superior intelligence. This is generally less important than it seems, but in a country which worships popular education and where all parents hope to give their children at least a better education than they have had, correct grammar and a speaking acquaintance with Robert Browning and Michael Angelo acquire a value out of all proportion to their power to function in life. When a likely young man comes courting who says, "You and me will go," and prefers the movies to Ibsen, it makes the young woman who aspires to culture question the long evenings of a lifetime.

This is especially true of the young woman who has risen intellectually and economically above her social class. Skilled preparation has given her an income superior to that of the men in the group where she was born; and she has been too busy studying and working to make social connections in the class where she thinks and works. The social emancipation of woman lags far behind her intellectual and economic freedom, so that these young women whom we are considering still move socially in their family planes. The men in that group are too ignorant and too poor to suit her; and the men with whom she works know her only as a stenographer, a teacher or a journalist. Our last census returns show how widespread this condition is, for except in New England, there is no excess of men in the cities. But in the countryside the excess of men is great, for they have been left behind in the struggle towards larger intellectual and economic independence which has swept the young women into the towns.

These women, possessing a fairly large intellectual and professional life, often move in a pitifully narrow social circle. The home, as a center for social activity, has been sadly squeezed in our emigration from the country into the city. Friends and acquaintances come less often than in the past to spend a few days with the family. Calls are more formal and brief. Acquaintances less often drop in for a meal; and meantime the older social conditions that limit a woman's initiative in making acquaintances still hold and are broken by young women only at considerable risk.

Even public meeting places are apt to be restricted to one sex. If the girl joins a club, it is a girl's club; if she joins the Y.W.C.A., the young men are a half mile away in the Y.M.C.A. When shall we be wise enough to turn both these institutions into Young People's Christian Associations? And meantime, while the hunting field is narrow the difficulty of selection has increased. A generation ago, as we have pointed out, a girl might hope to find a desirable mate among a dozen acquaintances. Now she needs to look over a hundred young men to find her own.

In the light of this analysis the wonder is not that we have so many unmarried women in America, but that we have so few. Nature has loaded the dice in favor of marriage and she generally has her own way. Many of these young women, however, will never marry. Nuns will continue to vow their virginity to the Celestial Bridegroom; reformers will spend their lives in securing social justice for their sisters and for their sisters' children; professional women will continue to seek fame and service; teachers will fight off the wars of the future, not with submarines and aeroplanes; but with ideas and ideals planted and nourished in young minds. Many other women, with no particular devotion to sustain them, will be held by the charm of pay envelopes and independent latch keys until it is too late; while the accidents of fate will leave many stranded in their struggle towards a complete life.

Meantime there can be no doubt that the most complete life a woman can live, at least between the ages of twenty-five and forty-five, is found in a marriage based on a deep and lasting love; and the same is no less true for a man. What, then, has our modern life to offer to celibate women as compensation for the life they do not attain?

There are at least certain negative values which come to them. In escaping vital experiences, the woman can at least recognize the fact that she also escapes the anxieties and troubles that are inseparable from family life. She will probably be lonely; but, on the other hand, when she wishes it she can be alone. In writing to a friend, who had lost his little daughter, Cicero says that all men would wish children were it not for the anxiety that they inevitably bring.

To put it more positively, the celibate woman retains her freedom of action. Through study, travel, art, science or society, she may reach a degree of self-realization not always attained by her sister who marries. Into her work she can carry much of the enthusiasm and devotion which, as wife and mother, she might lavish on husband and children.

The desire for service, which lies so deep in the nature of all good women, can often be more fully realized in a life of personal freedom than in one of marriage. At least there may be a different realization of very great value to the individual and to society. Such women as Clara Barton, Susan B. Anthony and Jane Addams have brought gifts of service to mankind far beyond what they would probably have given in their own homes. Each of these women probably recognized her personal loss; but many devoted wives and mothers have also recognized their loss through inability to enter into the wider service of public life. Unto no mortal is it given to live all the possibilities of life.

But more important than any of these compensations we have named is the power we all possess to live life vicariously. Our real living is never in the mere possession and use of things, but in what we think and feel about them. Lower animals live in facts; man lives in his ideas and ideals. All of life's values must be found on the way; when we arrive we are always in danger of becoming unconscious and so losing what we came to get. . . .

Locke's Beloved Vagabond lost all the facts of life, fame, money, the woman; but who that has wandered across the pages of romance with him does not envy him the keen appreciation of life, the realization of its realities, the high and compelling ideal, even against the background of poverty-stricken and often drunken facts. Jean Christophe lived all but his musical life vicariously. The woman he loved was another man's loyal wife; his children were born from other men's passions; his home was wherever he could feel the universe. He lived without the material realizations of life; but who of us would not desert unconscious wealth, houses and homes for such a conscious life?

The poet Dante illustrates in his own life the relative value of facts and dreams, of living life directly and living it vicariously, to a singular degree. He was married and had a family of children, but in all his voluminous writings there is no word of these facts of his daily existence. In his early youth he fell in love with Beatrice; we know very little about her; she married another man; and it is quite probable that Dante never even touched her hand; but she led him through Paradise. Since the poet's death, millions have read and have been shaped by the "Vitá Nuova" who have never even heard of the wife and children who were the facts of Dante's life.

If all this be true, then the modern woman who does not marry need not feel that life is closed to her, that having been denied the Garden of Realization she must stand before the gates and weep. When the angel with the flaming sword drove Eve out of Eden he opened the world of work and varied experience to her. Gifted with imagination and desire, she could create for herself new gardens of perfection; and if they were less real than those she had left, they were possibly more vividly realized than was the one where she had slumbered away the days of happiness.

Self-realization through vicarious living, this is the solution to a celibate life for the individual. Joan of Arc gave herself to religion and to her people. Madam Kovalevsky found at least relief for the letters that did not come in the honors that were lavished on her mathematical discoveries. Susan B. Anthony found her realization in the ideal life that was to come to all the women of the world; her sister, Mary Anthony, found a deep and rich realization in serving her better-known sister, who was to her all that home means to most women.

Thousands of our teachers are truer mothers to their children than are the mothers who bore them. In schools, libraries and social centers many fine women are to-day wedded to humanity; they are conceiving new ideals of social justice and are giving birth to opportunity for fuller living that shall bring conscious gladness to millions unborn.

For themselves and for all the higher purposes of civilization such lives may have great worth. Biologically they are lost; for the little strand of the stream of life that finds lodgment in them ceases with their death. This is a pity; and it is a pity, too, that back of the dream there should not be also the reality. One can not help feeling that the children of Dante and Beatrice would have added something of great worth to the fact of existence.

And that this may come in the future we must remodel our medieval institution of marriage. It must cease to be a political convenience or a religious sacrament and must become a biological truth. We must make it impossible for the state to sanction and for the church to sanctify the marriage of imbeciles and of old men and young girls. Women must be emancipated socially, as they have been emancipated intellectually and economically; and they must be given a larger and more direct share in choosing their life mates. We must put family finances on a basis of equal partnership that will attract self-supporting and self-respecting women. We must provide ample opportunities for young people to meet and know each other and we must recognize the fact that it is always a sin for men and women to live in the close companionship of marriage if they do not love each other.

INDEX

Abel, Mary Hinman 175-176
Adams, Abigail xxii
Addams, Jane 291, 327
aemenorrhoea xv, 266; also see menstruation
Allen, Grant xv, 1, 54
  "Plain Words on the Woman Question" 125-131; responses 132-136
Allen, N. 84-85
altruism 201, 202, 317, 318
American Equal Rights Association 195
American Home Economics Association 158
American Woman Suffrage Association 195
Angell, James Rowland 61
  "Some Reflections on the Reaction from Coeducation" 87-95
Angstman, Charlotte Smith xvi-xvii
  "College Women and the New Science" 174-180
Anthony, Susan B. 194, 195, 196, 197, 327, 328
Antisuffrage — see suffrage
Association for the Advancement of Women 196
Association of Collegiate Alumnae 58, 175, 176

Barnes, Earl
  "The Celibate Women of Today" 323-329
Barnes, Robert 266
Barton, Clara 291, 327
Bate, Mrs. Arthur 169
Beecher, Catharine, 55, 56-57, 156, 158, 161

Beecher, Henry Ward 52, 195-196
Billings, John S. 107, 108
biological determinism 1-53
birth control 105, 106, 118
birth rates xvi, 105-155; cause of the decline in 110, 116-118; importance of 126, 158; methods of computation 106-107; of black women 120; of college-educated men 110, 112-113; of college-educated women 61, 111, 114-116, 118, 131, 137-146; of immigrant women 106; of native-born women 106
Bissell, Mary Taylor 9-10
  "Emotions Versus Health in Women" 48-53
Blackwell, Alice Stone 198
Blackwell, Antoinette Brown 1, 8
Blackwell, Emily xv
  "The Industrial Position of Women" 271-280; response 281-285
Blackwell, Henry 195
brain size and intellectual capacity 7, 35-36, 41-43, 47, 59; differences between men's and women's 7, 33-34, 39-43; brain work, brain-forcing and its effect upon health and fertility 5, 37, 57, 78, 85
Bryn Mawr 55, 57, 58
Bushee, Frederick 109
Butler, Elizabeth Bearsdley 249

Camhi, Jane 202
Campbell, Helen 158, 179
Catt, Carrie Chapman 197, 199
celibacy 128-129, 323-329
Census statistics, on education 63; on

INDEX

Census statistics (*continued*)
employment 251-255, 256; on fertility 108
childbearing, childrearing—see motherhood
Clarke, Edward H. 57, 72, 77, 83
coeducation, arguments against 59-60, 74, 91-93, 100-103; arguments for 60, 94, 99-101; growth in 88; history of 59; and marriage 60, 93, 103-104; reaction against 60-61, 87, 101
college women, employment of 58-49; fertility of 61, 111, 114-118; involvement in domestic science 174-181; marriage of 58, 60, 61, 93, 103-104, 114-116, 138, 148
contraception—see birth control, birth rates
Coolidge, Mary Roberts Smith—see Smith, Mary Roberts
Cope, Edward D. 192, 199
"The Relations of the Sexes to Government" 210-216; responses 216-219
Cramer, Frank
"The Extension of the Suffrage to Women" 216-217

Darwin, Charles 1; classification of mental powers 7, 38-39; *Descent of Man* 3, 25-30, 37; male superiority 3-5, 27, 37; *On the Origin of Species* 2; theory of inheritance 3, 27; theory of natural and sexual selection 2-5, 25-27, 260
Degler, Carl N. 117, 156, 201, 202
divorce 106, 158, 185, 189-191, 192-193
Dods, Miss 169
domestic economy—see domestic science
domestic science, college courses 177, 179, 188; college women in 174-180; need for specialized training in 169, 171-173, 174, 177, 188, 191, 285; primary and secondary school courses 179
domestic science movement xvi, 157-191
domestic service 272-276
domesticity xiv, xxv, 71; status of 156-157, 164, 165; also see domestic science movement, homemaking
DuBois, Ellen Carol 195, 201
Duncan, Matthews 262

education for women 54-104; arguments for xxiii, 10, 45-46, 56, 60, 97-98, 161, 163; arguments against 10, 57, 59-61, 77, 147-152; effect upon health and fertility xv, xxiii, 57-58, 59, 72-74, 77-81, 83-85, 147-149; female seminaries 55, 56; for homemaking and motherhood xvi, xxiii, 55, 61, 71, 81, 127, 130, 134, 154, 163; history of 54-56; and marriage 58, 60, 61, 93, 103-104, 114-116; need for a special women's education 69-76, 81-86, 169, 171-173, 174, 177, 188, 191, 285; normal schools 5, 56; numbers of women attending institutions of higher learning 54-55, 62; purpose of 58, 70, 96-97, 135, 146, 154; also see coeducation, college women
Eastman, Elaine Goodale 310
Ehrenreich, Barbara 162-163
Eliot, Charles 110
Elliott, Ellen Coit
"Let Us Therewith Be Content" 316-322
Ellis, Havelock 13, 242
emotionality 9, 48-53, 265; also see hysteria, nervous disorders
employment (paid) of women 243-297, 301; arguments for 276-280; arguments against 247-248, 259-270, 292-297; danger of 247, 293-294; effect upon fertility 262, 294, 296;

exclusion from "male" professions 247; "female" professions 246; in clergy 262; in domestic service 272-280; in mills 246; in teaching 247, 286-291; in medicine 263; and menstruation and pregnancy 267, 269, 293; of college women 58-59; number of employed women 252; racial and ethnic division of 246, 252-253; and sexual differences 263-270, 293-295; sexual division of labor 245-246, 248, 256; status of 244, 249; wages 248, 249, 286-291

Engelmann, George J. 9, 110-111, 112, 137, 140, 141

English, Deirdre 162-163

equality, definition of 197; in education xxiii, 45-46, 54, 57, 74, 79, 97; in employment xxiv, xxvi; in intelligence 37-38, 45, 47; of men's and women's spheres xxv-xxvi, 156, 200; in political and legal rights xxiii, xxiv, 31-32, 200; in sexual differences 4-9, 197-199, 200, 201; in social status 33, 156-157

fatherhood, ideology of xxvi

Fernow, Olivia R.
"Does Higher Education Unfit Women for Motherhood?" 152-155

fertility — see birth rates

Fifteenth Amendment xxiv, 194-195, 203

Fourteenth Amendment xxiv, 194-195, 203

freedom xxv, 24, 49, 132, 321, 327

Freeman, Alice — see Palmer, Alice Freeman

Frederick, Christine 159-160

Gage, Matilda Joslyn 195

General Federation of Women's Clubs 196

Gilman, Charlotte Perkins xxi, xxv, 243-245

Gordon, Linda 117

Greeley, Horace 195

Grimke, Angelina and Sarah 194

Hall, G. Stanley 110, 113, 114

Hamman, Anne B. 298

Hardaker, M. A. xiv, 9
"Science and the Woman Question" 33-39; response 39-47

health of women 48-53; of college women 137-146; endangered by education 57-58, 77-86; endangered by employment 292-297

Heath, Mrs. Julian 159

Helmholtz, Hermann 6

Hewes, Amy 59

Hochbaum, Elfrieda — see Pope, Elfrieda Hochbaum

Hofstadter, Richard xxi

home economics — see domestic science, domestic science movement

homemaking, scientific 159, 169; status of 156-157, 159; time spent on 160-161; also see domestic science, domesticity

Howe, Julia Ward 196

Hunter, Robert 109

hysteria xv, 6, 48, 51, 211, 266; also see nervous disorders

immigrants, birth rate of 106; cause of race suicide 107-110

individuality xxvi, 298, 305-306, 321

infant mortality 172, 184-185

industrialization xxiv, 118; effect upon sexual division of labor xxiii, xxiv

Jacobi, Mary Putnam 9, 248

Jenkins, Therese A.
"The Mental Force of Woman" 217-219

Jordan, David Starr 170, 323
"The Higher Education of Women" 96-104

Keir, D. R. Malcolm
  "Women in Industry" 292-297
Klein, Viola 156
Kyrk, Hazel 160

Lerner, Gerda 156
Linton, Mrs. E. Lynn 230, 231
Livermore, Mary 196
Luce, Grace A.
  "Occupations, Privileges, and Duties of Woman" 236-238

Malthus 2
marriage 267, 313, 318, 319, 329; age at time of 114-116, 150, 154; effect upon fertility 262; foremost duty of women 55, 58, 126-136, 306; of American women 324; of college women 58, 60, 61, 93, 103-104, 114-116, 138, 148
Matthaei, Julie A. 156-157
Maudsley, Henry xiv, 50-51, 57, 260, 267-268, 269
  "Sex in Mind and in Education" 77-86
menstruation, significance for employment 248, 267, 293; relation to physical and intellectual ability 9, 248; relation to nervous disorders 267; also see aemenorrhoea
mental characteristics of women 5, 7, 8, 48-51, 47, 77-82; differences between women and men 4-5, 17-24, 34-38, 80-81, 90-91, 211-212, 260, 263
Minor v. Happersett 195
Mitchell, S. Weir 57, 58, 85
Morais, Nina xiv, 9
  "A Reply to Miss Hardaker on the Woman Question" 39-47
morality, difference between men and women 23, 213; superiority of women 31, 32, 163, 201, 219; inferiority of women 173
Morrill (Land Grant) Act 59
motherhood 308-311; ideal of xxiv-xxv, 10, 139, 157-302, 318; instinct for xxvi, 1
Mott, Lucretia 192, 194

National American Woman Suffrage Association 195
National Woman Suffrage Association 195
Nature xiv, 17, 21, 48, 52, 70, 78, 84, 85, 134, 135, 136, 151, 152, 197, 212, 220, 221, 223, 268; nature versus nurture xxvi
natural selection, theory of 2, 25, 27, 109, 199; also see Darwin, Survival of the Fittest
nervous disorders 9, 51, 150; also see hysteria, emotionality
New England Kitchen 176-178
"New Woman" 298-329
Nightingale, Florence 291
Nineteenth Amendment xxiv, 204
Norton, Alice P. 158, 179

Oakley, Ann 160
O'Neill, William L. 196

Palmer, Alice Freeman 57-58, 59, 60, 174, 178
Phillips, Wendell 195
Pope, Elfrieda Hochbaum
  "Women Teachers and Equal Pay" 286-291
progress, social 1, 5, 31, 109, 199, 200, 212, 282, 285, 312; racial 31, 211, 317

race suicide 106, 111, 150
reproduction xv, 57, 317; relation to hysteria 6; tension with individual development 6, 7, 8, 37-38, 45, 78, 148, 211; also see birth rates, menstruation
Repplier, Agnes 311
Richards, Ellen Swallow 158, 175, 176-178, 182
Robinson, Victor 105

Rockwood, Laura Clarke xvi, 163
"Food Preparation and its Relation to the Development of Efficient Personality in the Home" 181-191
Roosevelt, Theodore 61, 106
Rosenberg, Rosalind xiii

Sanitary Science Club 175
scientific homemaking—see homemaking, domestic science
Scudder, Lucy 194, 195
separate spheres, ideology of xxiii, xxiv, 157, 165, 200, 201, 282-284, 301; also see sexual division of labor
sex roles xiii, xxv
sexual differences xiii, 53, 78-81, 210-212, 235; and employment of women 259-270; also see brain size, emotionality, mental characteristics, reproduction
sexual division of labor xiii, 5, 81, 245-246, 248, 282
sexual selection, theory of 25-26, 260-262; also see Darwin
Simpson, J. Y. 266
Smith, A. Lapthorn xvi
"Higher Education of Women and Race Suicide" 147-152; response 152-155
Smith, Daniel Scott 117, 156
Smith, Gerrit 195
Smith, Mary Roberts xvi, xvii, 110, 116, 117, 179
"Education for Domestic Life" 170-173
Smith, Mrs. Burton
"The Mother as a Power for Women's Advancement" 307-311
Smith, Theodate L. 110, 113, 114
Smith College 55, 58
social-Darwinism xxi, 1, 109, 200
social progress—see progress
Spencer, Herbert xiii, xxi-xxii, 136, 260; on political rights of women 13; on reproduction 6, 7-8, 18, 30, 261; on sexual differences 1, 5-8, 17-24; on social progress 31; on tension between reproduction (genesis) and individual development (individuation) 6, 7, 18, 30, 37, 261
"Psychology of the Sexes" 17-24
Stanton, Elizabeth Cady 192, 194, 197, 198, 215; "Declaration of Sentiment" 192-193
Stone, Lucy 194, 195
Storer, H. R. 266-267, 269
suffrage for women 192-242, 316-317, 321-322; Amendment (Nineteenth) xxiv, 204; arguments for xxiii, 195, 197-203, 228-238; arguments against 10, 203, 213-215, 220-228, 241-242; and marriage 215; movement 194-197
Survival of the Fittest 2, 11, 17, 27; also see Darwin, Spencer

Talbot, George F.
"The Political Rights and Duties of Women" 220-228; responses 228-238
Talbot, Marion 61, 158, 175, 178
Tentler, Leslie Woodcock 248
Thomas, M. Carey 58
Thorndike, Edward 110, 112
Tilt, Dr. 266, 267
Tilton, Elizabeth 195
Tilton, Theodore 196
true womanhood, ideology of 301, 314
Tucker, George 107, 108
Tweedy, Alice B.
"Is Education Opposed to Motherhood?" 132-136
"Woman and the Ballot" 228-236; response 239-242

Van de Warker, Ely xv, 247-248
"The Relations of Women to the Professions and Skilled Labor" 259-270
Van Kleeck, Mary 58

Vassar 57, 58

Walker, Alexander 260
Walker, Francis A. 107-109
Walker, Kathryn 160
Welter, Barbara 301
White, Frances Emily 8
  "Women's Place in Nature" 25-32
Willard, Emma 56
Willard, Frances 196, 291
Willcox, Walter F. 107, 108
Wilson, Margaret Gibbons 162
Wishy, Bernard 161
Wollestonecraft, Mary xxii
woman question xxii, xxiv, 10, 61, 126, 129, 302, 307-308, 317
womanhood ideal of xxvi, 5, 127, 130, 135, 318; also see motherhood, "New Woman"

woman's intuition 1, 20, 38
woman's rights movement xxv, 127, 193, 194; also see suffrage
women's colleges 55, 57, 58-59, 98
women's movement xix, xxi, xxii, xxiii, xxv, 69, 127-129, 302
women's nature xiii, xiv-xv, xvi, xxii, xxvi, 1-53, 57, 77-86
women's sphere xxiii, xxv, 32, 71, 164, 318; also see separate spheres
Woodhull, Victoria 195-196
Wright, Carroll D. 58

Youmans, Edward Livingston xxi; editorials 69-76, 168-169, 281-285
Youmans, Eliza xxvii
Youmans, William Jay, editorials 158-169, 239-242, 281-285, 305-306, 312-315

ABOUT THE AUTHOR

A graduate of Harvard-Radcliffe College, Louise Michele Newman has taught expository writing and English literature at Mercy College for the past two years. She is now a University Fellow in the American Civilization program of Brown University.